Codes, Ciphers and Spies

John F. Dooley

Codes, Ciphers and Spies

Tales of Military Intelligence in World War I

John F. Dooley
Department of Computer Science
Knox College
Galesburg, Illinois, USA

ISBN 978-3-319-29414-8 ISBN 978-3-319-29415-5 (eBook)
DOI 10.1007/978-3-319-29415-5

Library of Congress Control Number: 2016933031

Published by Copernicus Books,
an imprint of SpringerNature.

Copernicus Books
SpringerNature
233 Spring Street
New York, NY 10013
www.springer.com

Printed on acid-free paper

This Copernicus imprint is published by Springer Nature
The registered company is Springer International Publishing AG Switzerland

For Diane and Patrick

*CI*TCO M*OE*AE HA*DS*R INT*E*T
YNFT*S* AIWDH *C*LNOO E*I*ETR
OW*P*SE LLS*H*O LDEM*E* FLWYA
*R*MIAJ N*S*IGR OL*S*LE IHY*P*I
NJNX

Preface

This book is about two different things. First, it started with the rediscovery of a series of a dozen articles written in 1927 on contract for *Collier's Weekly Magazine* about the cryptographic section of the Military Intelligence Division of the US Army during World War I by John Matthews Manly, a member of that division. These articles were never published, and they disappeared until recently, when copies were discovered in the William F. Friedman Collection at the George Marshall Foundation Research Library in Lexington, VA. The book describes how the articles were written, how they ended up in the Friedman Collection, and what they contain. The articles are presented; edited for grammatical, factual, and spelling mistakes (but spelling conventions from the 1920s are retained); and annotated to provide a context for their contents. The articles themselves contained no citations or bibliography, so these have been added where possible.

The second thing this book attempts to do is to put cryptology, particularly American cryptology, in the context of World War I. America was late in many ways in getting to the Great War. American cryptologists had to work very hard to catch up with their European counterparts who already had 3 years of experience in using code and cipher systems in a modern war by the time the Americans arrived in France in the summer of 1917.

The book is divided into four parts. Part I tells the story of the American Expeditionary Force (AEF), how it was organized and how it got to France, and gives us a peek into the military intelligence operations within the AEF during 1917 and 1918. For the entire 19 months that America participated in the Great War, the AEF was playing catch up to the Allies who had already been fighting for 3 years by the time that the first Americans arrived on the Western Front. The military intelligence organization was no different. A separate command from MI-8 (which handled domestic and diplomatic intelligence), the military intelligence unit in the AEF, designated G2-A6, had to be built from scratch and in the beginning was largely trained by their British and French counterparts. Manly provides us with insight into its operations and problems in three articles plus a separate essay to set the story for us.

Part II tells the story of MI-8 (Section 8, "The Code and Cipher Section," of the Military Intelligence Section of the Army General Staff), how it was organized and functioned, and how it dealt with domestic correspondence during the war. These articles focus on German espionage and civilian correspondence, including secret messages from German POWs interned in the United States. We learn how MI-8 acquired its intercepted messages, some techniques of decryption, and how MI-8 worked with the military intelligence counterespionage units within the Army.

Part III engages us with German efforts, largely through spies and sabotage, to hinder the supplying of American arms and ammunition to the Allies in the first 3 years of the war when the United States was still officially neutral. It's the story of German espionage and sabotage in the United States during the war. This includes a spy ring operating in New York, but run by German diplomats and embassy staff out of the German Embassy in Washington. It's the story of an unprepared German intelligence establishment trying to recruit competent spies and saboteurs in Germany and the United States and largely failing. Overall, the Germans were more an amateur or semipro team than an experienced group of professional intelligence agents. This story's central character is the infamous, but largely ineffective, female German spy Madame Marie de Victorica.

Finally, Part IV wraps up the narrative and brings the story back to John Manly.

Manly's articles provide us a window into his experiences in this environment, and the additional chapters attempt to flesh out the American experience in the war on both sides of the Atlantic. There is enough material in the various archives, even restricting ourselves mostly to cryptology, for several volumes. My hope is that this work will do justice to John Matthews Manly's contributions to the war effort and give the reader some insight into America's role in the last phase of the Great War.

Galesburg, IL, USA John F. Dooley

Acknowledgments

The copies of the Manly articles used here are in the public domain and are from Item 811 in the Friedman Collection at the George Marshall Foundation Research Library, Lexington, VA. They were ably scanned and transcribed by Elizabeth Anne King, Knox'13; any errors are mine.

I would like to thank the staffs at the Special Collections Research Center at the University of Chicago Library (which houses the papers of John Matthews Manly); the National Archives and Records Administration (NARA) at College Park, MD; and the George Marshall Foundation Research Library (where the William F. and Elizebeth S. Friedman papers are held) for their gracious help. Jeff Kozak and Paul Barron at the Marshall Library were particularly helpful and supportive. Rene Stein at the National Cryptologic Museum Research Library was, as always, very knowledgeable and helpful with my searches through the David Kahn Collection. I would also like to thank the research librarians at the Knox College Library, especially Anne Giffey and Laurie Sauer, for their kindness, patience, and help in digging up many obscure newspaper articles and books. Interlibrary loan is my savior. This research was funded in part by a grant from the Andrew Mellon Foundation and by grants from the Office of the Dean of the College at Knox College. As always, my wife, Diane, has been my inspiration, sounding board, and first and best editor.

Photo Credits

Figures 1.1 and 1.3 are from the John Matthews Manly Collection and are used with permission of the Special Collections Research Center at the University of Chicago Library. Figure 1.2 is used with permission of the New York Public Library. Figure 2.2 and Figs. 8.1 through 8.6 are used with permission of the American Battlefields Monument Commission. Figures 1.3, 2.3, 4.1, and 9.1 are from the William Friedman Collection of the George Marshall Foundation Research Library, Lexington, VA, and are used with permission. Figures 5.1 and 5.2 are copyrighted by Dr. Nicholas Gessler, Duke University, and are used with his kind permission. Figure 7.1 is from the David Kahn Collection at the NSA National Cryptologic Museum and is used with permission. Figures 13.1 through 13.5 and 14.1, 14.3, and 14.4 are in the public domain from the Library of Congress Prints and Photographs Division. All remaining figures are in the public domain and are from the National Archives and Records Administration, College Park, MD, or Wikimedia Commons.

Photo Credits

Contents

Chapter 1
Introduction

Abstract In 1927, John Matthews Manly, former cryptanalyst for the Code and Cipher Section of the US Army Military Intelligence Division during World War I, wrote a series of a dozen articles intended for *Collier's* magazine. These articles are published here for the first time. This chapter gives a brief bio of John Manly and a short introduction to the internals of codes and ciphers.

In 1927, John Matthews Manly, a college English professor who had served in the Code and Cipher Section of the US Army's Military Intelligence Division (MID) during World War I, decided to write a series of articles about MID and his experiences during the war. Manly found a magazine interested in his idea and proceeded to write a set of twelve articles. Manly's articles were never published and were subsequently lost for more than 80 years. This book is the story of those articles, published for the first time here. It is also the story of MID, of German espionage in America during the war, the military intelligence unit of the American Expeditionary Force (AEF), spies, coded letters, plots to blow up ships and munition plants, secret inks, arms smuggling, treason, and desperate battlefield messages. The articles give us a uniquely American perspective on the Great War and provide a look at what the war was like both on the domestic and Western fronts for the Americans who lived it. The story all begins with John Manly.

1.1 John Matthews Manly: Early Life

John Matthews Manly was born on September 2, 1865, in Sumter County, Alabama, the eldest of seven children born to Charles and Mary Esther Manly. He came from a very successful and politically active Southern family. His great grandfather Basil Manly was president of the University of Alabama, and his grandfather, Basil Manly, Jr., was a minister and president of Georgetown College of Kentucky. In 1861 Basil Manly delivered the inaugural prayer when Jefferson Davis became the Confederate president. Manly's father, Charles, carried on the pastoral and educational traditions of his family and was a Baptist minister and the president of Furman University in Greenville, South Carolina.

J.F. Dooley, *Codes, Ciphers and Spies*, DOI 10.1007/978-3-319-29415-5_1

Fig. 1.1 John Matthews Manly (Used with permission of U. of Chicago Library)

John Manly was a precocious child and student, completing a master's degree in mathematics at Furman University in 1884 at age 19. He then taught mathematics at William Jewell College in Liberty, Missouri, for 5 years before heading to Harvard University to pursue his Ph.D. in philology. As Harvard did not have a department of philology, Manly put together his own program, rounded up a dissertation committee, and earned his degree in 1890. He then went to Brown University to teach English and stayed there until 1898 when he was lured by the new University of Chicago's first president, William Rainey Harper, to head the English Department there. With just the hiatus of his service during World War I, Manly would remain head of the department until his retirement nearly 40 years later (Fig. 1.1).

1.2 Manly and Chaucer

John Manly is primarily known as one of the world's premier Chaucer scholars. His masterwork, the result of a more than 16-year collaboration with Dr. Edith Rickert (1871–1938), one of his Chicago colleagues, is the eight-volume annotated edition of Chaucer's *The Canterbury Tales*, *The Text of the Canterbury Tales*, published just before his death in 1940 (Manly and Rickert 1940). Dr. Rickert had passed away in 1938 and never lived to see their magnum opus published.

Edith Rickert was first Manly's student, then his colleague in military intelligence during the war and finally, starting in 1924, his colleague in the English Department at Chicago. Rickert graduated from Vassar College in 1891 and taught at various high schools and later at Vassar before pursuing her graduate studies in English at the University of Chicago in 1895. She received her Ph.D. in English and philology in 1899 (Manly was on her dissertation committee). In 1900 Rickert moved to England and began a career as a novelist and journalist. While in England she published four novels, several research papers, and numerous newspaper arti-

Fig. 1.2 Edith Rickert (about 1905) (The Miriam and Ira D. Wallach Division of Art, Prints and Photographs: Print Collection. The New York Public Library.)

cles and short stories. She also worked as a professional researcher for a number of scholars including John Manly (Fig. 1.2).

Returning to Chicago in 1909, Rickert worked first as a magazine editor and in 1914 began teaching part-time in the English Department at the University of Chicago. During America's participation in World War I, she worked with Manly in the Code and Cipher Section of the Military Intelligence Division in Washington, DC. Rickert returned to Chicago after the war and continued her teaching at the university; she was named an associate professor of English in 1924. During this period, she and Manly wrote four textbooks on English and American literature and English grammar. It was also at this time that she and Manly began their collaboration on Chaucer.

From 1924 on, Manly and Rickert would spend the first 6 months of every year in England, tracking down reference after reference to Chaucer and manuscripts of *The Canterbury Tales*. In the summer and fall, they would be back at Chicago, teaching and supervising a staff of primarily graduate students that would collate and index their findings from England. Their goal was to find and annotate every single version of *The Canterbury Tales* in existence with the object of creating an authoritative text of the work, an effort that Manly thought, "would necessarily require several years" instead took 16. Rickert foresaw this early on in their collaboration. As Manly said in the preface to the set, "At a very early stage in our undertaking [Rickert] felt the great complication and size of it and often asserted that we could never finish it if we worked like normal human beings." While somewhat controversial 75 years later, their eight-volume *The Text of the Canterbury Tales* is still viewed by many scholars as the definitive work on this classic (Fig. 1.3).

1.3 Manly and Cryptology

In addition to the academic side of John Manly, his avocation was cryptology. From the time he was a teenager he was interested in secret codes and ciphers. Manly visited the Riverbank Laboratories in Geneva, IL, as early as 1915 to talk to the

Fig. 1.3 John Manly, Edith Rickert, and colleague David Stevens on board the *Europa* in 1932 (Used with permission of U. of Chicago Library)

owner, Colonel George Fabyan, about the alleged authorship of Shakespeare by Francis Bacon and the Baconian biliteral cipher. He also consulted with other scholars on cryptologic texts. For many years Manly maintained an active and friendly correspondence with Herbert O. Yardley, his commander during the war and the founder of the first permanent cryptologic organization in the War Department. William F. Friedman, widely regarded as the father of modern American cryptology, and Charles Mendelsohn, a historian and a colleague during World War I, were both close friends of Manly's. Their correspondence touched on many topics including cryptology, the Voynich manuscript and Roger Bacon, and, of course, Chaucer.

1.4 Manly in MI-8

Upon America's entry into World War I, John Manly, then 51 years old, volunteered for service in the US Army. He visited Major Ralph Van Deman, the head of the Military Intelligence Section, as early as March 1917 to offer his services. Van Deman contacted Manly at the end of September, and he was inducted as a Captain on October 3, 1917, and later promoted to Major. He served in the Code and Cipher Section of the Military Intelligence Division, designated MI-8, under the leadership of Herbert O. Yardley.

At the time of the American entry into World War I, the US Army did not have a formal intelligence organization nor did it have an organization to intercept and break enemy code and cipher messages. With each war or military conflict that the United States would find itself ensnared, the Army would create an intelligence organization and relearn all the skills and lessons required. Finally, in 1903 with the creation of the General Staff, the Army formed a formal military intelligence organization. This organization was short-lived, however, and was subsumed under the War College in 1908 with its separate identity eliminated. This first Military Intelligence Section did not include any personnel whose job it was to break enemy code and cipher messages.

With the American declaration of war, Major Ralph Van Deman, considered the "Father of American Military Intelligence," convinced the War Department that the United States needed a separate intelligence unit if it were to be a full partner in the war in Europe. So in May 1917 the Military Intelligence Section of the General Staff was created with Van Deman as its first head.

Herbert O. Yardley was a code clerk in the State Department and had developed a familiarity with State Department codes and ciphers. He also taught himself how to break those same code and cipher messages, much to the chagrin of his superiors. In early June 1917, and with his superior's reluctant assent, Yardley approached Major Van Deman with a proposal to create a Code and Cipher Section within the brand new Military Intelligence Section (Yardley 1931, pp. 34–36). With Van Deman's enthusiastic approval, Yardley created the Code and Cipher Section and continued its work as a civilian after the war. Yardley was an organizational genius, a slick and astute salesman, and a self-taught cryptanalyst who built the first permanent cryptologic organization in the US Army. Starting with just himself and a single clerk, Yardley built an organization that ended the war with more than 165 personnel and five subsections to handle shorthand messages, secret inks, code and cipher creation, code and cipher solution, and training. After the war his joint War-State Department Cipher Bureau—the Black Chamber—was the only US government organization devoted to breaking code and cipher messages during the 1920s. The Cipher Bureau was credited with breaking the main Japanese diplomatic code in 1920 and provided the US State Department with decrypted Japanese telegrams during the Washington Naval Conference of 1921–1922 (Yardley, 1931, pp. 283–317). These decrypted telegrams gave the United States a bargaining advantage in the negotiations on naval warship tonnage during the conference. With the change in administrations in 1929, Yardley's Cipher Bureau lost its funding, and he turned to other pursuits, including writing fiction, running a restaurant in Washington, DC, and writing nonfiction magazine articles about cryptology (Fig. 1.4).

Manly was 24 years older than Yardley, and their relationship was close during the war and would remain close at least through the early 1930s. Manly was practically the only person who defended Yardley after the publication of Yardley's tell-all 1931 book *The American Black Chamber* made him a pariah in the American cryptologic community.

In October 1917, Manly joined a fast-growing group under then Lieutenant Yardley in MI-8 as the head cryptanalyst, chief instructor, and second-in-command. Manly had an impact from the very beginning. He recruited several of his colleagues and students from the University of Chicago, including Dr. Edith Rickert, for MI-8 and began solving German military and diplomatic code and cipher messages.

Manly's greatest coup during the war was the solution of the Pablo Waberski cipher in 1918, a story told well, if not completely accurately, in Yardley's book, *The American Black Chamber* (Yardley 1931, pp. 140–171). Waberski was a

Fig. 1.4 Lt. Herbert O. Yardley (*Public Domain*. From RG 457 National Archives (NARA), the Yardley Collection.)

German spy who crossed into the United States from Nogales, Mexico, with a lengthy cryptogram in his possession. He was captured in Arizona on February 1, 1918, just after crossing the border, and the cryptogram was sent to MI-8 in Washington, where it languished unsolved for several weeks.

In early May 1918, Manly and Rickert spent the better part of 3 days breaking the cryptogram—a double transposition cipher—and Manly was later called to testify at Waberski's trial at Fort Sam Houston in San Antonio. The solution of the cryptogram was the damning piece of evidence that convicted Waberski and earned him a death sentence (later commuted; Waberski was released from prison and deported to Germany in 1924) (Kahn 2004, pp. 41–43).

When Yardley was sent to France in August 1918 and later assigned to supervise the cryptographic section of the American delegation to the Peace Conference in early 1919, John Manly became commander of the Code and Cipher Section and oversaw its demobilization after the Armistice, returning to the University of Chicago in mid-1919.

1.5 Manly After the War

Manly returned to the University of Chicago and resumed his post as chair of the English department. He and Edith Rickert published several more books together, including textbooks on English and American literature and a series of grammars for elementary schools in addition to their collaboration on Chaucer. While the remainder of Manly's career was primarily focused on literature, he still found time to work in cryptology.

Like many American cryptologists of the day including both Yardley and Friedman, Manly became interested in the Voynich manuscript (Kahn 1967, pp. 863–872). The Voynich manuscript is a 240-page illustrated vellum codex that is written in an unknown language and alphabet. It is named after Wilfrid Voynich, a Polish book collector and dealer who acquired it from a Jesuit monastery outside Rome in 1912. Nearly every page is a combination of text and illustrations. Of the approximately 170,000 letters in the manuscript, an alphabet of 20–30 symbols would account for most of them. The vellum has been carbon-dated to the early to mid-fifteenth century, and the ink in which the text is written traced to a slightly later date. Professor William Newbold at the University of Pennsylvania announced a possible solution to the mysterious cryptogram in April 1921. He described the convoluted process requiring microscopes and a process of rearranging deciphered letters until they produced understandable Latin that he used to reach his decipherment. Manly began corresponding with Newbold and examining his claims and eventually came to the conclusion that Newbold's analysis was faulty and his decipherment incorrect. This led, later in 1921, to the publication by Manly of two papers on the Voynich, *Roger Bacon's Cipher Manuscript* (Manly 1921a) and *The Most Mysterious Manuscript in the World* (Manly 1921b). In these papers Manly laid out the various propositions about the manuscript and then analyzing Newbold's arguments and his process demolished them in their turn. Newbold would go on to publish a book on the Voynich manuscript in 1928 (Newbold 1928) in which he claimed to have deciphered the manuscript, and Manly would publish another paper shredding Newbold's techniques (Manly 1931).

1.6 The *Collier's* Articles

In the midst of his other scholarly work, Manly was also interested in telling the story of MI-8 during the war and of his own experiences as a member of the Military Intelligence Division. He was fully aware that much of what he could say was constrained by the secrecy demanded of everyone in MID with regard to their intelligence work during the war, but he thought he could tell enough of a tale to interest readers. In 1923 he contemplated writing an article for *Harper's Magazine* but eventually dropped the idea. Then in late 1926 the editor of *Collier's Weekly* magazine, Mr. William Chenery, wrote to Manly and suggested that Manly write a series of articles on the role of MI-8 during the war. Manly, who was about to head to England for his annual trip researching Chaucer, put Chenery off and said he would contemplate the series when he returned to the United States in June 1927.

Here is where our story really begins.

Neither Manly nor Chenery gave up the idea of the articles, and after some negotiations Manly eventually agreed to write a series of articles for *Collier's* on his World War I experiences (Manly 1927a). He was originally contracted to write up to six articles and finally between 7 and 12 articles, each 4,000 words, for *Collier's*, and was to be paid $2,000 for each article, a tidy sum for 1927 New York (Anonymous 1927).

In early September 1927 Manly went to New York, rented an apartment, engaged a secretary, and began to write. Herbert Yardley assisted him, although the extent of their collaboration is not completely known. By September 14, 1927, Manly had three articles on the Radio Intelligence Section in the American Expeditionary Force finished and sent them to *Collier's*. In a letter dated September 16, 1927

(Chenery 1927), Chenery indicated he was happy with the content of the articles but suggested they needed to be edited for the more casual style required for a general circulation magazine. By the end of September, Manly and Yardley had all 12 articles finished and were ready to get into the editing phase with an experienced magazine journalist that *Collier's* had engaged (Anonymous 1927, p. 6). Manly was also ready to be paid.

At this point, *Collier's* and Chenery began to back off. First of all, Chenery hadn't liked Manly's writing style in the articles he had seen—insisting it was too scholarly and not appropriate for the type of audience that *Collier's* catered to—and he assigned a freelance journalist named Davenport to rework the articles. Manly, Yardley, and Davenport had met in late September to discuss the content, and Davenport had gone off to rework the existing articles.

There was also an evolving dispute over the timing of payment for the articles. In the September 16, 1927, letter to Manly, Chenery had stated, "We pay on acceptance but in the case of a long series of articles are accustomed to space the payments over a certain period of time" (Chenery 1927, p. 1). Manly thought that he and Chenery had agreed that "pay on acceptance" meant "pay on delivery" and that Chenery had already agreed to "accept" the articles that Manly submitted (Anonymous 1927, p. 1). Every time during the latter half of September and the first half of October 1927 that Manly asked Chenery about payment, Chenery put him off (Anonymous 1927, p. 5, 7). Manly finally received a check for $2,000—one article—on October 10, 1927.

By late October—somewhere around the 20th—Chenery, Davenport, Manly, and Yardley met at *Collier's*, and Chenery informed Manly "…that the material was such that it was impossible for *Collier's* to accept the articles and offered to return the manuscripts of the twelve articles, with the understanding that Prof. Manly should keep the $2000.00" (Anonymous 1927, p. 7). After considerable discussion, on November 29, 1927, *Collier's* sent back the 12 manuscripts and John Manly kept the $2,000.

Apparently, John Manly sent the articles, unopened, to his attorney and at this point they disappeared. Manly never tried to get the articles published again.

John Matthews Manly passed away on April 2, 1940, having published his magnum opus on Chaucer but never having published a word on his experiences in MI-8 during World War I. Manly's younger brother Basil Maxwell Manly (1886–1950) was the executor of his estate, and it was apparently he who inherited the *Collier's* articles upon John Manly's death in 1940. Basil Manly was an economist and a longtime member of the Federal Power Commission. Rose Sheldon, in her annotated inventory of the William Friedman Collection (Sheldon 2000, p. 296), believes that Basil Manly was interested in seeing that the articles were published or included as part of a biography of his brother. This never happened.

Upon Basil Manly's death in 1950, the *Collier's* articles were apparently passed on to two US Army generals, Lt. General Alexander R. Bolling (see http://en.wikipedia.org/wiki/Alexander_R._Bolling) and Major General (GSC) F. L. Parks (http://en.wikipedia.org/wiki/Floyd_Lavinius_Parks), by an unknown source, possibly the executor of Basil Manly's estate. Bolling was then in command of the US Third

Fig. 1.5 William F. Friedman (Used with permission of George Marshall Foundation Research Library)

Army, which was the center of training for all new Army recruits. Parks was the head of the Army Information (read public relations) Department. These two gentlemen saw the value of the articles as historical artifacts and were interested in finding a permanent home for them (Anonymous n.d.). They proceeded to give the articles to William F. Friedman, then the head of the cryptographic division of the Armed Forces Security Agency (AFSA). Friedman evidently made a cursory examination of the articles, wrote up some brief notes, and then put them away, never to look at them again (Anonymous 1967) (Fig. 1.5).

In the late 1960s William and Elizebeth Friedman were making preparations to donate all their papers to the George Marshall Foundation Library, the *Collier's* articles among them. However, the Library does not hold the originals of the articles. Item 811 of the William Friedman Collection contains a set of photostatic copies of the articles along with a copy of the *Facts* memorandum that lays out the chronology of the creation of the articles during the last few months of 1927. None of the supporting documentation mentioned in the *Facts* memorandum or the originals of the articles themselves are present in the Friedman Collection, nor is their current location known. Neither the series of letters between Manly and Chenery nor the memoranda written by Herbert Yardley, described in the *Facts* memorandum, are in the Friedman Collection, nor are they in the John Manly Collection at the University of Chicago.

Curiously, the biggest story missing from the twelve articles is the Pablo Waberski story in which Manly played the leading role. There doesn't seem to be any reason to leave this story out. It has all the elements that *Collier's* may have been looking for—spies crossing the border, hidden messages, brilliant decryption work, and dramatic trial testimony. This story rates many pages in Yardley's *The American Black Chamber* (Yardley 1931, pp. 140–171), so why is it missing in these articles?

What is more, in the same folder in the Friedman Collection with the articles, there is a version of the story, apparently written by Manly after his testimony at Waberski's court martial in 1918. This 23-page document (included as Chapter

15 in the current volume) describes in detail how Manly broke the double transposition cipher message and also contains—apparently from memory—a transcript of Manly's testimony at Waberski's court martial at Fort Sam Houston in San Antonio, TX (Manly 1927b).

So this is the heart of our story. These 12 articles from 1927 give us a glimpse into German spies, US military intelligence, codes, ciphers, and secret inks and how America began to come of age in international relations, espionage, and signals intelligence as a result of World War I.

1.7 A Few Words on Codes and Ciphers

Secrecy in communications is known to have existed for close to 3000 years. As Kahn puts it, "It must be that as soon as a culture has reached a certain level, probably measured largely by its literacy, cryptography appears spontaneously – as its parents, language and writing, probably also did. The multiple human needs and desires that demand privacy among two or more people in the midst of social life must inevitably lead to cryptology wherever men thrive and wherever they write. Cultural diffusion seems a less likely explanation for its occurrence in so many areas, many of them distant and isolated" (Kahn 1967, p. 84).

Every discipline has its own vocabulary and cryptology is no different. This section does not attempt to be a comprehensive glossary of cryptology but rather gives the basic definitions and jargon.

Governments, the military, and people in business have desired to keep their communications secret ever since the invention of writing. Spies, lovers, and diplomats all have secrets and are desperate to keep them as such.

There are typically two ways of keeping secrets in communications. *Steganography* hides the very existence of the message. Secret ink, microdots, and using different fonts on printed pages are all ways of hiding the message from prying eyes. *Cryptology*, on the other hand, makes absolutely no effort to hide the presence of the secret message. Instead it transforms the message into something unintelligible so that if the enemy intercepts the message they will have no hope of reading it. *Cryptology* is the study of secret writing. A *cryptologic system* performs a *transformation* on a message—called the *plaintext*. The transformation renders the plaintext unintelligible and produces a new version of the message—the *ciphertext*. This process is *encoding* or *enciphering* the plaintext. A message in ciphertext is typically called a *cryptogram*. To reverse the process the system performs an inverse transformation to recover the plaintext. This is known as *decoding* or *deciphering* the ciphertext.

Steganography usually involves hiding the existence of the message physically in some innocent document or, recently, hiding it virtually in digital images. The most common form of steganography in history is the use of secret inks. The world of secret inks is divided into two types. *Organic inks* are those derived from the juices of fruits and vegetables—lemons, limes, oranges, and onions—and other

Table 1.1 The two dimensions of Cryptology

	Cryptography		Cryptanalysis			
Codes	1-part	2-part	Theft, spying	Probable word	Context	
Ciphers	Substitution	Transposition	Classical	Statistical	Mathematical	Brute-force
	Product cipher					

organic substances—milk, urine, blood, starches, etc. These inks can normally be developed using heat or water. *Sympathetic inks* are those that are normally derived from other chemicals and that must be extracted from compounds, including tannic acid, cobalt chloride, alum (aluminum potassium sulfate), iron sulfate, phenolphthalein, etc. Sympathetic inks require a separate chemical reagent as a developer.

The science of cryptology can be broken down in a couple of different ways; one is that it is concerned with both the creation of cryptologic systems, called *cryptography*, and with techniques to uncover the secret from the ciphertext, called *cryptanalysis*. A person who attempts to break cryptograms is a *cryptanalyst*. A complementary way of looking at cryptology is to divide things up by the types and sizes of grammatical elements used by the transformations that different cryptologic systems perform. The standard division is by the size of the element of the plaintext used in the transformation. A *code* uses variable-sized elements that have meaning in the plaintext language, like syllables, words, or phrases. On the other hand, a *cipher* uses fixed-sized elements like single letters or two- or three-letter groups that are divorced from meaning in the language. For example, a code will have a single *code word* for the plaintext "stop," say, 37761, while a cipher will transform each individual letter as in X=s, A=t, V=o, and W=p to produce XAVW. One could argue that a code is also a substitution cipher, just one with a larger number of substitutions. However, while ciphers have a small fixed number of substitution elements—the letters of the alphabet—codes typically have thousands of words and phrases to substitute. Additionally, the methods of cryptanalysis of the two types of system are quite different. Table 1.1 provides a visual representation of the different dimensions of cryptology.

1.8 Codes

A *code* always takes the form of a book where a numerical or alphabetic *code word* is substituted for a complete word or phrase from the plaintext. *Codebooks* can have thousands of code words in them. There are two types of codes, 1-part and 2-part. In a 1-part code, there is a single pair of columns used for both encoding and decoding plaintext. The columns are usually sorted so that lower numbered code words will correspond to plaintext words or phrases that are lower in the alphabetic ordering. For example,

```
1234  centenary
1235  centennial
1236  centime
1237  centimeter
1238  central nervous system
```

Note that because both the code words and the words they represent are in ascending order, the *cryptanalyst* will instantly know that a code word of 0823 must begin with an alphabetic sequence before "ce," thus eliminating many possible code word-plaintext pairs.

A 2-part code eliminates this problem by having two separate lists: one arranged numerically by code words and one arranged alphabetically by the words and phrases the code words represent. Thus, one list (the one that is alphabetically sorted) is used for encoding a message, and the other list (the one that is numerically sorted by code word) is used for decoding messages. For example, the list used for encoding might contain

```
artillery support  18312
attack             43110
company            13927
headquarters       71349
platoon strength   63415
```

while the decoding list would have

```
13927  company
18312  artillery support
43110  attack
63415  platoon strength
71349  headquarters
```

Note that not only are the lists not compiled either numerically or alphabetically, but also there are gaps in the list of code words to further confuse the cryptanalyst.

Cryptanalyzing codes is very difficult because there is no logical connection between a code word and the plaintext code or phrase it represents. With a 2-part code, there is normally no sequence of code words that represent a similar alphabetical sequence of plaintext words. Because a code will likely have thousands of code word-plaintext pairs, the cryptanalyst must slowly uncover each pair and over time create a dictionary that represents the code. The correspondents may make this job easier by using standard salutations or formulaic passages like "Nothing to report" or "Weather report from ship AD2342." If the cryptanalyst has access to enough ciphertext messages, then sequences like this can allow her to uncover plaintext. Still, this is a time-consuming endeavor. Of course the best way to break a code is to steal the codebook! This has happened many times in history, much to the dismay of the owner.

Codes have issues for users as well. Foremost among them is distributing all the codebooks to everyone who will be using the code. Everyone who uses a code must have exactly the same codebook and must use it in exactly the same way. This limits the usefulness of codes because the codebook must be available whenever a message needs to be encoded or decoded. The codebook must also be kept physically secure, ideally locked up when not in use. If one copy of a codebook is lost or stolen, then the code can no longer be used, and every copy of the codebook must be replaced. This makes it risky to give codebooks to spies who are traveling in enemy territory, and it also makes it very difficult to use codes in battlefield situations where they could be easily lost.

1.9 Ciphers

Ciphers also transform plaintext into ciphertext, but unlike codes, ciphers use small, fixed-length language elements that are divorced from the meaning of the word or phrase in the message. Ciphers come in two general categories. *Substitution ciphers* will replace each letter in a message with a different letter or symbol using a mapping called a *cipher alphabet*. The second type will rearrange the letters of a message but will not substitute new letters for the existing letters in the message. These are *transposition ciphers*.

1.10 Substitution Ciphers

Substitution ciphers can use just a single cipher alphabet for the entire message; these are known as *monoalphabetic substitution ciphers*. Cipher systems that use more than one cipher alphabet to do the encryption are *polyalphabetic substitution ciphers*. In a polyalphabetic substitution cipher, each plaintext letter may be replaced with more than one *cipher letter*, making the job significantly harder for the cryptanalyst. The cipher alphabets may be *standard alphabets* that are shifted using a simple key. For example, a shift of 7 results in

```
Plain:  abcdefghijklmnopqrstuvwxyz
Cipher: HIJKLMNOPQRSTUVWXYZABCDEFG
```

And the word *attack* becomes HAAHJR. Or they may be *mixed alphabets* that are created by a random rearrangement of the standard alphabet as in

```
Plain:  abcdefghijklmnopqrstuvwxyz
Cipher: BDOENUZIWLYVJKHMFPTCRXAQSG
```

And the word *enemy* is transformed into NKNJS.

All substitution ciphers depend on the use of a *key* to tell the user how to rearrange the standard alphabet into a cipher alphabet. If the same key is used to both encrypt and decrypt messages, then the system is called a *symmetric key system*.

Just like the security of a codebook, the security of the key is of paramount importance for cipher systems. And just like a codebook, everyone who uses a particular cipher system must also use the same key. For added security, keys are changed periodically, so while the basic substitution cipher system remains the same, the key is different. Distributing new keys to all the users of a cryptologic system leads to the *key management problem.* Management of the keys is problematic because a secure method must be used to transmit the keys to all users. Typically, a courier distributes a book listing all the keys for a specific time period, say, a month, and each user has instructions on when and how to change keys. And just like codebooks, any loss or compromise of the key book will jeopardize the system. But unlike codebooks, if a key is lost, the underlying cipher system is not compromised, and merely changing the key will restore the integrity of the cipher system.

While most cipher systems substitute one letter at a time, it is also possible to substitute two letters at a time, called a *digraphic* system, or more than two, called a *polygraphic* system. A substitution cipher that provides multiple substitutions for some letters but not others is a *homophonic* system. It is also possible to avoid the use of a specific cipher alphabet and use a book to identify either individual letters or words. This is known as a *book* or *dictionary cipher*. The sender specifies a particular page, column, and word in the book for each word or letter in the plaintext, and the recipient looks up the corresponding numbers to decrypt the message. For example, a code word of 0450233 could specify page 045, column 02, and word 33 in that column. Naturally, the sender and recipient must each have a copy of exactly the same edition of the book in order for this system to work. But carrying a published book or dictionary is significantly less suspicious than a codebook.

1.11 Transposition Ciphers

Transposition ciphers transform the plaintext into ciphertext by rearranging the letters of the plaintext according to a specific rule and key. The transposition is a *permutation* of all the letters of the plaintext message done according to a set of rules and guided by the key. Since the transposition is a permutation, there are n! different ciphertexts for an n-letter plaintext message. The simplest transposition cipher is the *columnar transposition*. This comes in two forms, the *complete columnar transposition* and the *incomplete columnar*. In both of these systems, the plaintext is written horizontally in a rectangle that is as wide as the length of the key. As many rows as needed to complete the message are used. In the complete columnar transposition, once the plaintext is written out, the columns are then filled with nulls until they are all the same length. For example, for the message "Second division advancing tonight," we can create a transposition table that looks like

```
s e c o n d
d i v i s o
n a d v a n
c i n g t o
n i g h t x
```

The ciphertext is then pulled off by columns according to the key and divided into groups of five for transmission. If the keys for this cipher were 321654, then the ciphertext would be

```
CVDNG EIAII SDNCN DONOX NSATT OIVGH
```

An *incomplete columnar transposition cipher* doesn't require complete columns and so leaves off the null characters resulting in columns of differing lengths and making the system harder to cryptanalyze. Another type of columnar transposition cipher is the *route transposition*. In a route transposition, one creates the standard rectangle of the plaintext, but then one takes off the letters using a rule that describes a route through the rectangle. For example, one could start at the upper left-hand corner and describe a spiral through the plaintext, going down one column, across a row, up a column, and then back across another row. Another method is to take the message off by columns but alternate going down and up each column.

Cryptanalysis of ciphers falls into four different but related areas. The *classical* methods of cryptanalysis rely primarily on language analysis. The first thing the cryptanalyst must know about a cryptogram is the language in which it is written. Knowing the language is crucial because different languages have different language characteristics, notably letter and word frequencies and sentence structure. The surest way to guess at the language in which a cryptogram is written is to identify the sender and receiver.

It turns out that if you look at several pieces of text that are several hundred words long and written in the same language that the frequencies of all the letters used turn out to be about the same in all of the texts. In English, the letter "e" is used about 13 % of the time, "t" is used about 10 % of the time, etc., down to "z," which is used less than 1 % of the time. So the cryptanalyst can count each of the letters in a cryptogram and get a hint of what the substitutions may have been. This type of *frequency analysis* can be used to determine what type of cryptologic system was used to create the cryptogram.

Beginning in the early twentieth century, cryptanalysts began applying *statistical* tests to messages in an effort to discern patterns in more complicated cipher systems, particularly in polyalphabetic systems. Later in the twentieth century, with the introduction of machine cipher systems, cryptanalysts began applying more *mathematical analysis* to the systems, particularly bringing to bear techniques from combinatorics, algebra, and number theory. And finally, with the advent of computers and computer cipher systems in the late twentieth century, cryptanalysts have had to fall back on *brute-force* guessing to extract the key from a cryptogram or, more likely, a large set of cryptograms.

References

Anonymous. 1927. "Manly vs. Collier's. Facts." Item 811. George Marshall Foundation Research Library. William Friedman Collection.

Anonymous. 1967. "Notes on Manly Colliers Articles." Item 811. George Marshall Foundation Research Library. William Friedman Collection.

Anonymous. n.d. "Undated Handwritten Notes Relating to the Manly Collier's Articles (Probably by Elizebeth Friedman)." Item 811. Friedman Collection, George Marshall Foundation Research Library, Lexington, VA.

Chenery, William L. Letter to John M. Manly. 1927. "Letter to John M. Manly," September 16. William Friedman Collection, Item 811. George Marshall Foundation Research Library.

Kahn, David. 1967. *The Codebreakers; The Story of Secret Writing*. New York: Macmillan.

Kahn, David. 2004. *The Reader of Gentlemen's Mail: Herbert O. Yardley and the Birth of American Codebreaking*. New Haven, CT: Yale University Press.

Manly, John M. 1921a. "Roger Bacon's Cipher Manuscript." *The American Review of Reviews*, July, 105–6.

Manly, John M. 1921b. "The Most Mysterious Manuscript in the World: Did Roger Bacon Write It and Has the Key Been Found?" *Harper's Monthly Magazine*, July.

Manly, John M. 1927a. "Articles for Collier's Magazine." Item 811. Friedman Collection, George Marshall Foundation Research Library, Lexington, VA.

Manly, John M. 1927b. "Waberski." Item 811. Friedman Collection, George Marshall Foundation Research Library, Lexington, VA.

Manly, John M. 1931. "Roger Bacon and the Voynich MS." *Speculum* 6 (3): 345–91. doi:10.2307/2848508.

Manly, John M., and Edith Rickert. 1940. *The Text of The Canterbury Tales*. Chicago: University of Chicago Press.

Newbold, William Romaine. 1928. *The Cipher of Roger Bacon*. Edited by Roland Grubb Kent. Philadelphia: University of Pennsylvania Press.

Sheldon, Rose Mary. 2000. *The Friedman Collection: An Analytical Guide*. George Marshall Foundation. Electronic. Lexington, VA: George Marshall Foundation Research Library. Retrieved from http://marshallfoundation.org/library/wp-content/uploads/sites/16/2014/09/Friedman_Collection_Guide_September_2014.pdf

Yardley, Herbert O. 1931. *The American Black Chamber*. Indianapolis, IN: Bobbs-Merrill.

Part I
The AEF

Chapter 2
The Americans Embark

Abstract When the United States declared war on Germany in April 1917, it was not prepared. In this chapter we discuss the creation of the American Expeditionary Force (AEF) and the reorganization of the American Army into a modern fighting force. With the advent of the first draft since the Civil War, the Army had to prepare for upwards of 2 million new men. This meant training camps, organizing transport, and raiding the existing Army divisions for officers and noncoms to train the new draftees. The chapter also talks about the creation of the first modern military intelligence units in the US Army, including the creation by Herbert Yardley of the first permanent cryptologic unit in the service.

2.1 America Stumbles into War

On April 6, 1917, the day war was declared, the United States was woefully unprepared to wage a modern war. Since the end of the American Civil War in 1865, the US Army had primarily done three things, police the South during Reconstruction, wage a one-sided war against the Native Americans with the ultimate purpose of moving them onto reservations, and patrol the US-Mexican border to prevent Mexican bandits and revolutionaries from crossing. The Army had fought one 3-month war against Spain in Cuba and the Philippines in 1898 and had put down an insurrection in the Philippines in 1899–1902. It had also invaded Mexico twice, once to occupy Veracruz over a fabricated insult in 1915, and a year later in the so-called Punitive Expedition, led by General John J. "Black Jack" Pershing in retaliation for an incursion into New Mexico by Mexican revolutionaries that led to several American deaths.

The Punitive Expedition, in particular, showed the many shortcomings of the existing American Army forces. In 1917 the Army, Navy, and Marines had fewer than 300 obsolete aircraft, none of them fit for combat. The Army had very few machine guns, half a dozen trucks, and few modern artillery pieces. The size of the regular US Army in April 1917 was about 128,000. The second leg of the US forces, the National Guard, had approximately 132,000 men in April of the same year, while the US Marine Corps had about 15,500 for a total combined force of around 275,000 (Ayres 1919, pp. 13–15). At that same time, the English had between 1.5 and 2 million men in the field, the French around 2 million, and the Germans nearly

J.F. Dooley, *Codes, Ciphers and Spies*, DOI 10.1007/978-3-319-29415-5_2

3 million on the Western Front. In April 1917 the Allies were in dire straits. Despite universal conscription the Allied Powers had suffered so many casualties in the nearly 3 years of conflict that they were having difficulty filling up their frontline forces with replacements. The long stretches in the trenches and the never-ending and fruitless offensives meant the troops were getting restless, desperate, and depressed. Later in 1917, a number of French and German regiments would mutiny and just refuse to fight. The Allied commanders pressured the Americans to send troops as soon as possible. The United States, despite a population of around 107 million, clearly could not field an army in the millions purely by volunteers. A new draft was needed.

The first selective service act since the Civil War was passed into law on May 19, 1917, and the first draft was held on June 5, 1917. After amendments, all men between the ages of 18 and 45 inclusive were eligible and had to register for the draft. During the 19 months of American participation in the war, 24,234,021 men registered for the draft, and 2,800,000 were inducted into the military (Ayres 1919, pp. 17–19). The draft was generally accepted in World War I (as opposed to the Civil War) because this time, the draft boards were local and were run by civilians, local citizens, not military men. There were also fewer exemptions, and men were not allowed to buy their way out of service or use substitutes, as was possible during the Civil War.

Training hundreds of thousands of new draftees took many of the noncommissioned officers away from the Regular Army. The noncoms were the only ones with sufficient experience to give the initial instruction to the draftees and to retrain the National Guard troops. Housing the new enlisted men was the first and most basic problem. A newly organized Cantonment Division began selecting sites and contractors to build brand new training camps for 16 National Army divisions and an equal number of National Guard divisions. In an extraordinary feat of organization and engineering, eight new training camps were two-thirds built and ready to house 400,000 draftees and National Guard soldiers by September 1917 (Hallas 2000, p. 22). Other supplies were also virtually nonexistent. “The flood of troops strained the supply of available equipment. A shortage of rifles saw some recruits drilling with wooden sticks. Hand grenades, machine guns, artillery pieces were all in short supply” (Hallas 2000, p. 24).

For tactical purposes, American forces were generally organized into divisions. An American division had roughly 27,000 enlisted men and 1,000 officers, about twice the size of the British and French divisions. To accommodate the expected number of draftees, the Regular Army would be expanded by volunteers and some draftees and would be divided into 20 divisions, numbered 1 to 20. The National Guard would provide 16 divisions, numbered from 26 to 42. The third leg, called the National Army, would be composed of an additional 18 divisions, numbered 76 to 93. Note that there are deliberate gaps in the numbering and some numbers are skipped. This was because the original War Department plan for the Army was to have 80 divisions in France by early 1919 and 100 divisions there by the end of the year. These distinctions (Regular Army, National Army, National Guard) were eliminated on August 7, 1918 in favor of a single, unified organization.

The first American forces, elements of the US 1st Division (later called the Big Red One), arrived in France in June 1917, preceded by General Pershing and his headquarters staff (Fig. 2.1). Initially, movement across the Atlantic was slow. The

Fig. 2.1 General John J. Pershing in 1917 (*Public Domain*. Reproduction Number: LC-DIG-ggbain-21134 (digital file from original negative); Rights Advisory: No known restrictions on publication.; Call Number: LC-B2- 3764-8 [P&P]; Repository: Library of Congress Prints and Photographs Division Washington, DC, USA http://hdl.loc.gov/loc.pnp/pp.print)

26th National Guard Division (the "Yankee" Division) sailed for France in September, as did the 42nd "Rainbow" Division. By March 1918, there were 318,000 American soldiers in France (Keegan 1999, p. 372). By April 1918, there were seven of the oversized American divisions in France, the equivalent of 14 British, French, or German divisions. The number would have been higher if not for the lack of transport ships to get the troops across the Atlantic. In February the British agreed to release a large number of transport ships for use by the Americans, and the number of American troops shipped to France accelerated. By June 1918 the number of American troops topped 650,000, and new troops were arriving at a rate of nearly 10,000 per day (Eisenhower 2001, p. 99). By August 1918, 1,300,000 men had arrived, and more were arriving at a rate of over 250,000 per month (Keegan 1999, p. 373). By November 1918, there were 42 American divisions in France, plus men in the Services of Supply bringing the total enlisted strength of the US ground forces to over 4 million, of whom 3 million were in France and the rest being trained in the United States.

As far as equipment was concerned, the United States was also deficient in 1917 and was forced to buy and borrow British and French arms, planes, tanks, and artillery until American industry ramped up in late 1918. By the end of 1918, the United States had 2,698 planes in service, of which 667, less than one-fourth, were of American manufacture (Ayres 1919, pp. 85–88). Of the almost 3,500 artillery pieces the American Expeditionary Force (AEF) used in France, only 477 were of American manufacture, and only 130 of those were used in combat (Ayres 1919, pp. 80–81). Despite possessing the world's largest automotive industry, the United States had to rely on French tanks for the operations of the AEF's Tank Corps, and in some instances British and French tank battalions supported US troops. 51,544 US-made trucks were sent to France, and another 50,000 or so trucks were purchased from the British and French. Of the American-made trucks, about 7600 were

"Liberty" trucks, the first standardized design for a military vehicle. All the Liberty trucks were designed by the War Department and had standardized parts and came in two sizes. The Standard A was a 1½ ton truck and the Standard B was either a 3 or 5 ton truck, both with a maximum speed of about 15 mph.[1] These trucks largely replaced horses, which were also in short supply, for short-haul work in getting supplies from railheads to the trenches; they were also used in the three American designated ports of St. Nazaire, Bordeaux, and La Pallice to move cargo from supply ships to the railheads (Ayres 1919, p. 62). American industry did a better job producing the infantry weapons. Barely 100,000 rifles were on hand for the Army's use when the war broke out; most of the American arms manufacturers were under contract to make arms for Britain and France. Two Army arsenals were producing the Model 1903 Springfield rifle and stepped up production during the war. Three private companies were producing the Lee-Enfield rifle for the British, and when they completed their contracts, they began turning out those rifles modified for the Springfield .30-06 cartridge, the standard American rifle ammunition. Since the Army had not purchased a large number of machine guns in the prewar period, the AEF was equipped almost exclusively with French Chauchat light machine guns until July 1918. Unfortunately for the Americans, the Chauchat was notoriously prone to misfiring because of dust and dirt and had a tendency to overheat easily. American industry, however, was able to come up with a new, more robust design for a light machine gun and by the end of the war had produced excellent weapons while ramping up production significantly. Beginning in July 1918, American units were being armed with Browning M1917 water-cooled machine guns and the soon to be famous M1918 Browning Automatic Rifle (BAR). The main advantages of the Brownings were their simplicity, and for the BAR, it's relatively lightweight. Both machine guns also used standard .30-06 Springfield rifle ammunition, greatly simplifying supply. By the end of the war, three American manufacturers had delivered 52,000 BARs (Ayres 1919, pp. 63–72).

2.2 The Americans Arrive

The American commander, Major General John J. Pershing, and his staff reached France in June 1917. The first American troops, the 16th Infantry Regiment of the US 1st Division, also arrived late in June 1917 (Fig. 2.2). The French were so desperate to see the Americans that a parade of freshly arrived American troops was organized through Paris on July 4, 1917. The parade and the 2nd Battalion, 16th Infantry, wound down the Champs-Élysées, over to Napoleon's tomb at Les Invalides and then across to the tomb of the Marquis de Lafayette in Picpus Cemetery. That is where a Lieutenant Colonel of the Quartermaster Corps, Charles E. Stanton, put the cap on the day with an impassioned "America has joined forces with the Allied Powers, and what we have of blood and treasure are yours. Therefore

[1] See https://en.wikipedia.org/wiki/Liberty_truck accessed on August 18, 2015.

Fig. 2.2 Soldiers of the American 1st Division arriving at the port of St. Nazaires, June 1917 (*Public Domain*. Used with permission of the American Battlefields Monuments Commission (ABMC) https://www.abmc.gov/sites/default/files/publications/AABEFINAL_Blue_Book.pdf)

it is that with loving pride we drape the colors in tribute of respect to this citizen of your great republic. And here and now, in the presence of the illustrious dead, we pledge our hearts and our honor in carrying this war to a successful issue. Lafayette, we are here!" (Coffman 1968, pp. 3–4).

The Allies were completely unimpressed with the American Army and thought that it would take too long for the Americans to organize and train themselves to become an effective modern fighting force. What the Allies wanted was basically untrained bodies that they could train, use as replacements for their own troops, and rush to the front lines to be cannon fodder (Coffman 1968, pp. 8–11; Eisenhower 2001, p. 12 and 16–17).

What the Americans wanted was a full partnership. Wilson and Pershing insisted on an independent American Army with its own section of the front and the ability to act independently or jointly with other Allied forces. What they eventually got was a compromise between these two positions, with an independent American Army having its own sector and also several American divisions integrated into British and French forces. By the late fall of 1917, four American divisions, two Regular Army divisions, the 1st (the *Big Red One*) and the 2nd, and two National Guard divisions, the 26th (*Yankee* Division) and the 42nd (*Rainbow* Division), were in France. Initial training for all divisions was in the United States and took 6 months, followed by advanced combat training in France by British and French instructors for an average of 2 months (Ayres 1919, p. 25). However, by spring 1918 the Allies were desperate for American help so training in the United States was shortened by about half to get soldiers over to France. This reduced training and particularly the lack of experience of the officer corps showed itself later in American behavior on the battlefield.

The first American troops, from the 1st Division (Regular Army), eased into the front lines east of Nancy in mid-October 1917. The 26th and 42nd Divisions followed the 1st in February 1918, and the 2nd Division moved up in March 1918 (Hallas 2000, p. 61). Typically Americans would take over from Allied troops in a "quiet" sector where there was little combat or movement for about a month before being moved into a more active sector of the front. The Americans did not have to wait long for the first casualties. The first American was wounded on October 23rd, and three American soldiers became the first killed in action during a German trench raid on November 3, 1917. Throughout the rest of the fall of 1917 and the winter of 1918, the Americans rotated in and out of the trenches and learned first the tedium and then the terror of being in the front lines. Throughout this period the Americans participated in trench raids and small actions in support of their French comrades. As the Americans began to be involved in more combat operations, the casualties started to mount (Hallas 2000, p. 69).

The relative calm of the winter of 1917–1918 came to a close quickly. The Russians had collapsed and in February 1917 the Czar abdicated. The Russian Provisional (Kerensky) government took over in March 1917 and declared that it would continue the war under its entente obligations. The Russian populace and, increasingly, the army were exhausted by the war and wanted nothing more than peace. The Germans allowed Vladimir Lenin to cross Germany from his exile in Switzerland in April 1917 in order to organize the workers' committees (Soviets) and work against the Provisional government.

On July 1, 1917, the Russians and Romanians launched a major offensive, known as the Kerensky Offensive, against the combined German-Austrian forces in the Western Ukraine. While the Austrians fell back, the Germans held and by July 23rd the Russian advance had collapsed. The Germans and Austrians then counterattacked and threw the Russians a further 250 km back into Russia. This disaster was the last effective Russian military action of the war. The Russian Army effectively disintegrated, and further German advances were only limited by their having overextended their lines of supply. The Kerensky Offensive was the last straw for most of the Russian populace, and anti-government riots began in major cities. The Bolsheviks took over Petrograd (St. Petersburg) and arrested the Provisional government leaders in early November 1917, putting in place their own soviet government. The Russians signed an Armistice on December 22nd and peace negotiations followed by the end of 1917. On March 3, 1918, the Russians and Germans signed the Treaty of Brest-Litovsk with the Russians basically capitulating to all the German demands. This freed up nearly 50 German divisions for action on the Western Front, and the Germans immediately began moving troops west[2] (Keegan 1999, pp. 375–381).

[2] See also https://en.wikipedia.org/wiki/Treaty_of_Brest-Litovsk. Accessed 18 August 2015.

2.3 American Military Intelligence Awakens

When it came to military intelligence, once again the United States was completely unprepared. At the beginning of the American entry into the war, there were exactly three Army officers trained in codes and ciphers—Parker Hitt, Frank Moorman, and Joseph Mauborgne—and, in a brilliant bit of Army wisdom, none of them were ultimately assigned to cryptographic duties (Gilbert 2012, p. 44). So in April 1917, the US Army military intelligence organization was once again starting from scratch.

Throughout the 140 years or so of the existence of the US Army prior to World War I, military intelligence had normally been a series of isolated and temporary stories of individual commanders who saw the need for knowing and understanding what the enemy was doing before battle. Beginning with George Washington commanders would create small organizations to manage communications, recruit spies, and infiltrate enemy lines to scout and gather intelligence. Amateurs who had no previous knowledge or experience with information gathering and intelligence analysis typically staffed these intelligence units. During the Revolutionary War, the War of 1812, the Mexican War, and the American Civil War, intelligence organizations would come into being and prove useful for the duration of the conflict. Once the war was over and most of the Army was demobilized, these organizations were disbanded. In 1860, just before the American Civil War, the War Department created the US Signal Corps under Major Albert J. Meyer to manage army communications. The Signal Corps was the first intelligence organization to remain in existence after the end of conflict. The US Army's Military Information Division (MID) was created in 1885 with its original function to acquire information about the armies of foreign nations. This led to the formation of the military attaché service in 1889. Military attachés were dispatched to foreign capitals, attached to the US Embassy there, and would report back to MID on military readiness and effectiveness of the host country's army. MID was charged with analyzing this data and reporting it back to the War Department. In 1903, the United States finally adopted the General Staff organization that all the other major powers had used for decades. Under this reorganization, MID was moved into the 2nd Division of the General Staff, known as G-2. This didn't last long, however. By 1908 the 2nd Division was merged into the Third Division—later called the War College Division—and MID lost its clout and its separate identity. So at the beginning of World War I in 1914, the United States, once again, had no real intelligence organization (Gilbert 2012, pp. 1–5). This state of affairs was finally rectified on May 3, 1917—nearly a month after war had been declared—when the War College Division created the Military Intelligence Section (MIS), and in June 1918 the name was changed to the Military Intelligence Division (MID) under Major Ralph Van Deman (Gilbert 2012, pp. 28–29). Van Deman, a Harvard graduate who also had both law and medical degrees, had been in the Army since 1891. He was in the first class of the Army War College in 1904 and had organized and run the Military Intelligence Division in the Philippines during the Philippine-American War (1899–1902). He was the ideal person to run the new MIS and would become known as the "Father of American Military Intelligence" (Gilbert 2012, pp. 11–13).

Van Deman lost no time in organizing the MID. Beginning with just a couple of enlisted soldiers and some civilians, Van Deman had, by the end of 1917, an organization of several hundred soldiers and civilians and a budget of over $1 million, modeled on the British military intelligence organization. By that time MID had five subsections: MI-1 Administration (Personnel and Office Management), MI-2 Collection and Dissemination of Foreign Intelligence, MI-3 Counterespionage (Military), MI-4 Counterespionage (Civilian), and *MI-8 Cable and Telegraph (Code and Cipher Section)*. By the end of the war, seven more sections had been added: MI-5 Military Attaches, MI-6 Translation, MI-7 Graphics (Maps), MI-9 Field Intelligence, MI-10 Censorship, MI-11 Passport and Port Control, and MI-12 Graft and Fraud (Gilbert 2012, p. 223).

On the AEF side, General Pershing also set up his staff organization along the lines of the British and French. As such, he had an Intelligence Section (G-2) headed by Major Dennis E. Nolan, a friend and contemporary of Van Deman's but with—at the time—much less experience in intelligence. Early on Van Deman and Nolan agreed that the AEF G-2 organization and MID would remain completely separate but would share as much information as possible. MID would also be charged with training many of the intelligence officers who would become part of the AEF. MID would concentrate on counterintelligence and domestic US matters, while AEF G-2 would focus on intelligence gathering and counterintelligence in France. Over the course of the war, the two organizations would work very closely together to the point of even sharing personnel (Gilbert 2012, pp. 31–33).

In addition, while the Navy had its own Office of Naval Intelligence, its job was mainly to observe other countries building programs and fleet maneuvers. For the duration of the war, the Navy and State Departments depended on the War Department's Military Intelligence Division (MID) to handle all interception and cryptanalysis of enemy cryptograms, including those in shorthand and in invisible ink.

As soon as the MID was set up, Van Deman began getting requests for decryptions of intercepted cablegrams, letters, and notes found on arrested aliens and Americans. Van Deman needed a cryptologic organization. Enter Herbert Yardley.

2.4 Herbert Yardley and MI-8

Herbert O. Yardley (1889–1958) was a mid-Westerner who had come to Washington in 1912 and was engaged as a code clerk at the State Department. Yardley was smart and ambitious and bored on the night shift at State. To while away the time, he started teaching himself the State Department diplomatic codes and ciphers and set himself the problem of decrypting random cablegrams that passed across his desk. Yardley quickly discovered two things. First, he was pretty adept at decrypting State Department-coded cablegrams, and second, the State Department codes and ciphers weren't very complicated and hence not very secure. His coup was when he was able to decrypt messages sent in the private cipher system used by President Wilson and his close friend and presidential advisor Colonel Edward House. At this point, in early 1917, Yardley wrote up all his notes and conclusions about the State Department systems into a 100-page memorandum and handed the memo to his boss (Yardley 1931, pp. 21–27). While his boss wasn't exactly pleased, he did understand that Yardley had talent and some good ideas. When the United States entered the war, Yardley set himself the goal of working on cryptograms for the War Department. Little did he know that at that time, there wasn't even one person in the War Department solving enemy cryptograms. By May 1917 Yardley had worked his way up to Major Van Deman's office and made his pitch to set up a cryptanalytic bureau within MIS. Not having any other better choices (in fact, having no other choices at all), Van Deman took Yardley up on his offer, commissioned him a First Lieutenant, and set him up in charge of a new subsection of MID, MI-8 the Code and Cipher Section (Yardley 1931, pp. 34–36).

Yardley got to work setting up MI-8 starting in June 1917. He started with two clerks and an avalanche of coded messages from the Army, the Navy, and the State Department. Yardley's first job was to find people in or out of the Army who had what he called "cipher brains," that peculiar twist of mind that allowed someone to see deeply complex patterns in encrypted messages and unravel them. This was not easy at first. Yardley's real coup was to hire Dr. John Matthews Manly, head of the English Department at the University of Chicago and an amateur cryptologist. Manly was commissioned a Captain and joined MI-8 in Washington in October 1917. It turned out Manly was a terrific cryptanalyst and a good organizer so Yardley placed him in charge of the cryptanalytic section of MI-8 and made Manly his second-in-command (Yardley 1931, pp. 38–39). Manly also brought along several of his colleagues from the University of Chicago, including Dr. Edith Rickert who would also prove to be an excellent cryptanalyst. She and Manly would together solve one of the most important cryptograms that MI-8 would see during the war, the Waberski cipher message. Eventually there would be six sections within MI-8: *code and cipher solution* (headed by Manly), *code and cipher creation*, *training*, *shorthand*, *communications*, and the *secret ink laboratories* (Yardley 1931, p. 47).

Early on, one of MI-8's most essential jobs was the training of cryptanalysts for both MI-8 and the AEF, but in fact there was no training program, no curriculum or training materials, and no one to do the training. MI-8 was in quite a fix. Van Deman and Yardley were saved by an offer from a civilian, George Fabyan, who was a wealthy textile businessman interested in making a name for himself. Several years earlier he had set up what was the country's first privately funded research laboratory at his Riverbank estate in Geneva, Illinois. One of the research areas that his staff worked on at the Riverbank Laboratories was cryptography. This was primarily because Fabyan was interested in proving that Francis Bacon had written the plays of Shakespeare, and he was convinced that there was a cipher in Shakespeare's *First Folio* that would prove it (Munson 2013, pp. 93–94). So Fabyan had several people on his staff that could solve cryptograms and create the training curriculum that the Army needed. Among these were a young couple, William F. and Elizebeth Smith

Fig. 2.3 Lt. William F. Friedman in 1918 (Used with permission of the George Marshall Foundation Research Library)

Friedman, who were destined to become the most famous pair of cryptologists in American history. So, in the fall of 1917, the Army contracted with Fabyan to solve cryptograms that would be sent to Riverbank and to set up a cryptologic training program at Riverbank. William Friedman designed the curriculum and taught most of the courses. Between November 1917 and March 1918, the Riverbank school trained 78 Army officers, the majority of whom joined the AEF in France. In March 1918 MI-8 took over training and moved the cryptology school to Washington (Barker 1979, pp. 3–8). Friedman was then free to join the Army; he was commissioned as a First Lieutenant in June 1918 and was assigned to the Radio Intelligence Section of Military Intelligence in the AEF, G2-A6, under Major Frank Moorman. Friedman would rise to head the cryptanalytic Code Solution section of G2-A6 for the remainder of the war. He was demobilized in the spring of 1919 (Clark 1977) (Fig. 2.3).

So by early 1918 both the Radio Intelligence Section of the AEF and the Code and Cipher Section of MID were set up, running, and cooperating. There were seven US Army divisions in France in various stages of training, four of them already moving into the frontline trenches. Pershing's headquarters at Chaumont was set up, and the Army Services of Supply was in the process of upgrading French ports, building hundreds of miles of railroad lines, and unloading tons of supplies each day. By the early spring of 1918, the AEF was in the line and engaging in independent combat operations.

References

Ayres, Leonard P. 1919. *The War With Germany: A Statistical Summary*. EBook. Washington, DC: Government Printing Office. https://archive.org/details/warwithgermanyst00ayreuoft.

Barker, Wayne G. 1979. *The History of Codes and Ciphers in the United States During World War I*. Vol. 21. Laguna Beach, CA: Aegean Park Press.

Clark, Ronald. 1977. *The Man Who Broke Purple*. Boston: Little, Brown and Company.

Coffman, Edward M. 1968. *The War To End All Wars: The American Military Experience in World War I*. New York: Oxford University Press.

Eisenhower, John S. D. 2001. *Yanks: The Epic Story of the American Army in World War I*. New York: The Free Press.

Gilbert, James L. 2012. *World War I and the Origins of U.S. Military Intelligence*. Lanham, MD: Scarecrow Press, Inc.

Hallas, James H. 2000. *Doughboy War: The American Expeditionary Force in World War I*. Boulder, CO: Lynne Rienner Publishers.

Keegan, John. 1999. *The First World War*. New York: Alfred A. Knopf.

Munson, Richard. 2013. *George Fabyan*. North Charleston, SC: CreateSpace Independent Publishing Platform.

Yardley, Herbert O. 1931. *The American Black Chamber*. Indianapolis, IN: Bobbs-Merrill.

Chapter 3
Overview of Cryptology and the Army

John Matthews Manly

Abstract This first, unnumbered, Manly article gives an overview of the "ears" of the armed services—the wireless interception services and the cipher bureaus. Manly discusses how wireless telegraphy transformed the communications of armies and the subsequent increase in the necessity of message secrecy and hence cryptology. Manly then gives a short history of codes and ciphers and motivates their use during the Great War. Untitled in the Friedman Collection, this article seems to belong at the beginning of the Manly sequence.

Everybody knows that during the World War, airplanes were the "eyes of the army.[1]" We have been told a thousand times how observers from both armies went out daily in high-speed airplanes to note the disposition of troops, the condition of the battle line, the location of camps and supply stations, points of strength and weakness, and evidences of intended attacks. Such information as these intrepid scouts brought back was of incalculable value, for men and material equipment are not more necessary for success in war than knowledge of the strength and plans of the enemy.

With the invention of these new eyes arose the need of interfering with their vision, and this expressed itself, as is well known, partly in attempts to put out the eyes and partly in devices for covering or distorting the things these eyes were sent to observe. Faster airplanes designed and armed for attack were the means of putting out the eyes. Smoke screens and all the thousand and one methods of camouflage were the means adopted to conceal and to distort. This whole story is too well known to need retelling, but no one, so far as I know, has told the story of the ears of the army or even asked what they were.

What were these "ears"? What new means of overhearing the enemy's talk and his plans were devised? And what were their effectiveness and use? The story of the developments in this field is perhaps as varied and picturesque and important as the story of the new eyes, and some day it will be told in full. In this series of articles, I purpose to tell only a small part of it but the part which perhaps is least well known.

[1] See http://query.nytimes.com/mem/archive-free/pdf?res=9B03E3DC1F3BE03ABC4B52DFB2668382609EDE retrieved June 24, 2014, and http://www.century-of-flight.net/Aviation%20history/airplane%20at%20war/Aerial%20Reconnaissance%20in%20World%20War%20I.htm retrieved June 24, 2014.

J.F. Dooley, *Codes, Ciphers and Spies*, DOI 10.1007/978-3-319-29415-5_3

Before we can answer the question, "What were the ears of the army?" we must answer a previous question "What was the voice?" or rather, "What were the voices?" for there were several of them. Some were old and familiar; others entirely new.

In the older days when a commanding officer wished to give orders to his men, he spoke to them face to face. Later, as fighting bodies became larger, he called a conference, disclosed his plans, and distributed his general orders. Where lack of time or other conditions made personal conference impossible, messengers had to be sent bearing oral or written messages; and written messages, it is well known, were carried not only by men but by dogs and carrier pigeons.

All these methods of communication still remain useful and necessary, but modern science, which has transformed the world, has perhaps nowhere made more radical changes in life than in the means and forms of communication in war. The four great inventions concerned in these changes are, of course, the telegraph, the telephone, the airplane, and the wireless telegraphy. Whether the wireless telephone can ever be of great service in war can well be doubted, but all the other great modern forms of communication played important parts in the Great War. Messages were sent by thousands over the telegraph and telephone wires which were laid with unexampled rapidity along the battle lines of the opposed armies; and speedy airplanes carried oral or written messages to points that had not been reached by these wires, and airplanes were called into service when for any reason the other means of communication failed or were insecure.

None of these methods, however, or perhaps all of them taken together, played so picturesque and important a part in the talk of the great armies as wireless telegraphy, for its voice was the loudest and most penetrating. Messages uttered in Berlin were received at Constantinople, at the remotest stations in the Black Sea districts, and at stations on the northern coast of Africa, and the answers could be plainly heard and understood. And down near the battlefronts were receiving and sending stations which kept the officers of the line in constant communication with the military and political chiefs seated at a great distance from them but able by this means to speak to them instantly and in spite of intervening obstacles.

Every new invention in war brings its counter-invention. To new guns and ammunition are opposed new methods of fortification and improvements in armor plate. To every move there is a countermove. In the old days military men attempted to hear what was said back of the enemy's lines with the aid of scouts and spies and the capture of messengers or messages. For the more modern means of communication, new methods of listening had to be devised.

First, perhaps, came the tapping of telegraph wires or, if this was impossible, the cutting of them to interrupt the talk. This was done in our own Civil War. The Intelligence Corps[2] of Generals Grant, Sherman, and McClellan and that of Generals

[2]For the Intelligence Corps of the two sides in the American Civil War, see for the North the Bureau of Military Information (BMI) for espionage and counterespionage and the US Military Telegraph Department (USMT) for military telegraphy and cryptography and cryptanalysis and for the South the Confederate Secret Service for espionage and counterespionage, the Confederate

Lee, Jackson, and Johnson each tried to attach wires and telegraph instruments to the wires carrying the messages of the opposed army, just as they attempted to intercept written orders and capture scouts and messengers.

With improvements in electrical science came the art of tapping telegraph and telephone wires not by actual contact but by induction. It was found that when a wire carrying telegrams or telephonic messages of the enemy could not be reached, the messages could, nevertheless, be obtained by laying down a wire in parallel in which by induction the electrical contacts made and broken in the other wire were exactly repeated. The currents set up by induction in the parallel wire were feeble, but electrical experts knew how to reinforce it and make it strong enough to operate a telegraph instrument or a telephone receiver and thus repeat the message carried by the other wire. This method of listening-in played a great part in the war and was of untold value as a source of information.[3]

With wireless telegraphy the situation was entirely different. In itself the wireless offered no guarantee or even possibility of privacy. The air was full of its voices. Messages were flying between regiments and battalions and army corps and their headquarters and commanders. Artillery batteries were receiving reports from meteorological stations on the conditions affecting the flight of shells and from airplane observers on the accuracy of their shots. Berlin was talking to army headquarters in France, in Bulgaria, and in Tiflis,[4] and to political agents in Tripoli and other distant lands. The allied stations at Chaumont, Souilly, Neufchatel,[5] and other places could hear all this and by tuning in properly could pick out from the confusion the particular sending station they wished to hear and record. Messages sent out from the great wireless station at Nauen[6] in Germany and intended for agents in

---

Army Signal Corps for cryptography and cryptanalysis, and the Confederate Military Telegraph for military telegraphy.

[3] In an analog telephone wire, a direct current carries the conversation down the wire. A magnetic field is generated around the wire at a 90-degree angle to the direction of the current. Placing an induction coil next to the telephone wire and within the magnetic field will induce a current in the second wire, reproducing the original signal without requiring any physical connection to the original telephone wire. The device as used during World War I depends on the original telephone wire being a single-wire system, using the ground itself as the return "wire." It is also very limited in range as the magnetic field drops off rapidly with distance from the wire.

[4] Tiflis is the modern city of Tbilisi, the capital of the Republic of Georgia.

[5] The American Expeditionary Force set up a number of wireless intercept stations beginning in October 1917. These included stations at AEF headquarters in Chaumont, Souilly, Neufchatel, and Gondrecourt, south of St. Mihiel. These stations, designated the Radio Intelligence Section under the Signal Corps, and run by the Second Field Signal Battalion, forwarded their German intercepts to the Radio Intelligence Service under Colonel Frank Moorman at Chaumont.

[6] The Nauen radio transmitting station is located about 23 miles west of Berlin and 16 miles northwest of Potsdam. It was the most powerful transmitter in Germany during World War I. Opened in 1906, the Nauen station was a research station of Telefunken, one of the first German radio companies. Georg Graf von Arco, the first chief technical officer of Telefunken, originally designed it. In 1913 the first high-power machine transmitter was installed in the station. This Alexanderson alternator transmitter enabled the station to operate at a very high power level and transmit continuous waves at enormous distances (more than 5000 km) using ultra-long wavelength amplitude

Tripoli were even caught and reported to Washington by our wireless intercept station in the forests of Maine.

Such a situation had never existed before in the history of the world. Two great armies were confronting each other in the field of battle, and each was giving the greater part of its most important messages and orders in such a way that its enemy could easily overhear them. Each was listening intently to everything that was being said, and each knew that the other was listening in the same way. But secrecy was no less important in this war than it had been in all that preceded it, nor were the precautions to insure secrecy taken with less constant and vigilant care.

Knowing that wireless telegraphy must be one of its chief means of communication and knowing also that the enemy could easily intercept all its wireless messages, each of the opposing powers from the very beginning of the war sets its best brains to devising means for preventing its intercepted messages from being read. To prevent them from being received and recorded was clearly impossible. The only means of rendering them intelligible to the persons for whom they were intended and unintelligible to the rest of the world was to use some form of code or cipher.

Simple codes and ciphers had been used in remote antiquity. They flourished to some extent during the Middle Ages and received a very considerable development in the wars and international intrigues that marked the sixteenth and seventeenth centuries. At the beginning of what is sometimes called the Modern Age, the experts of the Papal Chancery and those of the Republic of Venice were especially distinguished for their skill in cipher. But the knowledge and use of code and cipher spread rapidly. Henry VIII and Queen Elizabeth received confidential reports from their diplomatic agents in cipher. The ill-fated Mary Queen of Scots used cipher for communicating with those who were working and plotting to release her from imprisonment and accomplish her ambition to be Queen of England.[7] The instructions of Louvois,[8] a minister of Louis XIV, to his generals were commonly expressed in cipher. Napoleon and his generals used cipher and code, as did the patriots of the American Revolution.[9] In short military men and diplomats in every country of the world used secret writing. Notwithstanding all this, there was comparatively little development in the cryptographical arts between the sixteenth century and the

---

modulation. During World War I the Nauen station was operated by the German Navy and was the primary means of communication with the U-boat fleet.

[7] This was the famous Babington plot of 1586, organized by the Catholic Anthony Babington, which was uncovered by Elizabeth I's principal secretary and chief spymaster Sir Francis Walsingham. The plot determined to assassinate Elizabeth and put Mary on the throne of England. Walsingham had a double agent in Mary's retinue and had obtained a copy of the cipher system that Mary and the conspirators were using. When Mary dispatched a cipher message approving the details of the plot to assassinate Elizabeth, Walsingham sprung his trap; all the conspirators were caught, and Mary was convicted of treason and beheaded on February 8, 1587.

[8] Louvois was Louis XIV's Minister of State for War from 1666 to 1691. He is credited with increasing the size and reorganizing the French Army.

[9] Among the cipher systems used during the American Revolution were a homophonic substitution cipher designed by Charles Dumas and used by the Americans and a grille type cipher that used a cutout hourglass to reveal the secret message used by the British (Wilcox 2012, p. 14).

twentieth. One, and perhaps only one, great discovery in methods of solving unknown ciphers was made about the middle of the nineteenth century,[10] but apart from this the experts of the sixteenth century devised some systems that would have puzzled the experts of the nineteenth century, and some of them could have solved with all ease the best systems devised by their later brothers of the craft.

The reasons for this lack of development, in the first place, were the comparatively small number of cipher messages passing at any one time in earlier days and available for attack by enemy experts bent upon solving them and, in the second place, as a consequence of this, the small number of persons who had become expert in the subject. When armies were small and their commanders were in close contact with their men, there was small need of cipher messages, and only a few such messages would be sent in such a way that they could be intercepted by the enemy and attacked by its experts. It was a comparatively rare thing, therefore, that a cipher expert had an opportunity to attempt to decipher the messages of the enemy, and as a consequence, he usually had little material to work on. With the coming of wireless telegraphy in the World War, all this was changed. It was easy to intercept hundreds of messages and to record them in as accurate a form as that in which they would reach the persons for whom they were intended.

Now, it is a curious fact with regard to the clear forms of cipher that the more material there is available to the expert for study, the easier is it to solve and read the cipher. This seems not to be generally understood by the public, who think that the shorter a cipher is the more quickly and easily it ought to be read. As a matter of fact, the simplest cipher, if short enough, may be entirely unreadable by one who does not know the system or possess the key, whereas a message many times that length might be quickly and easily read by even a novice.[11] Of the older forms of cipher, there was probably none that could have been used with any degree of security where so much material was available for its solution as was furnished by the wireless operations in the war. As a matter of fact, this was not clearly understood at first, and both sides used systems that were easily read by their opponents. Something obviously had to be done. New systems had to be devised of such a nature that they could be easily and accurately written and read by those properly

---

[10] Manly is no doubt referring to the discovery of a general solution to Vigenère ciphers developed independently by Charles Babbage around 1854 and by Friedrich Kasiski in 1863. The Babbage-Kasiski method makes use of the fact that with a sufficiently long plaintext message, a Vigenère ciphertext will, with high probability, repeat a part of the key with an identical part of the plaintext and that this repetition will be a multiple of the key length. For example, say that the key is "barking" and at one point in the plaintext the word "regiment" lines up with the beginning of the key. If the word "regiment" appears again further down the message and if it once again lines up with the beginning of the key, then it must encrypt to the same sequence of ciphertext letters, and it must also occur at some multiple of the key length. Once the cryptanalyst can guess the key length, they can then separate the ciphertext into a set of monoalphabetic substitution ciphertexts and solve each individually.

[11] This fact is formalized in Claude Shannon's 1949 paper *Communication Theory of Secrecy Systems* as the *unicity distance* (Shannon 1949). It is the minimum length that a ciphertext must have in a particular cipher system in order to be uniquely decrypted. If a ciphertext is shorter than the unicity distance, it will admit multiple possible solutions.

instructed in their use but at the same time would offer the greatest obstacles to decipherment by any person for whom they were not intended.

This task was one of no small difficulty, for it was quite certain that by capture in a sudden raid or the infidelity of a trusted officer, any system that could be devised would soon be known to the enemy. A further requirement of any practical system, therefore, must be that messages written in it could not be read by a person familiar with the system unless he knew the particular way in which the system was used in these special messages.[12] This requirement soon led to throwing into the discard all the old systems and devising new ones, simpler, perhaps, in appearance but more effective in reality and better adapted to resist solution even under the enormously difficult conditions of modern times.

This work led naturally to subtler and more effective methods of decipherment by the experts on both sides that were attempting to break into the cipher system of the enemy and read their messages.

So, the Signal Corps with its stations for the intercepting of wireless communications was, in a certain sense, the ears of the army. They could and did intercept and record all the enemy's wireless messages, but having them so, they were much in the position of a person in a foreign land who hears people talking all about him but can understand no word of what they say. Cipher and code messages are to all intents and purposes an unknown form of language that must be mastered to be understood. The task of listening was the task of the Signal Corps, but the task of understanding what was said devolved upon the Code and Cipher Section, and it may, therefore, fairly be said that in the real and true meaning of the impression the code and cipher experts were "the ears of the army."

References

Shannon, Claude. 1949. "Communication Theory of Secrecy Systems." *Bell System Technical Journal* 28 (4): 656–715.

Wilcox, Jennifer. 2012. *Revolutionary Secrets: Cryptology in the American Revolution*. Ft. Meade, MD: Center for Cryptologic History, National Security Agency.

[12] This is known as *Kerckhoffs's principle*, named after Auguste Kerckhoffs, a Dutch-French cryptographer who proposed it in 1883. Simply stated, the principle is "assume the enemy knows the cipher system." This implies that the entire security of the system rests in the key. So keeping the keys to secure a system is the only way to protect the system from compromise. This leads to the classic problem in symmetric cipher system security—the *key management problem*.

Chapter 4
The AEF and Colonel Moorman

John Matthews Manly

Abstract The next three Manly articles are all about the cryptographic section of the American Expeditionary Force (AEF) in France during the war—G2-A6. This first article (Article I) contains several anecdotes about the AEF. It mentions Major (later Colonel) Frank Moorman as the head of the cryptographic section in France. The article also defines and describes the differences between codes and ciphers and gives some examples.

It was a steaming day in the summer of 1917[1] at G.H.Q. An officer who had come over with General Pershing was helping a newly arrived officer of the General Staff to find his quarters in the vast wooden structure, which had been hastily erected for the home of the head and heart of the American Army in France. The newly arrived officer was Lieutenant-Colonel (Frank) Moorman, who had been assigned to organize G2-A6, the Code and Cipher Section of G2, the Military Intelligence Division of the General Staff, and AEF.[2]

They entered a large room piled almost to the ceiling with bundles of typewritten sheets.

"Here," said the first officer, "are the wireless messages of the Germans intercepted by the Signal Corps. You are the code and cipher man. Let us hear what is in them as soon as you can."

"Of course," he added hastily, seeing the amused and bewildered look on the face of Colonel Moorman, "we do not expect you to read them all at once. Just pick out the most important ones and let us have what they say" (Fig. 4.1).

This incident illustrates the ideas and feelings of the average fighting officer on the American front toward the code and cipher work. Colonel Moorman, to whom I am indebted for this incident, says that at first our officers had been very skeptical about the value of the code and cipher men and the trustworthiness of the information which they derived, or pretended to derive, from reading the messages of the enemy, just as the British officers had been skeptical on the same points in 1914.

[1] Manly's original date is 1918, but this is incorrect. Moorman was one of the first staff officers to arrive in France after the declaration of war in April 1917. He was reassigned from Coastal Artillery to G2-A6 on July 28, 1917. General order No. 8 of July 5, 1917, established the AEF General Headquarters staff, including the cryptographic unit G2-A6.

[2] In July 1917, Frank Moorman was a Major in the Regular Army.

J.F. Dooley, *Codes, Ciphers and Spies*, DOI 10.1007/978-3-319-29415-5_4

Fig. 4.1 Lt. Colonel Frank Moorman, 1918 (Used with permission of the George Marshall Foundation Research Library)

Fortunately for the code and cipher work of AEF, the first message giving definite information of an attack that was deciphered by G2-A6 was sent up to headquarters in time to make preparations to meet it. Skepticism of the value of code work could not last long after the decipherment of such messages as the following (here given in translation):

> After artillery preparations at 6 A.M. Extend attack in cooperation with 7th Army Corps.
>
> 14th Infantry Division. Corps Headquarters wishes wire entanglements placed in front line at once. Corps Headquarters 7.
>
> Schmettow.[3] Position on right flank 52, 14D, reestablished. Enemies yielding over south slope of heights.
>
> Corps Headquarters. Enemy is retreating. The division is advancing on the whole front.

The great importance attached by the Germans themselves to code and cipher work is indicated by certain wireless cipher messages intercepted between the Political Section of the General Staff and German agents.

> Very secret. Cessation of hostilities between Turkey and Entents (sic) is imminent and consequent discontinuance of U-boat communication with Tripoli is in immediate prospect.[4] The entire German personnel of the Mirr Expedition, including Newmann and Wali are to ship at once on U-boat C-73 and the following boat U-73 for return to Germany. Continuation follows.
>
> 32471. Make full use of U-boat C-73 for transportation of German personnel. Whether wireless station should be turned over to the Turks or destroyed is left to your judgment. Bring with you cash, important documents and easily transportable articles of value. Untransportable documents and propaganda material are to be destroyed.
>
> In connection with telegram 32471 of October 30: On departure from Tripoli destroy all codes and ciphers.

[3]A German army corps, designated *Cavalry Corps Schmettow*, was created in August 1916 and participated in the Romanian Campaign under the command of Generalleutnant Eberhard Graf von Schmettow. It was later designated as 65th Corps in January 1917 and ended the war on the Western Front.

[4]The Ottoman Empire and the Allies signed the Armistice of Mudros on October 30, 1918. All hostilities in the Middle Eastern theater ceased at noon the following day. The Allies began occupying Constantinople on November 13, 1918, and this effectively began the partitioning of the Ottoman Empire.

The order for the destruction of cipher and code material is given in greater detail in the following cipher message, officially known as PQR:

> Please carefully and immediately burn without remainder and destroy the ashes of, all papers connected with the war, the preservation of which is not absolutely necessary, especially papers now in your hands or reaching you hereafter which have to do with the Secret Service and the service of the representatives of our General Staff and Admiralty Staff (strictest silence concerning the existence and activity of these representatives is to be observed now and for all future time, even after the conclusion of peace) which might be compromising or even unpleasant for us if they came to the knowledge of our enemies, who are still endeavoring to obtain possession of such papers.
>
> Lists, registers, accounts, receipts, account-books, etc., are especially included in these papers, as well as correspondence with this office by letter and telegraph on the subjects mentioned.
>
> Cipher-books, codes and cipher-keys and directions that are still in use are excepted for the present, and most particular attention must be paid to keeping them in absolute safety.
>
> Please report in writing en claire the execution of this order so far as it relates to papers now on hand and then burn this so-called order for burning, which, for further reference, I herewith designate as PQR, and the contents of which together with this designation you will please retain in memory.

The activities of some of the most famous German agents in the Western world will be told in later articles.

The British had learned the lesson of the trustworthiness and value of code and cipher work in connection with the great German withdrawal on the Flanders front in 1916.[5] The British code men had informed them that the withdrawal was to occur, but the higher officers refused to believe until they woke up one morning and found the Germans gone. After this both the British and French believed thoroughly in the value of code decipherment and tried to get for this work the best analytical brains that could be found.

One of the most brilliant demonstrations of the tactical value of the information often carried in cipher messages was given during the first 2 days of June 1918, when certain messages deciphered by Captain Painvin of the French Cipher Bureau warned the Allies of an impending attack north of Montdidier and enabled them to thwart it.[6]

"The success of the British and French had been so great," says Colonel Moorman, "that the officers of the A.E.F. who, of course, were kept thoroughly informed of what the other Allies were doing, came to believe that the solution of code and cipher messages was accomplished by a kind of magic and took no time or labor."

---

[5] Manly is probably talking about the phased German withdrawal along the northern and central Western Front beginning February 21, 1917, back to the newly created Hindenburg line. The German withdrawal to their new defensive positions was completed before the British offensive at Arras on April 9, 1917, and their general offensive in Flanders from June 7 to November 10, 1917.

[6] This is likely the Noyon-Montdidier offensive launched by the Germans on June 9, 1918, as part of their overall series of spring offensives. The message Painvin solved was in the ADFGVX cipher system. Painvin had partially solved the ADFGX system over the course of March through April 1918. The ADFGVX system was introduced on June 1, 1918, just before the German offensive. See Article III for more details on Painvin's solution of ADFGVX.

As a matter of fact, the art of code and cipher construction and the equally important art of code and cipher attack were—like so many other arts and sciences connected with war work—in a very elementary state at the beginning of the war and developed with phenomenal rapidity. When the advance guard of the American Army arrived in France, both the art of constructing ciphers and that of attacking them had attained a subtlety and effectiveness unexampled in the history of the world.

It had long been known to American cipher experts like Colonel Hitt, Colonel Moorman, and Colonel Mauborgne[7] that the British Army had used a cipher system known as the Playfair system. Although all three of these officers had solved problems in this cipher, it had long been regarded by the British as impossible of solution by enemy attack. Colonel Moorman, therefore, asked one of the first British cipher experts with whom he came into contact whether the British Army was still using the Playfair system.

"No," said the expert, "we found a method to solve it in 30 min.[8]"

This collapse of an impossibility into a job for 30 min measures the development that had taken place in code and cipher work between the summer of 1914 and that of 1917.

For a time, at the beginning of the war, the art of attacking ciphers far outstripped that of constructing them. The result of this was that on the Western Front, the Germans used cipher for only a few months, after which they abandoned it entirely for the use of trench and field codes.[9] And when the Americans arrived in France, German communications on the Western Front were carried on entirely in these codes. Cipher was not used again until the great offensive in March 1918, as will be told later on.

A word or two explaining the difference between code and cipher may be necessary. Laymen use the terms interchangeably and indiscriminately, but to the expert they mean entirely different systems of secret writing. To put it briefly, a code is an artificial language made up entirely of groups of letters or figures expressing in general either single words or phrases and sentences. For convenience in spelling out proper names and other words not provided for in the code vocabulary, certain

[7] At America's entry into World War I, Parker Hitt, Frank Moorman, and Joseph Mauborgne were the three Regular Army officers with the most experience in cryptology and particularly in cryptanalysis. In fact, they were practically the only officers with any experience in cryptanalysis in the Army. All three had been trained (and been instructors) at the Signal Corps School at Fort Leavenworth, KS. The standard US Army cryptanalysis text, *Manual for the Solution of Military Ciphers*, was written by Parker Hitt (1916). Mauborgne perfected Gilbert Vernam's one-time pad cryptographic system into a truly unbreakable cipher. He would go on to become the Chief Signal Officer of the US Army from 1937 to 1941.

[8] Major (then Lieutenant) Joseph Mauborgne wrote a pamphlet laying out the general solution to the Playfair cipher in 1914 while a student at the Army Signals School in Fort Leavenworth, KS (Mauborgne 1914).

[9] This is a mistake. The Germans used first double transposition ciphers and later polyalphabetic ciphers throughout 1914–1916. The Germans did not begin using trench codes until early 1917 (Kahn 1967, pp. 314–316).

groups are also set aside to indicate single letters and syllables or groups of letters. Cipher systems are, on the other hand, special methods of disguising the forms of an existing natural language. A substitution cipher is merely a new way of spelling an already existing language. New values agreed upon between the correspondents are temporarily assigned to the characters—letters, figures, or other symbols—used in re-spelling the words of the language chosen as the basis for communication. A transposition cipher disguises the normal spelling by a systematic disarrangement of the order of the normal letters.

The following examples will show how the same message may be encoded in a tri-numeral code and enciphered by three different methods of encipherment:

Code example

Code groups	Meaning
028	After
135	Artillery preparation
160	At
742	6 AM
391	Extend
163	Attack
236	Cooperation with
885	7th Army Corps

The message would be sent in the following manner: 028 135 160 742 391 163 236 885.

4.1 Ciphers[10]

Here is the same message enciphered by the American Army Cipher Disk with the key of E. More complicated ways of using the disk are discussed in a later article.

```
After artillery preparation at six am extend
EZLAN ENLWTTANG PNAPENELWQR EL MWH ES AHLARB
attack cooperation with seventh Army Corps
ELLECU CQQPANELWQR IWLX MAJARLX ENSG CQNPM
```

[10] In modern parlance, this is a *monoalphabetic substitution cipher*. The US Army Cipher Disk used during World War I was made of a heavy cardboard stock with two disks of different diameters and a spindle in the center around which the disks could rotate. The outer disk contained the 26 uppercase letters in the order A B C D E F G H I J K L M N O P Q R S T U V W X Y Z and the inner disk contained the lowercase letters in the order z y x w v u t s r q p o n m l k j i h g f e d c b a. To encipher, one aligns the "a" on the inner disk with a key letter on the outer disk (in Manly's example, the "a" aligns with the "E") and then uses the inner disk letters for the plaintext and the outer disk letters for the corresponding ciphertext. So the first word in Manly's example, "after," becomes EZLAN. See Figs. 5.1 and 5.2 in Chap. 5 for the Army Cipher Disk.

Divide in groups of five letters and send thus:

```
EZLAN ENLWT TANGP NAPEN ELWQR
ELMWH ESAHL ARBEL LECUC QQPAN
ELWQR IWLXM AJARL XENSG CQNPM
```

Here is the same message enciphered by a single transposition. The series of numbers forming the transposition key is written at the top of the columns. The message is written in horizontal lines.

5	3	8	9	4	6	7	1	10	11	2	12
a	f	t	e	r	a	r	t	i	l	l	e
r	y	p	r	e	p	a	r	a	t	i	o
n	a	t	s	i	x	a	m	e	x	t	e
n	d	a	t	t	a	c	k	c	o	o	p
e	r	a	t	i	o	n	w	i	t	h	s
e	v	e	n	t	h	a	r	m	y	c	o
r	p	s									

To send the message, read down vertical columns in order of numbers in the key and divide in five-letter groups.

```
TRMKW RLITO HCFYA DRVPR EITIT
ARNNE ERAPX AOHRA ACNAT PTAAE
SERST TNIAE CIMLT XOTYE OEPSO
```

To convert a simple transposition cipher into a double, the process of transposition is repeated. This may be done either with the same key or with a different one. In the following example, a different key is used. The message obtained by the first encipherment is written horizontally.

6	8	4	7	12	11	2	10	3	5	9	1
t	r	m	k	w	r	l	i	t	o	h	c
f	y	a	d	r	v	p	r	e	i	t	i
t	a	r	n	n	e	e	r	a	p	x	a
o	h	r	a	a	c	n	a	t	p	t	a
a	e	s	e	r	s	t	t	n	i	a	e
c	i	m	l	t	x	o	t	y	e	o	e
p	s	o									

The letters are then taken vertically to obtain the final form of the message:

```
CIAAE ELPEN TOTEA TNYMA RRSMO
OIPPI ETFTO ACPKD NAELR YAHEI
SHTXT AOIRR ATTRV ECSXW RNART
```

The value of either a code or a cipher system depends, of course, upon the ease and accuracy with which it can be written and read and upon the capacity of individual messages for resisting decipherment even by an expert who knows the system in a general way and has plenty of messages to work on. Many years ago, Edgar Allan Poe declared that human ingenuity cannot concoct a cipher that human ingenuity cannot resolve.[11] Whether this is true or not has long been a subject of discussion. Many systems that seem absolutely indecipherable on the basis of a single message yield readily to attack when a large number of messages are available for study. In connection with the operations of an army, innumerable messages must, of course, be sent, and the radio makes them accessible to the enemy as well as to the person for whom they are intended. The requirements for a good army cipher or code are, therefore, very exacting.

4.2 Codes

The solution of an enemy's codes is a very different problem from the solution of his ciphers. When a cipher system has been solved, the reading of individual messages is usually simple and rapid, but even when the general system of a code is known, a new code in the same system can be worked out only very slowly and on the basis of the solution of a large number of individual messages. Fortunately, accidents or carelessness on the part of the enemy often gave invaluable assistance.

Colonel Moorman relates an interesting instance of this, which occurred early in March of 1918, just before the big drive.[12]

> An offensive had been looked for, and everyone was counting on it. About the tenth of March, the Germans put into service along the entire Western front a new code different from any they had been using. We called it the tri-numeral code. Our code men were very eager to solve it because of the coming drive, and both the British and French urged us to put our best men on the job. While we were at work, we picked up a message in the old code which said, 'Previous message not understood. Repeat in old code.' Our whole staff then set to work checking over the messages and finally found two sent by this station to which this one was addressed, which checked exactly in number of code words and in general form. We assumed that they were different forms of the same message and this soon proved to be true.
>
> Lieutenant Georges Painvin, the chief code expert of the French, an analytical genius of the highest order, was a regular wizard in solving codes. On the basis of this single message he worked out a complete system of this new code by the end of the second day, and before the Germans themselves were familiar with it our men were reading it as easily as they were.[13]
>
> It is too much to say that the solution of this code changed the result of the war, but it undoubtedly cost the lives of many German soldiers and saved the lives of many of the Allies. That foolish mistake of one German officer certainly cost them dear.[14]

---

[11] Poe, *The Gold Bug*, "…it may well be doubted whether human ingenuity can construct an enigma of the kind which human ingenuity may not, by proper application, resolve (Poe 1843)."

[12] The first German Spring Offensive began on March 21, 1918.

[13] This new tri-numeral trench code, one of a series called the *Schlüsselheft* (codebook), was introduced at midnight on March 11, 1918. The initial break in the *Schlüsselheft*, using the two messages mentioned by Manly, was made by American cryptanalyst Lieutenant Hugo Berthold early in the morning of March 11. He shared his break with the British and French cryptanalytic organizations, and Painvin completed the break into the superenciphered code (Kahn 1967, pp. 335–336).

[14] The Germans began using trench codes in the spring of 1917. Eventually, their standard trench code grew to consist of about 4000 groups of numbers. In March 1918, the Germans introduced a new trench code, the *Schlüsselheft* (codebook), that was used through their spring offensive. The codebook itself was never changed, but the superencipherment key was changed about twice a

Happy accidents like this cannot be counted on, but sometimes the hand of fate can be forced. It occurred to some ingenious mind that this same situation could be created again by cooperation between the code section and the artillery. The most authentic version of the story comes from an artillery officer.

> We had learned," he says, "that a new code would go into effect at a certain hour and that the books would not be distributed until just before they were to be used. We figured that a certain station could be isolated by artillery fire and prevented from receiving its book. A continuous heavy fire was, therefore, kept up for many hours so that the messenger who was to deliver the books could not get through. The next morning a message arrived in the new code, but this station was unable to read it and asked for a repeat in the old code. Our men picked up the message in both codes and by comparing them, got a big jump in the solution of the new code.

In some versions of the story, the artillery is said to have surrounded the German station with a box barrage and kept it up for 24 h.

But carelessness was not the only human weakness by which our code men profited for the solution of German codes. For 4 months, the German code work was in the charge of a Lieutenant Jaeger, who insisted on signing every message with his name. To spell out this name, four code groups corresponding to j, ae, g, and er were used. Of course, our men soon came to know that the last four code groups of a message would mean these letters. As the code books were constructed on a system which was already known from the old codes, the knowledge of the groups for spelling Jaeger's name gave not only these meanings in the new code, but the thirty or forty others which were connected with them by the system. Vanity of vanities! Lieutenant Jaeger's name finally disappeared from the messages, owing perhaps to the discovery of this and his assignment to other duty. In any case, as Colonel Moorman remarks, "His loss was keenly felt by the American code men."

These incidents show vividly how quick our men were to profit by the mistakes of the enemy. We cannot believe that the Germans were any slower in this matter than we were, for unfortunately it was difficult—or speaking more accurately, impossible—to impress upon our fighting men the necessity for exercising the utmost care in communications and following exactly the instructions that were given them for the use of code messages.

The most ludicrous instance of this, perhaps, occurred when an officer in command of one of our stations sent a message by wire to another station in plain

---

month. At the same time, on March 6, 1918, the Germans introduced the ADFGX cipher invented by Colonel Fritz Nebel that French Captain Georges Painvin later broke. The ADFGX cipher was a combination substitution and fractionating transposition cipher that replaced letters in the plaintext with pairs of letters from the set ADFGX using a modified Polybius square and then broke up and transposed the resulting ciphertext based on a key. Painvin solved the ADFGX cipher in special cases by April 1918. Right before their offensive in June 1918, the Germans introduced a variation on the ADFGX cipher, the ADFGVX cipher that allowed them to encode all 26 letters of the Roman alphabet and the ten numerals, 0–9 in the square. Painvin also broke this cipher system, but only sporadically ((Kahn 1967 pp. 333–336) (*Schlüsselheft*) and (Kahn 1967, pp. 339–347) (ADFGVX)). A general solution for the ADFGVX cipher was not completed until 1933 when William Friedman, Frank Rowlett, Abraham Sinkov, and Solomon Kullback of the US Army's Signal Intelligence Service (SIS) published a solution (Rowlett et al. 1934).

English. He thought he had disguised it successfully because instead of dividing it into words, he had separated the letters into groups of five to make it resemble the five letter groups of our old army code, which was then in use. When charged with this negligence by his commanding officer, he said, "I don't understand how to use that code book. Anyway, what's the difference? The Germans haven't any brains." It is difficult to say what might have been—or what perhaps were—the results of this bit of ignorance and carelessness.

Unfortunately, this carelessness was not limited to the use of codes. Men on the front were working under conditions absolutely unparalleled in human history, and it was extremely difficult for them to realize that arrayed against them were the ablest brains in the whole German world of science and business, making use not only of the latest refinements of modern inventions, but even introducing new applications of science that had not yet been developed or applied in civilian life. How this sometimes worked out is strikingly shown by the dangers of unguarded conversations over the telephone, as told by an officer of the Signal Corps.

> "We knew," said he, "that messages over a telephone or telegraph wire could be picked up by induction, so we laid down induction wires paralleling the telephone wires of the Germans, and with proper instruments were able to hear all conversations passing over the German wire. Much valuable information was obtained in this manner. Sometimes definite plans were overheard. More frequently the information was not direct but was obtained by inference from remarks that the speakers themselves must have felt would be of no service to us even if overheard. For instance, we would hear German soldiers in a certain camp making inquiries as to the location of dressing stations, supply camps, and the like. This, of course, made it quite certain that they were strangers to the station and consequently that the station was being relieved. In one case a listener reported that one division had been relieved and another had taken its place because he recognized that the new voices had a softer accent."

Of course, while we were listening in on the talk of the Germans, they were equally busy in their attempts to hear what we were saying to one another and in making inferences from what they heard. Anyone familiar with the freedom with which the men in our camps talked to one another and exchanged in confidence information they had picked up or guesses they had made concerning the movements and plans of our troops will readily believe that more than once, to our serious cost and injury, the Germans profited by such indiscretions.

References

Hitt, Parker. 1916. *Manual for the Solution of Military Ciphers*. Fort Leavenworth, KS: Press of the Army Service Schools.

Kahn, David. 1967. *The Codebreakers; The Story of Secret Writing*. New York: Macmillan.

Mauborgne, J. O. 1914. *An Advanced Problem in Cryptography and Its Solution*. Army Service Schools Press. William Friedman Collection. George Marshall Foundation Research Library. http://marshallfoundation.org/library/digital-archive/advanced-problem-cryptography-solution/.

Poe, Edgar Allan. 1843. The Gold-Bug. *The Dollar Newspaper*.

Rowlett, Frank R., Solomon Kullback, and Abraham Sinkov. 1934. *General Solution for the ADFGVX Cipher*. Washington, DC: U.S. Army Signal Intelligence Service.

Chapter 5
Cryptology at the Front and at Home

Abstract World War I saw an explosion of work in cryptology. The advent of radio—wireless telegraphy—made it crucial that armies be able to hide the contents of their communications from the enemy. Both sides also struggled with the problem of creating systems that were secure, but also usable under battlefield conditions and by soldiers who had minimal training in cryptology. This chapter explores the ways in which both sides created new systems to hide their plans from the enemy.

As Manly mentions in Chap. 3, intelligence was a long time coming to armies in combat. Until World War I, most European and the American armies did not have permanent intelligence organizations, relying instead on ad hoc groups that were created on the whim of a particular commander in a particular situation. The advent of the General Staff system in the last half of the nineteenth century in Europe and the start of the twentieth in America began the permanent inclusion of intelligence groups as an influential part of planning and execution, formalizing this work and providing a reporting structure for intelligence officers. The bedrock of all intelligence work is to find out the enemy's order of battle, that is, the location, strength, and organization of all its troops before the beginning of an engagement. Military intelligence is acquired in three different ways:

> "First, combat units, directly by reconnaissance and indirectly by capturing prisoners and documents, acquired most of the army's tactical intelligence and frequently helped with larger concerns. Second, agent networks traced enemy troop movements… Finally, three specialist organizations under military discipline provided intelligence through technical means. … aircraft photographs on the enemy's rear trench lines, flash spotters and sound rangers on the locations of its artillery pieces for counter-battery fire… and lastly, intelligence acquired from the enemy's communications, dealt with a wider issue, the enemy's intentions and order of battle at operational and strategic levels" (Ferris 1988, p. 25).

This last form of intelligence is known as *signals intelligence*."Signals intelligence involved the interception of messages; traffic analysis, or the inferences derived from the observation of the procedures and patterns of communications circuits; and the solution of codes and ciphers. Most of these techniques became sophisticated only during the Great War itself, which was the dawn of modern signals intelligence" (Ferris 1988, p. 25).

Before the advent of the telegraph in the 1830s, the telephone in the 1870s, and the coming of radio communications—wireless telegraphy—in the late nineteenth

J.F. Dooley, *Codes, Ciphers and Spies*, DOI 10.1007/978-3-319-29415-5_5

century, signals intelligence focused on the capture of mounted messengers or runners and the interception of messages they carried. These messages, if in code or cipher, were then cryptanalyzed to discover the enemy's intentions. The invention of the telegraph increased the ability of opposing forces to acquire messages via physically tapping the telegraph lines of their opponents. The telephone also allowed for tapping of wires and increased the volume of information that could be intercepted. The problem with both the telegraph and the telephone though was that the interception of enemy communications often did not survive the beginning of combat. In World War I, as soon as a battle began, telephone and telegraph lines near the front were typically severed, and communications between an army's headquarters and its frontline troops were relegated to the use of mounted messengers or runners, many of whom would be captured or killed. Thus commanders had a much more difficult time controlling their troops after the beginning of battles. The greatly increased size of modern armies only served to make this problem much worse.

Radio, however, had the ability to fix this communication problem and also provide an opportunity for a giant expansion of signals intelligence. By 1910, most of the European powers were actively exploring the use of radio in their militaries. The French, in particular, were working on using radio closer to the front lines and were preparing to intercept and cryptanalyze radio messages in wartime. While radio helped solve the problem of communicating between frontline troops and headquarters further back, there was a glaring problem with radio as well. Telegraph and telephone messages are carried over a single wire from point to point, but radio is inherently a broadcast medium. So whatever message a regimental headquarters sent back to its division was heard not just by the division radio operators but also by every radio receiver tuned to the same frequency and within the broadcast range of the sender. This was the signals intelligence opportunity provided by armies using radio communications; interception was much easier.

In the fall of 1914, as the Imperial German Army moved into Belgium and France and as the French and British withdrew deeper into French territory, the two armies had two different responses to the problem of communications. While both the British and French used radio to some extent—and to a larger extent than expected before the war—they were falling back into their own territory and had the use of the existing telephone and telegraph infrastructure behind their new lines. The Germans, on the other hand, were advancing into new territory, their lines of communication becoming increasingly longer. The Imperial German Army advanced into territory whose infrastructure was being at least partially destroyed by combat. They typically could not use the French telephone and telegraph lines and so depended on radio as they first advanced on Paris, then retreated, and finally raced the British toward the English Channel. Neither side was prepared cryptologically for the unprecedented use of radio in fall 1914. Both sides were using field ciphers during this time, and it turned out that field ciphers take a long time to construct a message and were very prone to errors in the creation and transmission of messages via radio. "Since all (the) ciphers in use were cumbersome and speedy transmission was the order of the day, they continually sent crucial messages in clear" (Ferris 1988, p. 26). Sending messages via radio in plaintext rather than enciphering them just gives the enemy all the information they need for free.

By December 1914, the opposing lines on the Western Front had stabilized, and both sides were digging in. The French and British updated the telephone and telegraph lines on their side of the Front, and the Germans strung new telephone and telegraph lines throughout Belgium and northern France. The use of radio diminished substantially between the end of 1914 and mid-1916. But landlines could still be tapped and listened in on. Also, because frontline telephone and telegraph lines would be cut by artillery fire and runners would be captured or killed, radio was the only reliable form of communications once a battle started. Because of problems like this, by the summer of 1916 radio was again in common use, and also by this time both sides were producing radio sets in a smaller size and in a quantity to move radios down to the battalion level. From this point on, radio interception and the cryptanalysis of intercepted messages would be one of the primary sources of intelligence for both sides on the Western Front.

During the course of the war, both sides used three different types of cryptologic systems to mask their radio, telegraph, and telephone messages. The first was cipher systems, where individual letters were either replaced by different letters (substitution), or moved to different parts of the message according to some rule (transposition). These cipher systems were more prevalent in the early part of the war, from 1914 through mid-1916, although there were cipher systems in use all the way through to the Armistice. Second, one-part code systems were used. Codes typically replace larger syntactic blocks, including letters, words, phrases, and sometimes complete sentences, with code words. These code words are usually sequential and matched with the alphabetically arranged plaintext blocks. One-part codes use a single table arranged alphabetically where each plaintext element is paired with either a numeric or letter code word. This sequential matching of code words and plaintext elements is a weakness of one-part codes. Because of this weakness, one-part codes are nearly always then superenciphered using a simple substitution or transposition cipher to add an extra layer of security. Lastly, both sides used two-part codes where two tables are used, one for encipherment and one for decipherment. This allows the code words to be randomly selected and makes for much more difficult cryptanalysis than with one-part codes. It also eliminates the need for superencipherment and makes the code easier to use in the field. The physical codebook, however, is twice as long.

5.1 Allied Codes and Ciphers in France

France During the war, the French used two different cipher systems, a polyalphabetic cipher with a running key using mixed alphabets and an irregular columnar transposition cipher (Kahn 1967, p. 312). The polyalphabetic cipher system used by the French was close to a Vigenère system. The Vigenère cipher is a polyalphabetic system that uses a table of shifted direct standard alphabets and encrypts letter by letter using a short, repeating keyword. It is, unfortunately, one of the simplest polyalphabetics to solve.

Table 5.1 shows what is now known as the Vigenère tableau.

The top row of the table is the plaintext alphabet, and the leftmost column is the key alphabet. In this system, of course, both correspondents must know the keyword. The encipherer takes the next letter from the keyword to select the row to use. The plaintext letter is selected from the appropriate column of the top row and the intersection of the row, and the column is the ciphertext letter. If the key is PARIS and the plaintext is "Attack along the Marne..." then for the first few letters we would get

```
Key:     P A R I S P A R I S P A R I S P A R I
Plain:   a t t a c k a l o n g t h e m a r n e
Cipher:  P T K I U Z A C W F V T Y M E P R E M
```

There are two significant problems with the standard Vigenère system. First, since we are using shifted standard alphabets, once a cryptanalyst correctly guesses a single letter in one of the alphabets, all the rest of the letters for that alphabet are also known. And second, because the keyword repeats and is typically short (almost certainly less than 26 letters), any cryptogram of any length is susceptible to analysis using the Babbage-Kasiski method which looks for repeated pieces of plaintext enciphered with the same portion of the keyword. These two weaknesses make the standard Vigenère useful only for very short messages with longer keys and whose value is measured only in hours.

The French improved on the standard Vigenère in two ways. First, instead of using shifted standard alphabets, they used mixed alphabets. This requires the cryptanalyst to find each letter in each alphabet used in a cryptogram in order to find the plaintext of the cryptogram. This isn't much of an improvement, but it will slow the cryptanalyst down somewhat. An example of a Vigenère tableau using mixed alphabets is in Table 5.2.

Second, the French abandoned the short key used in the standard Vigenère and used a running key. This technique begins with a regular keyword, but instead of repeating the keyword, the user then uses the plaintext as the rest of the key. For example, in the above cryptogram, instead of using the repeated key PARISPARISPARISPARI, with a running key we would use PARISATTACKALONGTHE as the key. Again, this adds an extra bit of complexity to the system and increases the cryptanalysts effort in finding a solution, but it does not make the cryptogram insoluble. Once the cryptanalyst has solved the first few letters of the cryptogram, they have the key for the next few letters.

The second cipher system used by the French was an irregular columnar transposition cipher that they used from 1914 through 1916. In an irregular transposition cipher system, one or more of the letters in a column are skipped or are taken off differently than the rest of the letters in the column, resulting in a jump in the order of the letters in the resulting cryptogram. In their irregular columnar transposition, the French accomplished this by taking several of the columns off via diagonals.

Table 5.1 A modern Vigenère tableau

	a	b	c	d	e	f	g	h	i	j	k	l	m	n	o	p	q	r	s	t	u	v	w	x	y	z
A	A	B	C	D	E	F	G	H	I	J	K	L	M	N	O	P	Q	R	S	T	U	V	W	X	Y	Z
B	B	C	D	E	F	G	H	I	J	K	L	M	N	O	P	Q	R	S	T	U	V	W	X	Y	Z	A
C	C	D	E	F	G	H	I	J	K	L	M	N	O	P	Q	R	S	T	U	V	W	X	Y	Z	A	B
D	D	E	F	G	H	I	J	K	L	M	N	O	P	Q	R	S	T	U	V	W	X	Y	Z	A	B	C
E	E	F	G	H	I	J	K	L	M	N	O	P	Q	R	S	T	U	V	W	X	Y	Z	A	B	C	D
F	F	G	H	I	J	K	L	M	N	O	P	Q	R	S	T	U	V	W	X	Y	Z	A	B	C	D	E
G	G	H	I	J	K	L	M	N	O	P	Q	R	S	T	U	V	W	X	Y	Z	A	B	C	D	E	F
H	H	I	J	K	L	M	N	O	P	Q	R	S	T	U	V	W	X	Y	Z	A	B	C	D	E	F	G
I	I	J	K	L	M	N	O	P	Q	R	S	T	U	V	W	X	Y	Z	A	B	C	D	E	F	G	H
J	J	K	L	M	N	O	P	Q	R	S	T	U	V	W	X	Y	Z	A	B	C	D	E	F	G	H	I
K	K	L	M	N	O	P	Q	R	S	T	U	V	W	X	Y	Z	A	B	C	D	E	F	G	H	I	J
L	L	M	N	O	P	Q	R	S	T	U	V	W	X	Y	Z	A	B	C	D	E	F	G	H	I	J	K
M	M	N	O	P	Q	R	S	T	U	V	W	X	Y	Z	A	B	C	D	E	F	G	H	I	J	K	L
N	N	O	P	Q	R	S	T	U	V	W	X	Y	Z	A	B	C	D	E	F	G	H	I	J	K	L	M
O	O	P	Q	R	S	T	U	V	W	X	Y	Z	A	B	C	D	E	F	G	H	I	J	K	L	M	N
P	P	Q	R	S	T	U	V	W	X	Y	Z	A	B	C	D	E	F	G	H	I	J	K	L	M	N	O
Q	Q	R	S	T	U	V	W	X	Y	Z	A	B	C	D	E	F	G	H	I	J	K	L	M	N	O	P
R	R	S	T	U	V	W	X	Y	Z	A	B	C	D	E	F	G	H	I	J	K	L	M	N	O	P	Q
S	S	T	U	V	W	X	Y	Z	A	B	C	D	E	F	G	H	I	J	K	L	M	N	O	P	Q	R
T	T	U	V	W	X	Y	Z	A	B	C	D	E	F	G	H	I	J	K	L	M	N	O	P	Q	R	S
U	U	V	W	X	Y	Z	A	B	C	D	E	F	G	H	I	J	K	L	M	N	O	P	Q	R	S	T
V	V	W	X	Y	Z	A	B	C	D	E	F	G	H	I	J	K	L	M	N	O	P	Q	R	S	T	U
W	W	X	Y	Z	A	B	C	D	E	F	G	H	I	J	K	L	M	N	O	P	Q	R	S	T	U	V
X	X	Y	Z	A	B	C	D	E	F	G	H	I	J	K	L	M	N	O	P	Q	R	S	T	U	V	W
Y	Y	Z	A	B	C	D	E	F	G	H	I	J	K	L	M	N	O	P	Q	R	S	T	U	V	W	X
Z	Z	A	B	C	D	E	F	G	H	I	J	K	L	M	N	O	P	Q	R	S	T	U	V	W	X	Y

Table 5.2 A Vigenère tableau using mixed alphabets

R	B	S	L	N	M	V	J	E	U	H	G	I	X	F	K	A	O	Z	C	D	P	Y	Q	T	W
B	A	U	I	H	E	N	D	S	Q	K	G	F	Y	L	O	R	T	W	M	V	J	X	C	P	Z
Y	H	T	C	W	Q	N	Z	P	L	J	V	R	X	K	E	M	F	A	I	D	O	G	B	U	S
U	R	L	X	F	G	Q	C	Y	T	P	H	I	S	J	O	N	A	V	Z	K	B	D	E	M	W
V	Q	P	H	G	R	U	I	S	Y	N	D	F	A	X	M	L	J	Z	B	E	K	T	O	C	W
L	N	I	B	X	G	Y	R	D	P	K	T	E	J	A	O	V	Z	H	U	M	W	F	S	Q	C
P	G	M	A	O	E	R	U	D	I	L	S	T	N	H	X	C	B	J	F	Z	K	V	W	Y	Q
O	I	C	J	E	Q	D	W	G	U	L	X	F	V	H	N	T	K	S	Z	R	P	Y	A	M	B
K	O	E	Z	J	L	M	X	U	A	F	C	R	D	P	Y	N	G	H	B	S	Q	I	T	W	V
D	R	B	A	T	O	G	V	I	Y	Q	M	J	C	F	W	P	X	L	U	E	S	N	H	Z	K
H	A	D	T	Y	M	I	O	F	Q	S	R	K	U	L	C	E	G	V	N	X	Z	P	B	J	W
X	E	Q	B	D	I	R	U	M	A	N	P	V	W	H	G	Z	T	F	K	J	O	Y	C	L	S
N	U	Y	O	D	T	K	J	X	W	P	Q	R	Z	F	V	E	H	M	S	C	G	I	L	B	A
C	R	V	O	J	M	W	I	Y	T	B	S	K	Z	X	G	D	P	U	H	A	E	N	Q	L	F
M	X	H	O	T	N	U	K	V	E	P	S	D	Z	Y	J	L	I	B	G	A	C	F	W	Q	R
W	V	K	Q	F	E	R	M	U	H	A	X	N	P	B	O	Z	L	D	T	J	S	Y	I	C	G
E	O	B	F	M	V	T	K	R	X	J	I	S	Q	U	P	A	Z	C	Y	L	W	H	N	D	G
T	S	A	W	M	O	H	R	I	L	N	C	D	X	J	P	Q	E	B	G	F	U	Z	K	Y	V
Q	Z	N	D	Y	P	W	K	O	R	M	U	C	F	V	T	J	L	X	A	B	S	E	H	I	G
A	Q	E	K	H	G	Z	D	Y	M	O	L	C	J	U	T	W	F	B	S	P	V	R	I	X	N
S	G	Z	A	B	D	F	L	T	R	V	I	M	P	K	H	O	X	W	Q	Y	C	J	E	U	N
Z	A	P	J	V	G	D	O	C	T	L	N	F	W	U	Y	B	Q	E	K	I	H	M	X	R	S
J	D	W	U	P	X	E	B	L	G	Y	V	O	H	Z	F	K	S	I	Q	A	N	T	M	R	C
F	J	C	E	M	D	Q	G	X	K	A	S	V	U	H	R	O	L	B	N	T	P	I	Z	Y	W
I	U	T	R	S	Y	J	L	O	H	K	N	V	G	C	F	A	B	M	P	Z	E	X	W	Q	D
G	U	W	R	D	Z	E	B	X	N	A	Q	I	M	C	K	Y	H	F	V	S	O	J	T	P	L

The French used a standard rectangular transposition block of N columns, where N is the length of the key. But instead of just taking the letters off by columns as is normal in a columnar transposition, the French added another step. Before taking the columns off, they instead took several diagonals of letters off, beginning with a particular column and running off down the table and to the right. They then did the same thing, starting with different columns, but taking the diagonals off down and to the left. Finally, they then used the key and took the rest of the letters off by columns, but skipping the letters that had already been chosen via the diagonal step; this is the interrupted part. By skipping those letters, the transposition cipher was more difficult for the enemy to decrypt (Kahn 1967, pp. 312–313).

As an example, say the message the French wanted to send is *Bombardment to begin at five am. First three battalions to step off at six am. Second regiment to step off at eight am*, the key is BIBLIOTHEQUE, and we take off from the rightward diagonals beginning at 6, 8, and 11 in that order and from the leftward diagonals from 10, 5, and 2. Our columnar table then looks like Table 5.3.

And we first take off the letters for the right diagonals and get OEIAFNTA BISLARPX DTHSII. Taking the left diagonals we get (skipping letters we've already removed) NFTCTT MAPSEF MBA. Finally we take off the columns in keyword order from 1 through 12, again, skipping all the letters we've taken off by the diagonals. The final message, broken up into five-letter groups is OEIAF NTABI SLARP XDTHS IINFT CTTMA PSEFM BABOB SAMFT ENENT ETEEO XFXOD TMATM AAROT IGTEE RNLOO GIRSE MTFSH TVETG O.

The resulting message is more difficult to solve than a regular columnar transposition, but as Kahn says, "The diagonals break up the columnar segments that the cryptanalyst juxtaposes and adjusts to solve uninterrupted columnar transpositions. But the diagonals constitute segments of their own, and the columns, though fragmented, keep their constituent letters together instead of scattering them, as does the double transposition" (Kahn 1967, p. 313).

Great Britain Like the French, at the beginning of the war, the British used cipher systems that they'd developed before the conflict began. The British also used two cipher systems during this period. The first was a polyalphabetic substitution cipher with a running key that was nearly identical to the French.

Table 5.3 French irregular transposition table

1	6	2	8	7	9	11	5	3	10	12	4
B	I	B	L	I	O	T	H	E	Q	U	E
b	o	m	b	a	r	d	m	e	n	t	t
o	b	e	g	i	n	a	t	f	i	v	e
a	m	f	i	r	s	t	t	h	r	e	e
b	a	t	t	a	l	i	o	n	s	t	o
s	t	e	p	o	f	f	a	t	s	i	x
a	m	s	e	c	o	n	d	r	e	g	i
m	e	n	t	t	o	s	t	e	p	o	f
f	a	t	e	i	g	h	t	a	m	x	x

The second and more common cipher used by the British from 1914 to 1916 was the Playfair. Sir Charles Wheatstone, the physicist, mathematician, and engineer, invented the British system, known as the Playfair cipher, in 1854. It acquired its name from Baron Lyon Playfair, who spent years popularizing the cipher and attempting to get the British government to adopt it. The British Army finally adopted the Playfair in the 1890s as their field cipher. It saw its first use during the Boer War (1899–1902) and was used as the field cipher down to the company level during the first years of World War I (Kahn 1967, pp. 198–202; Bauer 2013, pp. 166–178).

The Playfair cipher is a *digraphic substitution cipher*, one that encrypts two letters at a time. Every plaintext digraph is encrypted into a ciphertext digraph. It is based on a five by five Polybius square that uses a keyword to map 25 of the 26 letters of the Latin alphabet (I and J are either mapped together in a single cell, or J is just dropped). The keyword is dropped in row by row, deleting any repeated letters, and then the rest of the alphabet is filled in to complete the square. For example, if the keyword is MONARCHY, then the Playfair square looks like Table 5.4

Messages are enciphered according to the following rules:

1. Break up the plaintext message into two-letter groups. Break up any double letters (like SS or LL) by inserting a null letter (like Q or X or Z) between the repeated letters. If the message has an odd number of letters, just add a null to the end.
2. Encipher each two-letter group separately.
3. If the two letters in a group are in the same row, then encipher the group by taking the letter immediately to the right of each letter in the group. So if the square in Table 5.4 is used and the plaintext pair is HY, then the ciphertext is YB. If you run off the right side of the square, just loop around to the beginning of the row.
4. If the two letters in a group are in the same column, then encipher the group by taking the letter immediately below each letter in the group. So in Table 5.4, if our plaintext is CL, then the ciphertext is EU. If you run off the bottom of the square, just loop around to the top of the column.
5. If the two letters are in different rows and columns, then you "complete the rectangle" by first going across the row where the first letter is, to the column that contains the second letter, and using the letter you find at the intersection as the cipher letter. Do the same thing for the second letter. So in Table 5.4, if our plaintext is MG, then the ciphertext is NE, in that order.

Table 5.4 Example of a Playfair cipher square

M	O	N	A	R
C	H	Y	B	D
E	F	G	I/J	K
L	P	Q	S	T
U	V	W	X	Z

Table 5.5 Playfair square using the keyword FRIEDMAN

F	R	I	E	D
M	A	N	B	C
G	H	K	L	O
P	Q	S	T	U
V	W	X	Y	Z

Deciphering is just the inverse of enciphering.

Say we want to send the message "*flee, all is discovered*" using a Playfair cipher with the keyword FRIEDMAN. Then the Playfair square will look like Table 5.5.

The first thing we do is divide up our plaintext into digraphs, making sure to break up any repeated letters with nulls

```
FL EX EA LX LI SD IS CO VE RE DX
```

We now use the rules above to encrypt each digraph separately

```
Plain:   FL EX EA LX LI SD IS CO VE RE DX
Cipher:  EG IY RB KY KE UI NX OU YF ID IZ
```

And finally we break the ciphertext up into five-letter blocks for transmission

```
EGIYR BKYKE UINXO UYFID IZ
```

5.2 Cryptanalyzing a Playfair Cipher

David Kahn gives an excellent description of the difficulties of solving a Playfair cipher:

> In the first place, the cipher's being digraphic obliterates the single-letter characteristics – *e*, for example, is no longer identifiable as an entity. This undercuts the usual monographic methods of frequency analysis. Secondly, encipherment by digraphs halves the number of elements available for frequency analysis. A 100-letter text will have only 50 cipher digraphs. In the third place, and most important, the number of digraphs is far greater than the number of single letters, and consequently the linguistic characteristics spread over many more elements and so have much less opportunity to individualize themselves. There are 26 letters but 676 digraphs; the two most frequent English letters, *e* and *t*, average frequencies of 12 and 9 percent; the two most frequent English digraphs, *th* and *he*, reach only 3.25 and 2.5 percent. In other words, not only are there more units to choose among, the units are less sharply differentiated. The difficulties are doubly doubled. (Kahn 1967, pp. 201–202)

This is not to say that Playfair cipher messages are unsolvable; they are eminently solvable. For long Playfair ciphertexts, or when one has a large number of cipher messages, one can resort to digraph frequency analysis. Otherwise, luck, careful observation, and a deep understanding of how the cipher works are the best methods. To complicate matters a bit, the British, in their version of the Playfair, used a randomized key square instead of using a simple keyword to begin creating the square. This made the cryptanalysis more difficult, but only by a little. US Army Lt. Joseph Mauborgne was the first to publish a general solution to a Playfair cipher (Mauborgne 1914). Fortunately for the British, this solution was not well publicized until much later.

While they did continue to use cipher systems to some extent, both the British and the French switched to the use of trench codes beginning in mid-1916.

5.3 American Codes and Ciphers in France

Just as with everything else involved in getting an American Army to France, the Americans were learning about cryptology all through 1917 and the first part of 1918. The Americans entered the war just as the Allies were switching from ciphers at the front lines to trench codes. By the time the AEF had a cryptanalytic section—the Radio Intelligence Section (RIS), known as G2-A6—in place in the fall of 1917, the use of ciphers by the Allies on the Western Front was practically eliminated. Nonetheless, the Americans did contemplate the use of some ciphers during the war, if only briefly. Initially, the Americans proposed to use the Playfair cipher system previously adopted by the British. But by this time, even the British were admitting that the Playfair was difficult to use in the field and was insecure, so this idea was dropped early on. The Americans then decided to use their own Army Cipher Disk as their frontline cipher system. This was even less secure than the Playfair. The Cipher Disk was two circular pieces of celluloid attached with a pin through the center. The outer circle was divided into 26 cells, and each cell was inscribed with a letter of the alphabet. The inner circle, which was designed to rotate, was also divided into 26 cells and the cells inscribed with a reversed alphabet. See Figs. 5.1 and 5.2 for an illustration. The Cipher Disk was used by setting a key letter on the inner circle opposite the A on the outer circle. Plaintext letters were then read on the inner circle, and the corresponding ciphertext was read from the outer circle. The key letter was not changed for the entire message, making this a monoalphabetic substitution cipher system and so insecure as to be essentially worthless as a field cipher. Even if the key letter was changed during a single message, the cipher was then just equivalent to a Beaufort cipher, a simple polyalphabetic cipher that by this time was equally easy to decrypt. Wiser heads in the American intelligence community prevailed, and the Army Cipher Disk was not used during the war (Friedman 1942, p. 4 and 31).

By the time the American G2-A6 cryptologic organization was functioning late in 1917, the Allies had transitioned from field ciphers to field codes, known as trench codes. The Signal Corps' Code Compilation Section commanded by Captain Howard Barnes developed the American trench codes. Because the American Army had never had a field code before, Barnes was forced to start from first principles and also to plead with his British and French Allies for help, which was slow in coming. His goal was to create a simple, easy to use, field code that was also secure. According to Barnes,

Fig. 5.1 The front of the American Army Cipher Disk (Photo by Dr. Nick Gessler; used with permission) (Copyright Dr. Nick Gessler, Duke University; used with permission)

The data on the subject of codes was most limited in scope. Previous to this war the United States Army had never had a codebook, properly so called, for field service, and had had recourse to the cipher disk or short-lived emergency codes. Moreover, the Army was confronted with a foe who had profited not only by their own experiences of 3 years but the mistakes of the Allies which they had observed through their interception of wireless messages and the information gained from captured code books. At first the British and French were rather reluctant to disclose the systems which they had adopted for their codes, but eventually copies of obsolete editions were turned over to this Section for reference and study. With this meager data the compilation of a front-line code was begun. The fundamental principle upon which the books were founded was a complexity sufficient to delay solution with a simplicity sufficient to afford ease of operation.

The first American Trench Code, a small book consisting of some 1,600 words and phrases, was intended for distribution down to and including companies actually in line. Accompanying it were certain tables containing a distorted alphabet. It was proposed to change these tables at frequent intervals and thus delay the solution of intercepted messages. This Trench Code was never in fact actually delivered to the front line, and went no farther down than regimental headquarters because of the danger of capture. An edition of 1,000 was printed. (Friedman 1942, p. 9)

Fig. 5.2 The back (including instructions) of the American Army Cipher Disk (Photo by Dr. Nick Gessler; used with permission) (Copyright Dr. Nick Gessler, Duke University; used with permission)

Why was the American Trench Code never "actually delivered"? It turns out it wasn't very secure. The code was a one-part code that used a monoalphabetic substitution cipher as its superencipherment step to make code messages more secure. The code words were divided up into several sections to make encoding a message easier. In early May 1918, as the Trench Code was being prepared for distribution, the AEF's Assistant Chief Signal Officer, Major Parker Hitt (Hitt would later become the Chief Signal Officer for the American First Army), decided he wanted to test the security of the system. Hitt then went to Major Frank Moorman, head of G2-A6, and asked if one of his cipher experts would attempt to break through a set of coded messages and decode them. Lieutenant J. Rives Childs, who had been in France only since February 1918 and was the head of the Cipher Solutions Section of G2-A6 (despite having only a few months experience as a cryptanalyst), was then given a copy of the trench code and a set of 44 cryptograms that were encoded using the trench code and then enciphered using the superencipherment alphabet, called the "encoding table" and keys. Childs was not given the details of the superencipherment nor was he given the encoding table or keys. His job was to strip the superencipherment from the messages and then read them using the codebook. Less than a day later, Childs returned to Moorman with all 44 messages solved and returned to plaintext (Childs 1978, pp. 205–206; Kahn 1967, p. 327; Friedman 1942, p. 10).[1] This was the death knell for the first American Trench Code.

Barnes then turned to creating a new code. This time it was to be a two-part code with no superencipherment. The two-part code would have the advantage of being easier to both encode and decode in the field, and it eliminated the need for the second step of superencipherment. Its main disadvantage was that with sufficient traffic, something that would be available in the run-up and initial phases of an offensive, the Germans would be able to begin to pick apart the codebook. A long offensive also increased the chances that the Germans would capture a copy of the codebook. Because of the disadvantages, Barnes committed to creating, printing, and distributing a new version of the code every 10 days to 2 weeks, something unheard of before. The result was the spectacularly successful *River* series of codes, beginning with the delivery of the *Potomac* code on June 24, 1918. The first edition was 2,000 copies and contained codewords for 1,800 words and phrases. Each page contained about 100 codeword/plaintext pairs, and with both the encoding and decoding tables, null codewords, instructions, and blank pages for notes, the code was squeezed into just 47 pages and was designed to fit in a pocket. All the River series codes used a special "typewriter" font that made them easier to read in the field. The code was released down to the battalion level. *Suwanee* followed *Potomac* in an edition of 2,500 copies on July 15.

After this half a dozen more River code versions followed before the Armistice, including *Wabash*, *Mohawk*, *Allegheny*, *Hudson*, *Colorado*, and *Osage* in increasing numbers of copies. Beginning with *Allegheny*, the codewords made up of letters were replaced with codewords formed from numerals. In addition, starting with *Allegheny*, an Emergency Code List insert was added to the back and front flaps in order to speed up message creation even more (Fig. 5.3).

When the American 2nd Army was created in September 1918, Barnes decided to create separate codes for each Army, and the *Lake* series was born. The first *Lake* series code, *Champlain*, was issued on October 7, 1918, followed by *Huron* on October 15, and then *Seneca* (Fig. 5.4). At the time of the Armistice on November 11, the *Niagara* code was being printed and the *Michigan* and *Rio Grande* codes were being developed. Altogether in the 5-month period from June 24 through November 11, the Code Compilation Section developed, printed, and released 14 different trench codes. While they were doing this, they also developed three different frontline codes designed to be used at the company level, and a 38,000 codeword staff code for AEF headquarters use (Friedman 1942, pp. 17–19). Three different times in the 5 months codebooks were captured by the Germans, but each time Barnes released and distributed a new code within a very few days. His system was an enormous success.

[1] See also the story in Manly's Article II in Chap. 6.

SECRET **EMERGENCY CODE LIST**

To be used only with the "Huron Code."
To be Issued down to companies.
To be used only for communications within divisions.
To be completely destroyed, by burning. when in danger of capture or after a new code has been issued.

Precede Every Message in This Code by "RO"

About to advance...SP	AB...Gas is being released
Ammunition exhausted...BX	AF...Trenches
Are advancing...XP	AG...At
At...AG	AP...Objective reached
Attack failed...FS	AV...Enemy fire has destroyed
Attack successful...XA	AW...Relief being sent
Barrage wanted...BD	AX...Captured
Be ready to attack...SM	AZ...Look out for signal
Being relieved...ZB	BD...Barrage wanted
Captured...AX	BF...Right
Casualties heavy...BJ	BJ...Casualties heavy
Casualties light...SF	BM...Using gas shells
Center...XY	BP...left
Enemy...PF	BS...Enemy trenches
Enemy barrage commenced...SB	BX...Ammunition exhausted
Enemy fire has destroyed...AV	BY...Wire entanglements destroyed
Enemy machine gun fire serious...ZF	CA...Our
Enemy trenches...BS	CB...Situation serious
Everything O. K...CZ	CM...Message not understood
Everything quiet...FC	CP...Need water
Falling back...SX	CX...Raiders have left
Gas is being released...AB	CZ...Everything O. K.
Have broken through...PG	FA...How is everything
How is everything...FA	FB...Recall working party
Increase range...XG	FC...Everything quiet
Left...BP	FM...Stopped
Look out for signal...AZ	FS...Attack failed
Machine gun ammunition needed...XB	FX...Using high explosive shells
Message not understood...CN	FY...Tank stuck
Message received...ZP	FZ...Not ready
Near...SA	PB...Trenches have been occupied
Need water...CP	PF...Enemy
Not ready...FZ	PG...Have broken through
Objective reached...AP	PM...Strong attack
Our...CA	PO...Rush
Our artillery is shelling us...PV	PV...Our artillery is shelling us
Raiders have left...CX	PX...Reinforcements needed
Recall working party...FB	SA...Near
Reinforcements needed...PX	SB...Enemy barrage commenced
Relief being sent...AW	SC...Troops
Relief completed...XP	SF...Casualties light
Rifle ammunition needed...SZ	SM...Be ready to attack
Right...BP	SP...About to advance
Rush...PO	SX...Falling back
Situation Improving...ZX	SZ...Rifle ammunition needed
Situation serious...CB	XA...Attack successful
Stopped...FM	XB...Machine gun ammunition needed
Stretcher bearers needed...ZJ	XF...Are advancing
Strong attack...PM	XG...Increase range
Tank atuck...FY	XP...Relief completed
Trenches...AF	XY...Center
Trenches have been occupied...PB	ZB...Being relieved
Troops...SC	ZF...Enemy machine gun fire serious
Using gas shells...BM	ZJ...Stretcher bearers needed
Using high explosive shells...FX	ZP...Message received
Wire entanglements destroyed...BY	ZX...Situation improving

Fig. 5.3 AEF emergency code list (*Public Domain*. From Friedman document on American Field Codes at NARA)

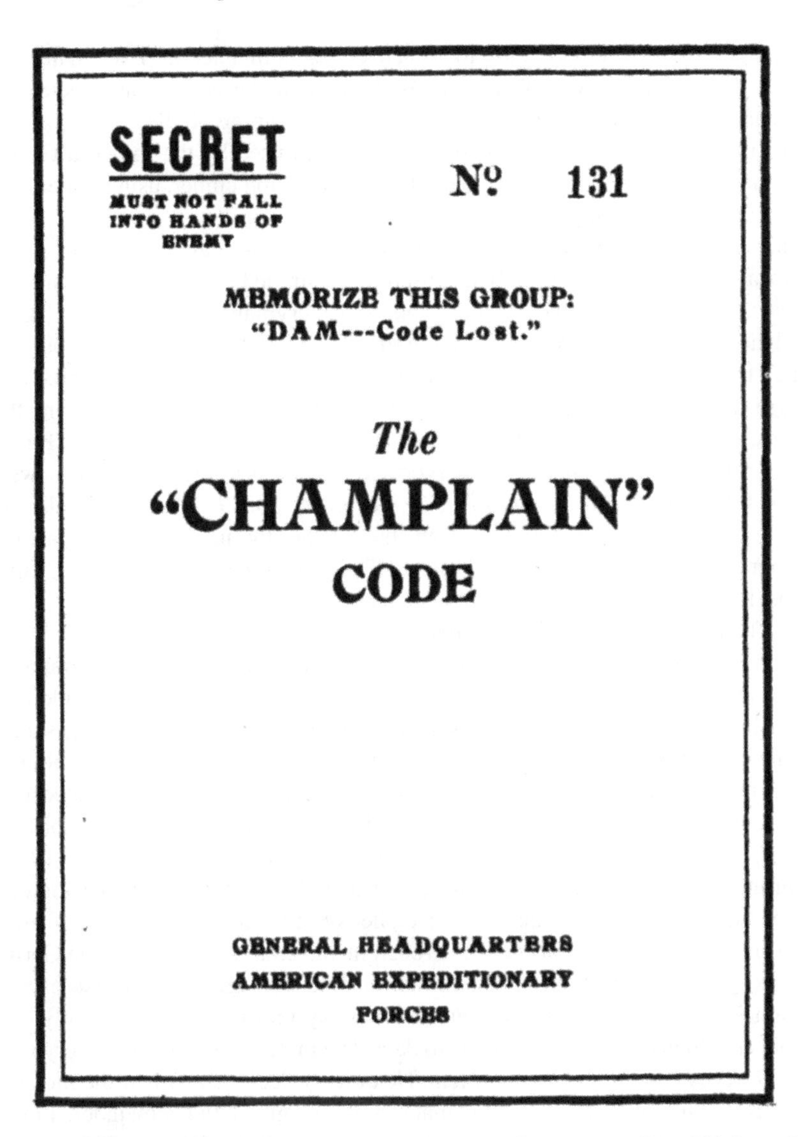

Fig. 5.4 Cover of the Champlain Lake Series codebook (*Public Domain*. From Friedman document on American Field Codes at NARA)

5.4 German Codes and Ciphers in France

Just like the Allies, the Imperial German forces in France used a combination of ciphers, one-part codes with superencipherment and two-part codes. Unlike the Allied systems, however, nearly all the German cipher and code systems were solved by the cryptanalytic bureaus of France, Great Britain, and the United States during the war. The Germans also used cipher systems far longer than the Allies, not switching most of their systems to trench codes until later in 1917. They also continued to release new cipher systems throughout the entire war, whereas the Allies had basically stopped using ciphers by the beginning of 1917.

The most commonly used type of cipher system used by the Imperial German forces on the Western Front was the double transposition cipher, using either a single key or, less frequently, two keys.

The first double transposition cipher used by the Germans starting in August 1914 was called the ÜBCHI cipher. It used just a single key and the Germans only changed the key about every 2 weeks. The French cryptanalysts broke ÜBCHI and found the current key on October 1, not in time for the Battle of the Marne, but early enough to help with the Race to the Sea. The Germans changed the key three more times in October and November, never suspecting that the French were reading their messages, and each time the French broke the cipher and retrieved the key. Finally, when the news of the French cryptanalytic successes was printed in a Paris newspaper, the Germans changed their cipher on November 18, 1914. It didn't really help because while seemingly complex from the outside, the new cipher was actually cryptographically weaker than the ÜBCHI. The new cipher, called the ABC cipher by the French, used a Vigenère tableau but always used the same keyword—ABC. The resulting ciphertext was then disarranged using a single columnar transposition. The single columnar transposition meant that a regular anagramming technique could be used to break the transposition. And the fact that the Germans were using three consecutive shifted standard alphabets to encipher messages meant that the ciphertext equivalent of a plaintext letter was at most two cipher alphabet letters away, making it much easier to recognize digraphs in the transposition. French cryptanalyst Lt. Colonel Anatole Thévenin broke the ABC cipher by December 10, but the Germans continued to use the ABC cipher until May 1915.

During 1915, the Germans continued to use cipher systems of increasing complexity. Mostly, these were transposition cipher systems, or a combination of substitution and a single transposition. The French managed to keep up with the German changes. In late 1915, though, just as radio traffic was again on the rise over the Western Front, the Germans introduced a new system that was seemingly more complicated than any of their systems to date. The French threw their best cryptanalyst—Georges Painvin—at the problem. He quickly decided that half the messages were fakes, but for 2 weeks could make nothing out of the remainder. Finally, Painvin and Major Adolphe Olivari broke the system (Kahn 1967, p. 307). It was a combination cipher, like the ABC, and they named it the ABCD cipher system. This time the Germans used four alphabets, but with the twist of using the Vigenère tableau with an interrupted key. Recall that the major weakness of the Vigenère cipher is that the key repeats at regular intervals. If, instead, the encipherer breaks the key at random places and starts over again, the regularity is interrupted, and cryptanalytic techniques like the Babbage-Kasiski method are harder to use. For example, if the key is BIBLIOTHEQUE, the encipherer can use the key as follows: BIBLIOTH BIBL BIBLIOTHEQU BIBLIOT BI, etc. The sender and receiver of the cryptogram just have to agree on where the interruptions are. One technique for this is to encipher a sentinel letter, say "J" at the point where the key will be interrupted. The receiver will decipher a letter, get a "J," and know to start the key over again at the beginning (Gaines 1956, p. 143). The French broke the ABCD cipher within a few weeks of its introduction, and the Germans withdrew the cipher in April of that year.

The Germans continued to use ciphers intermittently during the war, with very limited effect. They focused on transposition ciphers and ciphers that were a combination of substitution and transposition, but they did use a couple of pure substitution ciphers. One example of a bad use of substitution was the "777" cipher, so called because all messages in this cipher began with 777. This cipher was used only briefly in 1918 between General Kress von Kressenstein in Tiflis (now the capital of Georgia, Tbilisi) and Berlin. This cipher was a mixed alphabet monoalphabetic substitution cipher. The French and Americans broke it immediately and so did the Germans. This resulted in the now famous response to General Kressenstein "The cipher prepared by General von Kress solved here at once. Its further use and operation is forbidden. (Signed) Chief Signal Officer, Berlin" (Childs 1919, p. 1).

The German cipher system in the longest use was the *Wilhelm* or *Füer GOD* cipher mixed alphabet polyalphabetic substitution system introduced in 1916 and used until the Armistice. It was known as *Füer GOD* because most of the messages in it were from the main German radio station at Nauen outside of Berlin to a receiver whose call letters were G O D. This cleverly devised system used a modified Vigenère tableau of 22 cipher alphabets and a set of 30 keys of lengths between 11 and 18 letters that were monoalphabetically enciphered and rotated through messages, with each message using a different key in order. The Vigenère tableau looks like Table 5.6.

These alphabets are not very well mixed, which will give the cryptanalyst an advantage. Note that there are repetitions in some of the rows, for example, row L has two Vs, and row S has two Ys. This may be a transcription error on Childs' part, or it may be intentional. Also note that there are groups of letters that nearly always appear together. For example, ABCDEF appear together several times, and ABCD appear together even more frequently. Also, GHJK appear near each other often, as do LMNOP. All of these constitute weaknesses in the cipher. The 30 keywords were 30 common German words or phrases that were then enciphered using a monoalphabetic substitution before being used to pick the key alphabets from the tableau. The keys were changed for each message, but there were only 30 of them, so they would rotate fairly frequently. The 30 keywords and their translations are:

1. SPRINGBRUNNEN (Fountain)
2. VOLLWICHTIGES (many important things)
3. PROBIERSTEIN (touchstone)
4. MARIONETTENSPIELER (puppeteer)
5. BESTIMMUNGSORT (destination)
6. FAHRTUNTERBRECHUNG (stopover)
7. WAGENWECHSEL (change carriages)
8. WOLLENSIEFAHREN (do you want to move)
9. NOCHETWASMILCH (something else milk)
10. ESSCHMECKTMIR (I like it)

Table 5.6 The alphabets for the *Wilhelm* cipher system

	A	B	C	D	E	F	G	H	I	J	K	L	M	N	O	P	Q	R	S	T	U	V	W	X	Y	Z
A	S	Q	R	Y	V	X	U	Z	T	W	B	D	C	A	E	J	H	K	I	F	G	P	M	O	N	L
B	L	O	P	N	M	Q	S	R	T	U	V	Z	X	Y	W	C	A	B	H	E	D	G	J	F	K	I
C	P	O	N	M	R	T	S	Q	W	Y	U	X	Z	V	C	A	B	E	D	F	J	G	K	H	I	L
D	I	F	H	J	G	N	K	L	M	P	O	T	S	R	Q	V	Y	U	X	Z	W	D	B	C	A	E
E	X	U	V	Z	Y	W	A	C	B	E	D	G	I	H	J	F	K	M	O	N	L	T	R	S	Q	P
F	U	X	Z	W	Y	V	A	E	B	C	F	D	I	H	G	J	N	K	M	L	S	P	O	R	T	Q
G	A	C	D	B	H	J	F	I	G	E	M	N	L	K	O	T	R	S	T	Q	Y	Z	V	U	X	W
H	B	A	D	C	F	G	E	I	H	J	N	O	K	M	L	S	R	P	T	Q	W	X	V	Y	U	Z
I	T	R	S	Q	Y	W	X	Z	V	U	E	B	A	C	D	K	F	J	I	G	H	M	L	P	N	O
J	L	M	O	N	T	Q	R	P	S	Z	X	U	Y	V	W	B	A	C	D	E	G	J	H	F	K	I
K	M	O	K	N	L	Q	S	R	P	W	Z	T	V	U	X	Y	D	B	A	C	E	F	J	G	I	H
L	I	E	H	F	G	L	O	M	J	K	N	Q	P	T	R	S	X	V	Y	U	Z	V	B	A	D	C
M	H	F	I	G	N	M	J	K	O	L	Q	P	S	R	V	T	Z	U	W	X	Y	B	E	D	C	A
N	C	D	A	B	G	H	E	J	F	I	K	M	P	O	L	N	T	R	Q	S	X	U	Z	W	V	Y
O	E	C	D	B	A	F	J	I	G	H	L	K	O	N	M	S	P	Q	T	R	Z	U	Z	V	W	Y
P	R	Q	P	S	Z	W	T	V	U	X	Y	D	B	C	A	G	I	E	J	H	K	F	O	N	L	M
Q	V	Y	X	U	Z	W	C	A	B	E	D	I	H	G	F	L	K	N	M	J	Q	O	T	P	S	R
R	B	A	C	H	D	J	F	E	G	I	L	O	N	P	K	M	S	Q	R	U	Z	T	Y	V	W	X
S	Q	Y	Z	V	X	A	B	C	E	F	D	M	J	I	G	K	A	P	L	N	S	R	O	Y	U	T
T	E	D	I	G	H	F	L	M	K	P	O	N	R	Q	J	S	U	X	T	Z	W	V	Y	C	A	B
U	R	T	S	W	V	Y	Z	U	X	F	A	C	B	E	D	J	K	I	G	H	O	N	M	P	Q	L
V	M	O	L	N	P	S	R	Q	X	T	Y	W	Z	U	V	A	D	C	B	H	F	I	K	E	J	G

Childs (1919, p. 2)

11. KORKZIEHER (corkscrew)
12. ZUCKERZANGE (sugar tongs)
13. SCHWIEGERVATER (father-in-law)
14. WIELANGEDAUERTES (how long does it take)
15. WIEVIELKOSTETES (how much does it cost)
16. FREMDENBUCH (visitors' book)
17. BENEDIKTINER (Benedictine)
18. SELTERWASSER (seltzer water)
19. BEIWEMKAUFENSIE (who is buying them)
20. PAPIERHANDLUNG (stationery business)
21. PFERDEVERMIETER (horse hire)
22. GOLDARBEITER (gold workers)
23. HANDLUNGSGEHILFE (clerk/executive assistant)
24. HANDSCHUHMACHER (glover)
25. INSTRUMENTENMACHER (instrument maker)
26. KAMELTREIBER (camel driver)
27. RADIERGUMMI (eraser)
28. BESORGEDIEPFERDE (get the horses)

29. DUNKELKAMMER (darkroom)
30. RECHTSGELEHRTER (legal scholar/lawyer)

And the reconstructed substitution alphabet is

```
A B C D E F G H I J K L M N O P Q R S T U V W X Y Z
O H N R G S Q F E - A U I P B V - C J K M T L - - D
```

While this enciphering of the keywords seems to add more complexity, it really only adds an extra step for the encipherer. Because the encipherment is monoalphabetic, it retains all the language characteristics of the original words, and those will be reflected in which of the mixed alphabets in the Vigenère tableau are selected. Luckily for the Germans, the Wilhelm cipher was used infrequently to communicate from Berlin to German outposts in North Africa and was not of any real strategic importance.

Many of the problems that the Germans had with their transposition ciphers were operator errors; notably an operator would fail to do the second transposition and just send out a message with a single transposition. These were significantly easier for the Allies to decrypt, and they gave away the keys for the correctly enciphered messages. See Childs' memorandum *German Military Ciphers* for a detailed discussion on decrypting transposition ciphers (Childs 1919, p. 3–12). The most famous and most difficult cipher system created by the Germans was the ADFGVX cipher released on March 10, 1918, and used for the rest of the war. ADFGVX is a combination substitution and transposition cipher. See Chap. 7 for a detailed discussion of the ADFGVX cipher.

While the Allies began using trench codes in 1916, it took the Imperial German Army until early 1917 to switch most of their communications close to the front line from ciphers to codes. Like the Allies, the Germans began with a simple, small code that could be printed in just a few sheets of paper. This *Befehlstafel* ("command table") only used two-letter code words and was in use in various editions from March 1917 till March 1918 (Kahn 1967, p. 315). In June 1917, the Germans released the first of their *Satzbuch* (sentence book) trench codes for use down to the regiment level, restricting the *Befehlstafel* to the battalion level and below. The *Satzbuch* was a two-part code of about 2,000 codewords in its first edition and was later expanded to over 4,000 codewords. Its codewords were all three letters long, and they all began with either K, R, or U. Hence, the Allies called it the KRU code. The Germans replaced the *Satzbuch* with new editions about every 30 days. Eventually, this was down to a replacement schedule of about every 2 weeks. To avoid capture of a codebook, *Satzbuch* codes were not allowed within the so-called danger zone, that is, within three kilometers of the front line (Friedman 1919, pp. 3–4). Over time, the Germans expanded the *Satzbuch* code, allowing the use of initial letters S and A as well as K, R, and U; the Allies then called the code the KRUSA code. Various editions of the KRUSA code were used by the Germans up to the Armistice. The Allies succeeded in solving many *Satzbuch* and *Befehlstafel* messages, often because of an operator error or because an operator would repeat a message in two different editions, giving the Allied cryptanalysts an immediate entry into the new code.

On March 5, 1918, just 5 days before the ADFGX cipher was released, the Germans replaced the *Befehlstafel* with a new three-number code called the *Schlüsselheft*. The *Schlüsselheft* was designed only to be used in the 3 km danger zone near the front, and it was the only code allowed in the danger zone. It was a one-part code of only 1,000 three-digit groups and was always superenciphered; the Allies called it the three-number code. The superencipherment was done via an enciphering table, the *Geheimklappe*, distributed with the code (Kahn 1967, p. 315; Friedman 1919, p. 66). While the code itself was never changed or replaced (Friedman 1919, p. 70), the *Geheimklappe* was replaced regularly (daily in the last month of the war) until the Armistice under the assumption that if the codebook itself was captured, there was still some security in the superencipherment. The superencipherment via the *Geheimklappe* was clever. Two tables were used, one for enciphering (the *Verschlüsselungstafel*) and one for deciphering (the *Entschlüsselungstafel*). Only the first two digits of each code word were enciphered. Encipherment was done by splitting those two digits into row and column indexes of the 10 × 10 *Geheimklappe* tables and using the two-digit number at the intersection as the new prefix for the codeword. Decipherment worked the same way. See Fig. 5.5 for examples of the *Geheimklappe* enciphering and deciphering tables (Barker 1979, pp. 220–221).

The first break in the *Schlüsselheft* came on the very night it was released. In the early morning hours of March 11, 1918, Captain Hugo Berthold of the AEF Radio Intelligence Section, Code Solving Section, was given several messages in the new code. Among them was one particular message:

```
ÄN v X2 (Souilly 0040) 0025 CHI-13
845 422 373 792 240 245 068 652 781 245 659 659 504
```

This message is obviously in code and in a new three-number code that Berthold had never seen before. On the very same telegram sheet, Berthold also saw the following message:

```
X2 v ÄN (Souilly 0052) 0025 CHI-13
OS RGV KZD
```

This message is in the KRUSA code (the Three-Letter code) that Berthold could read at least part of. The interesting part of the message is the OS, which was the German abbreviation for OHNE SINN, meaning "makes no sense." Berthold also discovered that the code group RGV means "old." Now, what Berthold had was a message from radio station ÄN at 0040 sent to station *X*2 in the new code. Twelve minutes later, at 0052, *X*2 sends a message to ÄN saying "Your message 0025 makes no sense…old…." So what Berthold needs is the decipherment for the last KRUSA codeword, KZD. He guesses that it means something like "resend," making the return message something like "Your message 0025 makes no sense. Resend in old code." Berthold immediately begins looking through the other telegram sheets

Fig. 5.5 Geheimklappe enciphering and deciphering tables (From Friedman document on German Codes in WWI at NARA. *Public Domain*)

Verschlüsselungstafel.

	0	1	2	3	4	5	6	7	8	9
0	23	48	60	05	78	35	58	64	29	52
1	20	77	33	59	21	70	02	40	63	08
2	11	49	01	69	47	41	79	74	22	42
3	32	76	39	18	75	30	09	51	80	65
4	61	19	43	81	06	56	73	62	10	28
5	85	50	24	88	31	84	27	90	55	57
5	03	91	96	53	68	16	44	89	15	87
7	97	25	71	04	95	34	14	37	93	38
8	26	72	54	92	13	83	45	00	66	67
9	86	12	98	36	99	46	82	17	94	07

Entschlüsselungstafel.

	0	1	2	3	4	5	6	7	8	9
0	87	22	16	60	73	03	44	99	19	36
1	48	20	91	84	76	68	65	97	33	47
2	10	14	28	00	52	71	80	56	49	08
3	35	54	30	12	75	05	93	77	79	32
4	17	25	29	42	66	86	95	24	01	21
5	51	37	09	63	82	58	45	59	06	13
6	02	40	47	18	07	39	88	89	64	23
7	15	72	81	46	27	34	31	11	04	26
8	38	43	96	85	55	50	90	69	53	67
9	57	61	83	78	98	74	62	70	92	94

to see if he can find another message from ÄN to *X*2. Almost immediately, Berthold finds another message in KRUSA from ÄN to *X*2:

```
ÄN v X2 (Souilly 0057) 0025 CHI-14
UYC REM KUL RHI KWZ RLF RNQ KRD RVJ UOB KUU UQX UFQ RQK
```

So station ÄN had responded at 0057 to station *X*2 using the same message number (0025) and the partially broken KRUSA code, with a message almost exactly the same length (14 code groups instead of 13) as the original message. Berthold was then able to create a mapping between known codewords in the KRUSA code and most of the equivalent codewords in the new three-number code (Friedman 1919, pp. 70–72). The extra codeword was UOB, which was a null. His results were immediately sent to the French and British cryptanalysts who by the next day had recovered the *Geheimklappe* enciphering tables. Two weeks later, on March 25, the French captured a *Schlüsselheft* codebook. From that time forward, all the Allies had to do was recover the superencipherment keys when they changed and they could then read all the *Schlüsselheft* messages.

The trench codes of both sides were effective means of keeping messages secure, provided the messages were kept short, the instructions in how to use the code were followed carefully by the operators, and the total number of messages was kept low. The German KRUSA was particularly effective, but only up to the point where a sector of the line became active. When an offensive was being prepared and during the duration of the offensive, the amount of wireless traffic increased dramatically. This gave the cryptanalysts more material to work with, and it increased the probability of error on the part of the operators. These times allowed the Allied cryptanalysts the opportunity to solve more codewords. However, the Germans countered this opportunity by changing the KRUSA codes fairly frequently, forcing the Allies back to the beginning (Friedman 1919, p. 65).

The three-number code was breakable, but the number of messages and the daily key changes stretched the Allied cryptanalysts to the limit. Friedman has this to say about the American efforts against the three-number code:

> In the 244 days from March 12 to November 11, 1918, there were intercepted 952 stenciled pages of text intercepted opposite our forces, an average slightly under 4 pages per day. The majority of the messages were short, from 2 to 15 groups.
>
> During the first weeks of the code about one-half of the messages were sent in the base (with no superencipherment). This proportion fell off steadily so that during the last months the number of messages in the base constituted less than 5 percent of the total.
>
> The messages in the base were decoded as they came into our Radio Intelligence Offices at the First and Second Armies. Enciphered messages could be decoded only as keys in which they were enciphered were worked out. It is no exaggeration to say, however, that over 50 percent of the enciphered messages were deciphered and decoded, either in whole or in part…
>
> The outstanding fact in a study of the Three-Number Code wireless messages is that nothing of importance was sent by wireless when there was time to send it by messenger. It is, however, equally clear, that when there was not an opportunity to communicate by messenger, the most important messages could be sent in the Three-Number Code.
>
> The Three-Number Code had many excellent features. The Secret Key was a small slip of paper pasted into the back cover of the codebook by a narrow strip of edging on either the top or left hand side. When in use it projected from the codebook, so that both enciphering and deciphering tables showed clearly. When not in use it folded back, so that it was entirely within the book. There was no danger of it being lost. When capture was imminent, the key could be torn out and thrown away or destroyed. It could be burned in one-hundredth the time it would have taken to burn the codebook itself. If captured, it could easily be replaced. It could be changed just as often as deemed desirable.
>
> The capture of the codebook was foreseen when the system was adopted. The method of enciphering adopted was not sufficient to protect the contents of all the messages from decipherment. But the idea of a code book with a key which can be frequently changed, easily destroyed when in danger of capture, and quickly replaced by a new key if captured, was excellent.

The uniformity of the system along the entire front was desirable. It minimized confusion. Further, the fact that the base of every code book was exactly the same as the base of every other code book together with the fact that every unit of any size was known to have a copy, enabled any unit to communicate with any other unit in a code, which, though simple would nevertheless prevent the contents of the message from being exploited by the enemy troops opposite for at least several hours.

These advantages of uniformity were counterbalanced by too long an adherence to the same edition of the code. (Friedman 1919, pp. 82–83)

References

Barker, Wayne G. 1979. *The History of Codes and Ciphers in the United States during World War I*. Vol. 21. Laguna Beach, CA: Aegean Park Press.

Bauer, Craig P. 2013. *Secret History: The Story of Cryptology*. Boca Raton, FL: CRC Press.

Childs, J. Rives. 1919. *German Military Ciphers from February to November 1918*. 1016. Paris: United States Army Expeditionary Force. Friedman Collection. National Archives, College Park, MD.

Childs, J. Rives. 1978. "My Recollections of G.2 A.6." *Cryptologia* 2 (3): 201–14. doi:10.1080/0161-117891853018.

Ferris, John. 1988. "The British Army and Signals Intelligence in the Field during the First World War." *Intelligence and National Security* 3 (4): 23–48.

Friedman, William F. 1919. *Field Codes Used by the German Army during the World War*. 209. Washington, DC: War Department. Friedman Collection. National Archives, College Park, MD.

Friedman, William F. 1942. *American Army Field Codes in the American Expeditionary Forces during the First World War*. Washington, DC: War Department, Office of the Chief Signal Officer. https://www.nsa.gov/public_info/_files/friedmanDocuments/Publications/FOLDER_267/41784809082383.pdf

Gaines, Helen Fouché. 1956. *Cryptanalysis; a Study of Ciphers and Their Solution*. New York: Dover Publications.

Kahn, David. 1967. *The Codebreakers; The Story of Secret Writing*. New York: Macmillan.

Chapter 6
American Codes and Ciphers in France

John Matthews Manly

Abstract This Manly article (Article II) contains more AEF anecdotes and a discussion of general methods of breaking codes and ciphers. It includes discussions of the weaknesses of the Playfair cryptographic system, the US Army Cipher Disk, and the Vigenère polyalphabetic system. This article repeats the story of Lt. J. Rives Childs and the US Army trench code. That story is told in more detail (and less accuracy) in Yardley's *American Black Chamber*. This article also confirms the story from Ronald Clark's biography of Lt. William Friedman breaking the Pletts cipher machine in 1918. The article ends with a cliffhanger about the ADFGX cipher.

One of the best stories of the "electric screen"[1] concerns a French corporal assigned to duty at one of the AEF stations. While the electric screen was in operation, there were few German conversations to report. This station, however, turned in one morning many pages of German conversation giving information of considerable importance, including details of the relief of stations, the movements of troops, and preparations for an impending attack. Inquiry revealed the fact that the French corporal, having got tired of doing nothing and having lost interest in the pursuit of trench rats and smaller game, had taken about a mile of fine wire, crossed No-Man's Land, made his way through the German trenches, located a telephone central, and attached his thin wire to one of the principal trunk lines. He then crawled back to his station and for more than 4 h copied every word that passed over the trunk line. Incidentally, he examined one of the German listening-in sets and brought back with him a sample of the induction wire used with it. When asked why he did not bring the instrument itself, he replied, "There were several German soldiers sitting around and I didn't feel like making any disturbance" (Moorman 1920b, pp. 1039–1044).

Colonel Moorman says that in the early stages of the war, it was customary to assume that increased activity on the part of the German radio stations indicated preparations for increased activity of the fighting troops (Moorman 1920b, p. 1041). To defeat this inference, the Germans adopted the plan of sending "fakes," or meaningless messages, in periods of inactivity, for the purpose of maintaining a fictitious

[1] "Electric screens" were normally electric generators set up behind the trench lines that would generate electrical "noise" in order to prevent the enemy from listening in on telephone traffic using induction cables and listening posts (Moorman 1920b, p. 1042).

J.F. Dooley, *Codes, Ciphers and Spies*, DOI 10.1007/978-3-319-29415-5_6

appearance of activity (Moorman 1920b, pp. 1040–1041). During the interval between the adoption of each new code and the time when we really began to get solutions of it, this faked activity was very deceptive.

> We then began", says Colonel Moorman, "to study with greater care than before the identity and location of the enemy radio stations, with a view to aiding in the determination of the character of the messages. With the love of system characteristics of the Germans, the fictitious messages were all sent in the same manner—that is, across divisional boundaries. If, therefore, we could determine from the call letters of the sending and receiving stations that they were in different divisional areas, we could safely assume that the messages were meaningless and the activity fictitious. On the other hand, when we had learned to read the messages, we were aided in determining whether sending and receiving stations were in the same or different divisional areas by the meaning or lack of meaning of the messages passing between them. (Moorman 1920b, p. 1041)

The instances in which our code men were able to cooperate with other branches of the service were many. Some have already been mentioned. Another type of service of immediate value is described by a wireless operator. It occurred in connection with the work of enemy airplanes in assisting their artillery.

The general practice was this: As soon as the observing plane was well clear of the ground, it would try out its radio apparatus by giving its own call letters and those of the receiving station. Our code men kept a record of each plane for each day. If a plane continued its work and carried out its observation in spite of interference, its space for the day would be colored red. The space of the plane whose observer was easily frightened, or got excited and failed to signal the shots, would be colored yellow. Other colors indicated other characteristics. This record was hung up in view of the radio operator; and when a plane went up, he could tell from its call and a glance at his chart whether or not it was worth attention. If its space was colored red, obviously the location of the battery working with it became a matter of great importance. The identity of the battery could generally be determined from the call letters; and it was often possible not only to warn our own men of the direction from which battery fire might be expected but in some cases even to lay down a counterfire and prevent the attack (Moorman 1920b, p. 1041).

The high value that information derived from deciphered messages concerning the situation and plans of the enemy had for officers in charge of field operations is sufficiently shown by some of the messages quoted in our previous article from the experience of both the British and the AEF. But decipherments that are achieved only after the events to which they refer have already occurred are obviously of little or no value to men of action. Such decipherments may have a scientific interest and—as we shall see later—may ultimately have some practical application and value, but for use in operations, speed in decipherment is not only important; it is absolutely vital.

Speed in deciphering code and cipher messages may be attained in various ways. I have already cited instances—such as those involving the repetition of a message in two codes and Lieutenant Jaeger's unwise vanity concerning his signature—in which the work of our code experts was greatly facilitated by accident. Fortune also sometimes favored them in other ways. Occasionally, a trench or field code was captured in some sudden descent upon an enemy's station. Once, a code for the direction of airplane activities was procured in an interesting manner. On the night of September 15–16, 1918, one of the giant airplanes of the enemy was brought down, and although it burst into flames and the codebook it contained was burned to a cinder, experts were able to handle the burned leaves with such success that a considerable part of the code could be read, and from this it was possible to work out the rest of it.

But such occurrences are infrequent and cannot be depended on for success. The scientific methods of penetrating the secrets of an unknown code or cipher are the only resource that can be depended upon for regular and satisfactory results.

Codes and ciphers capable of resisting the present methods of an enemy attack are remarkable examples of human skill. They are constructed with a clear recognition of the fact that a system is not a good one unless it makes possible the sending of messages that resist attack even when the system itself is known to the enemy.[2] A perfect system would produce messages which could not be deciphered by the inventor of the system any more easily than by other competent experts.[3]

Such requirements were too exacting for the ciphers and codes in use at the beginning of the war. The British, as we have seen, were obliged to abandon the Playfair system, which the subtlety of modern methods of attack had reduced from impregnability to incapacity for resisting attack more than 30 min. They had also rejected a cipher machine, the Pletts Machine—a greatly improved form of a machine originally regarded as producing ciphers that were absolutely indecipherable—because Lieutenant Friedman, of the Riverbank Laboratory and later of the AEF code section, was able to decipher it and work out a general method of attack on the basis of only five[4] messages.

The situation in the American Army was no better. The official cipher was known as the Army Cipher Disk. When used with a keyword—as was customary—this disk produced a cipher identical with that produced by the Beaufort method of using the famous Vigenère table, a form of cipher which for nearly 200 years had rejoiced in the title of "Le chiffre indéchiffrable." Such Army experts as Parker Hitt, Moorman, and Mauborgne had pointed out its lack of security and demonstrated

[2] This is *Kerckhoffs's principle*.

[3] The one-time pad is the only known such system. It was originally invented by Gilbert Vernam of AT&T in 1917 and was perfected by Major Joseph Mauborgne, US Army, in 1918. In his seminal paper, *Communication Theory of Secrecy Systems*, Claude Shannon calls this "perfect secrecy" (Shannon 1949).

[4] In Clark (1977), the number of messages is given as six. Kahn (1967) gives the number as five. The Pletts device was a cipher disk that was a modification of the Wheatstone cipher device. It utilized two mixed alphabets on a pair of rotating disks with a different number of symbols on each disk (so that the multiple of the number of ways of selecting the alphabets was large). In 1917, the British were considering it as a replacement for the Playfair as their tactical cipher system. William Friedman broke the five messages and uncovered the keys in approximately 3 hours. The British subsequently abandoned their idea of using the Pletts in the field (Kahn 1967, p. 372; Clark 1977, pp. 57–60).

that it could be deciphered by an expert almost as easily as a single-alphabet substitution cipher.[5]

At one time, it was proposed to use the Cipher Disk with a running key, that is, to take as the key a continuous passage in some book agreed upon—beginning, for example, at the top of a certain page and enciphering each successive letter of the message by the successive letters of the chosen key. This method, originally regarded as absolutely impervious to attack, was solved by Captain Yardley on the basis of two messages; and Captain Yardley and I later devised a method whereby with four messages a clerk could solve them on a Burroughs adding machine.[6]

It became increasingly clear that a practical system, one that would resist the subtlety of modern attacks, must be devised by persons expert in methods of attack. But for a time, this was not recognized or admitted except among the small body of cipher experts. On August 5, 1918, General Kress von Kressenstein,[7] who was in charge of field operations in the neighborhood of Tiflis, began to send highly important messages to the Berlin Headquarters in a cipher that he clearly regarded as indecipherable. On August 8, an intercepted wireless message from the Chief Intelligence Officer in Berlin to General von Kress read as follows:

> The cipher prepared by General von Kress was solved here at once. Its further use in operations in forbidden.

This was not surprising, as it had also been solved at AEF G.H.Q.[8]

With equal simplemindedness, the American Commission to Russia[9] provided its own cipher without consultation with cipher experts. When its messages were submitted to MI-8 at Washington, they were found to be in a fine, old-fashioned system, using various keywords of 10–12 letters—a system so simple that when the messages were assigned as problems to a class in cipher in MI-8 which had had only

[5]The Beaufort cipher is similar to the Vigenère polyalphabetic cipher system. It uses the same tableau as the Vigenère, but uses a slightly different algorithm for converting the plaintext characters into the ciphertext characters. The Beaufort is a reciprocal cipher system; the algorithm for encryption and decryption is the same. In the Vigenère, the encryption and decryption algorithms are different. In operation, the US Army cipher disk was identical to the Beaufort cipher system.

[6]There does not appear to be any other record to substantiate this claim. William Friedman did devise a general solution for auto-key ciphers (Friedman 1939).

[7]General Kress von Kressenstein (1870–1948) was a German general who spent most of the war commanding Ottoman troops in Palestine. He was assigned to the Ottoman Army in 1914, first as a military engineer and then as chief of staff and later commander. He commanded the Ottoman troops defending Gaza in 1917 and later commanded the 8th Ottoman Army. After the December 5, 1917, Armistice of Erzincan that removed the Ottoman Empire from the conflict, Kressenstein was transferred to the command of the German Caucasus Expedition. On May 28, 1918, the Treaty of Poti granted German protection to the new Democratic Republic of Georgia. Kressenstein marched the 3,000 men of the Expedition to Tbilisi (Tiflis) in June 1918 and commenced operations against the new Turkish Army. He was relieved on October 21, 1918.

[8]This is the so-called 777 cipher. The cipher was a monoalphabetic cipher and was solved by Lt. J. Rives Childs around August 5, 1918 (Kahn (1967), p. 337; Childs 1978, p. 211).

[9]On March 3, 1917, Czar Nicholas II abdicated and a Provisional government was formed. President Wilson organized the American Mission to Russia, led by former Secretary of War Elihu Root, in May 1917 with the objective of keeping Russia in the war on the side of the Allies.

Fig. 6.1 Lt. J. Rives Childs and Captain Herbert Yardley in Paris, 1918 (*Public Domain*. From U.S. Army History web site. https://www.ikn.army.mil/apps/MIHOF/biographies/Yardley,%20 Herbert.pdf)

3 days of instruction, they were solved in a single afternoon by even the slowest members of the class. Our codes also were not up to date. The Army code had been in use since the Spanish-American War[10]; and the trench codes prepared for use on the Western Front were not devised by men familiar with modern code attack.

In order to demonstrate the dangers arising from the imperfections of our field code, Lieutenant Childs of the cipher section was assigned the duty of attacking our messages just as if he were an enemy expert. He was supplied with our code messages just as an enemy expert would have been supplied with them by his radio operators and was to attempt to decipher them without the aid of the codebook, just as the enemy expert would. He was then to tabulate the information thus gained, in order to show what the enemy might easily learn—and probably was learning—by deciphering our messages.[11] (Fig. 6.1)

---

[10] It is not clear what army code Manly is talking about here. The standard War Department Telegraph Code 1915 was the high-level code used by the AEF until it was replaced in 1918. The 1915 code was not considered secure, even in 1915.

[11] Childs did not break the code. He broke the superencipherment of the proposed trench code in May 1918 (Kahn 1967, p. 327). Childs was working on the original Army Trench Code, which was a one-part code (just a single set of code words, usually created sequentially, used for both encryption and decryption). Messages created using this code were then enciphered using a monoalphabetic cipher in order to improve security. Childs had access to the codebook and knew the system of superencipherment used. Childs proved that the superencipherment was too weak. This story is also misreported in Yardley's book *The American Black Chamber*. (See the discussion in Chap. 5.)

The experiment took place at the time of the great attack on the St. Mihiel salient. Childs was able to penetrate the secrets of our code and to tabulate in his report the whole American battle order, giving a description on the troops engaged and their disposition on the front.[12]

The care necessary in all communications on the front is well illustrated by another incident that occurred at the same time. In order to give Lieutenant Childs the same facilities for attacking the code that the German experts would have, he was supplied with reports of telephone conversations overheard on the induction wires described in our articles of last week. Among these conversations was that of a certain operator, who reported by telephone that some of his switchboard wires had been broken by the passage of tanks and heavy artillery, which had been moving into a small wood near him all night, and that an attack would take place the next morning. As he gave the location of his station, it is clear that if the Germans were listening in upon us, as we were upon them, they may well have obtained from his conversation information of the highest military value.

The security of a code or cipher system depends not only upon its merits but also upon the intelligence and care with which it is used. A single message carelessly enciphered or incorrectly handled may give enemy experts just the clues needed for a successful attack upon the system, for these experts, like eagles, are continually sweeping the air in search for their prey. They hover about, as it were, awaiting the fatal moment when some member of the enemy's staff shall make a mistake; and then they swoop down upon him.

On July 3, 1918, the Germans introduced for communication with General Kress von Kressenstein, whose headquarters were at Tiflis, a new cipher, known to the Allied code men as ALACHI, from the word with which every message began. When Childs first got one of these, he examined it and said, "Look here what they have sent! Can you imagine their doing anything so foolish? This is nothing but a plain transposition cipher. I will just sit down here a moment and work it out."

Three weeks later he had changed his mind, for he was still sitting and working. He had discovered that it was not a plain single transposition cipher but a highly complicated double transposition, and he was still laboring over every message in the hope of finding a method of solution. On July 24, the first of the messages intercepted that day resisted his attacks, as before; but when he examined the second, he found that something had happened. The operator at Constantinople had forgotten to make the second transposition and had left the message a plain single

[12] Childs' experiment was conducted in May 1918 and had nothing to do with the St. Mihiel assault, which happened September 12–16, 1918. By the time of the St. Mihiel assault, the American forces were using the "River" series of codes rather than the original Army Trench Code. The River series (the first code was named *Potomac*) was a two-part code that used two lists, one for encryption (and sorted by the plaintext words and phrases to be encoded) and one for decryption (and sorted by codeword). This system provided much better security than the original trench code. The first River series code was published on June 24, 1918. See Chap. 5 for the discussion of American trench codes.

transposition.[13] It yielded readily to attack and not only confirmed the analysis of its character made by Childs but gave a transposition key which enables him to read not only that message but the one that had just preceded it and other double transposition ciphers written with the same key.[14]

For the amusement of readers who would like to see how such a cipher is solved, I will print the message and add a few remarks on the method of solution:

```
OSM v COS ALACHI 152
RRSCH NSEAT NWENT URZAF LBDLN IEAEE FAIRL
ROMGH NENMF NNSIU ZTDLI ORFAJ ENFJN IAUBT
OETRC AERIS RKRAD FAIIT LLEUA HBHRS ENGIO
VOALT RRJBU IEIGT TETLN NEWEL ITAEZ FPKET
LNITP SGMHB AT
```

If this is a columnar transposition cipher, it will have been put up in a grille, somewhat like that illustrated in our first article. For example,

```
2 5 10 1 8 4 9 3 6 7
t h i s i s a s i n
g l e t r a n s p o
s i t i o n c i p h
e r
```

The encipherment consists in taking the letters vertically in the order of the numbers heading the columns: STITG SESSI SANHL IRIRP NOHIR OANCI ET. [Manly now switches back to the original challenge cipher. Ed.]

The simplest method of solving such a cipher as this is based upon the fact that in the German language, the letter C is almost always followed by an H or a K. In this message, there are two Cs and five Hs. If we can find the H that goes with each C, the two or three letters preceding the Cs and Hs ought also to pair off, so as to form parts of German words. Let us try this:

```
(First set) RO RE RA RS  (Second set) EO EE EZ ES

RM RU RH RG              TM TU TH TG
SG SA SB SM              RG RA RB RM
CH CH CH CH              CH CH CH CH
HN HB HR HB              AN AB AR AB
NE NH NS NA              EE EH ES EA
SN SR SE ST              RN RR RE RT
```

Upon examination of the paired columns of the first set, we are at once struck with the last two pairs of letters—NA and ST—in the last column of pairs. NA suggests the word NACH and almost cries aloud to be matched with one of the columns of pairs in the second set. Experiment shows that the best matching of all the letters

[13] The ALACHI cipher messages Childs was looking at went first from Berlin to Tiflis and then from Tiflis to Constantinople. It was the German operator in Tiflis who erroneously used just a single transposition on July 24 to give Childs the opening he needed (Kahn 1967, pp. 338–339).

[14] This story is told in Childs (1919) and in Moorman (1920a).

is afforded by the second of these columns, for it gives not only the good syllable BRA but also the syllable or word STAB:

```
RS
RG
SMEE
CHTU
HBRA
NACH
STAB
```

In this particular case, we can see at a glance that the columns may be seven letters long and cannot be longer, for the letters before C and after H determine the beginning and the end of our first column of pairs. Usually we should count the intervals of the letters in our group CHTU and then try to find a number that would exactly or nearly divide them all. From C to H, the count is 145 letters; from H to T it is 76; and from T to U it is 35. But 35 is 7×5, and 76 is one less than 7×11, and 145 is two less than 7×21. The message contains 152 letters, which is two less than 7×22. So the message was put up in twenty-two columns, of which the first twenty have seven letters each and the last two have six each.

We can now cut up the message into runs of six or seven letters and try to match these runs as columns, keeping a sharp look out for combinations—such as ZU. EI.EN.ER.ST.—that suggest German words. But as we do not know where the two short columns of six have been placed, we must be prepared to find some columns beginning one or two letters earlier than would be the case if all the columns had seven letters each. When the columns are properly cut and matched, they will give a message in German that may be translated thus:

> Lt. Col. Baron von der Goltz and Major Count Graf Jolfsmeel with staff, will arrive in Braila on July 26th; they will then continue their journey to Tiflis. General Staff, Political Section, Berlin.

The cipher experts of all the Allies had been working for 3 weeks on this cipher; and when Childs at once telegraphed his discovery to them, he received the next day a message of congratulation from Captain Painvin, the great genius of the French Cipher Bureau.

Good fortune of this sort comes seldom, even to the man who is alert to seize the opportunity and profit by it. Such errors were not uncommon on the part of the operators at Constantinople, Poti, Wicoliev, and Sebastopol, but they were never committed by the highly trained staff at Berlin.

Scientific attack on ciphers of the modern type is not a matter of lucky guessing or of unaided analysis, but of keen scientific research, often based upon an enormous amount of preliminary drudgery. Much purely clerical and mechanical work is involved in the preparations for attack upon a system and often also in the solution of particular messages, after the system is thoroughly understood. Two things are necessary for success: plenty of ciphers and plenty of men.

Colonel Moorman has an amusing anecdote in this connection. When he arrived at AEF GHQ to organize code and cipher work, one of his first problems was to get adequate personnel. The Signal Corps, which had been piling up intercepted messages until there was a whole room full of them, would say, "You don't need any more of these messages; you have more now than you can handle."

Then Colonel Moorman would reply, "We have more men coming from Washington, and we need plenty of messages."

When the men came, General Nolan, Chief of G2, would say, "What do you need with all these men?" and Colonel Moorman would reply, "We have to have all these men—look at all these messages!" (Moorman 1920a, p. 7).

As a matter of fact, a plentiful supply both of messages and of men was necessary for successful attack upon the German codes and ciphers, and Colonel Moorman was right in demanding more of both. But he could not foresee that among the men to be sent from Washington in response to his call would be Lieutenant J. Rives Childs, who later revealed himself as the Painvin of the AEF Cipher Section, and Lieutenant Sellers, who was the first American to solve a German field and trench code.

Colonel Moorman's call reached Washington early in October 1917. The cipher and code work there had just been organized by Colonel R. H. Van Deman, as part of the work of the Military Intelligence Division, which his foresight, energy, and enthusiasm were creating.

When war was declared, the military intelligence staff of the United States Army consisted of Van Deman, who was then a major, Lieutenant A. B. Cox—later Lieutenant Colonel Cox—and one stenographer. With characteristic thoroughness, Colonel Van Deman organized, as rapidly as he could secure the necessary authority and the necessary men, all the branches of a well-equipped intelligence division. For the code and cipher work, he began with two civilians, who were immediately commissioned as officers. It was necessary to take civilians, because all the officers of the Regular Army who were known as experts in this field had been assigned to other posts.[15]

In June 1917, Colonel Van Deman asked for the services of H. O. Yardley, who for several years had been engaged in code work in the Department of State and in connection with this had developed an interest in cipher and a knowledge of it. At the end of September, he notified John M. Manly, a professor of English in the University of Chicago, who had volunteered his services in April, that if he would make immediate application, he would be commissioned for the code and cipher work. These two officers and two clerks constituted the whole of the Code and Cipher Section in Washington at the beginning of October 1917. When Colonel Moorman's call for more men arrived, it was obviously necessary to secure them by a special effort.

A number of young officers who had enlisted in various branches of the service had, as it happened, been selected for military intelligence work and had been sent to M.I.D., then located at the Army War College. It was found that four of them were ready to go into the code and cipher work and to take the necessary training.

[15]The two civilians were Herbert O. Yardley, then a code clerk at the State Department, and Professor John M. Manly, chair of the English Department at the University of Chicago. The only three Army experts were Parker Hitt, Frank Moorman, and Joseph Mauborgne (Yardley 1931).

Captain Yardley and Captain Manly improvised a course of instruction and for a month or so conducted classes in methods of cipher construction and attack.[16]

But meanwhile it had been decided that the Code and Cipher Section of M.I.D. should be a Central Cipher Bureau, to which should be sent for decipherment all messages and documents supposed to be in code or cipher that were obtained anywhere in the United States by any of the divisions of the Army, by the Department of Justice and other government departments, or by any of the civilian organizations that were cooperating with the government. The volume of messages which came to MI-8, therefore, soon became so great that, in spite of the addition of a cipher expert from the Department of Justice and the constant addition of officers and clerks as rapidly as suitable candidates could be found and trained, it was impossible to give sufficient attention to the special training of these officers for the AEF code and cipher work.

At this juncture arrived a welcome and generous invitation from Colonel George Fabyan, of Riverbank, Illinois. Colonel Fabyan had long been interested in Baconian ciphers and had organized a staff for work in that field. Immediately upon the declaration of war, he had offered the services of himself and his staff to the government and had sent two members of the staff—later commissioned as Captain Powell and Lieutenant Friedman—to the Army service school at Fort Leavenworth, to learn what could be learned about codes and ciphers from the Army officers stationed there.[17]

Pending the organization of an official Code and Cipher Section, a number of code messages and a larger number of ciphers had been sent to Colonel Fabyan from Washington and deciphered by him and his staff.[18] When, therefore, he learned that it was necessary to give instruction in code and cipher work to officers for the AEF, he placed Riverbank and all its facilities at the service of the government for this purpose. The four young officers in Washington were immediately sent there for further instruction; and later 100 officers were detailed to Riverbank for the same purpose.

---

[16] These four officers were Lts. J. Rives Childs, Lee Sellers, Robert Gilmore, and John Graham. They were given their cryptographic instruction in late fall of 1917 at Riverbank Laboratories by William and Elizebeth Friedman, and not by Yardley and Manly at MI-8. See Childs (1978). MI-8 began cryptographic instruction in early spring of 1918.

[17] This sentence is not correct. Powell did attend the Army Signal School and then returned to Riverbank as a US Army Captain. Friedman was not trained at the Signal School at Fort Leavenworth. Friedman was self-taught in cryptology and, with Powell and Elizebeth Smith Friedman, trained approximately 80 Army cryptanalysts at Riverbank Laboratories in three different classes from the fall of 1917 until April 1918 when MI-8 took over the training function (Finley 1995, p. 136).

[18] Fabyan volunteered the services of Riverbank in March 1917. Captain Joseph Mauborgne visited Riverbank and subsequently (April 11, 1917) wrote a recommendation that the Riverbank resources be used. The cryptanalysts at Riverbank, notably William and Elizebeth Friedman and Dr. (later Captain) J. A. Powell, decrypted a number of cipher messages forwarded to them by the War, State, and Postal departments from the summer of 1917 until spring of 1918 when MI-8 took over all decryptions.

About the middle of December 1917, it was decided that training of the four young officers was sufficiently advanced to justify their transfer to the AEF, but owing to delays in transportation they did not arrive at G.H.Q. until about February 1, 1918. This was within a few weeks before the great drive of March, the preparations for which resulted in an enormous increase in German cipher messages on the Western Front.

Lieutenant Childs visited the British Cipher Bureau in London and the French Bureau in Paris and was quickly inducted by Captain Brooke-Hunt and Captain Georges Painvin into the knowledge of German ciphers that these two great experts and their staffs had up to that time. To one acquainted, as Lieutenant Childs was, only with the forms and principles of cipher that had been in use before the outbreak of the war, the knowledge revealed to him by the British and French experts must have seemed marvelous indeed. But greater surprises still were in store for him; and one of them came a very few weeks later when the great German General Staff suddenly sprung upon the astonished experts of the British and French bureaus a new cipher, unlike any that had ever before been used.

Rumors were circulating on all sides concerning the great drive that the Germans were preparing for the spring. Lieutenant Childs was at the Cipher Bureau of the French War Office in consultation with Captain Painvin when there were brought in copies of wireless messages consisting entirely of the five letters, ADFGX. Captain Painvin looked at the telegrams for a moment and threw up his hands with a gesture of dismay!

References

Childs, J. Rives. 1978. "My Recollections of G.2 A.6." *Cryptologia* 2 (3): 201–14. doi:10.1080/0161-117891853018.

Clark, Ronald. 1977. *The Man Who Broke Purple*. Boston: Little, Brown and Company.

Finley, James P. 1995. *U.S. Army Military Intelligence History: A Source Book*. Fort Huachuca, AZ: U.S. Army Intelligence Center.

Friedman, William F. 1939. "Military Cryptanalysis, Part III. Simpler Varieties of Aperiodic Substitution Systems." Washington, DC: War Department, Office of the Chief Signal Officer.

Kahn, David. 1967. *The Codebreakers; The Story of Secret Writing*. New York: Macmillan.

Moorman, Frank. 1920a. "Wireless Intelligence." Presented at the Meeting of Officers of the Military Intelligence Division, Washington, DC, February 13.

Moorman, Frank. 1920b. "Code and Cipher in France." *Infantry Journal* XVI (12): 1039–44.

Shannon, Claude. 1949. "Communication Theory of Secrecy Systems." *Bell System Technical Journal* 28 (4): 656–715.

Yardley, Herbert O. 1931. *The American Black Chamber*. Indianapolis, IN: Bobbs-Merrill.

Chapter 7
Painvin Breaks a Cipher

John Matthews Manly

Abstract Manly's Article III takes up the ADFGX (later ADFGVX) story. It contains a brief description of the ADFGX and ADFGVX cipher systems and contains a discussion of Painvin's solution. There is also a discussion of the "Wilhelm" cipher used by the Political Section of the German Army for propaganda purposes. Finally, there is a general discussion of the different types of systems used in different parts of the German Army.

At a crucial moment, on the eve of the long-heralded German push of March 1918, the wonderfully organized German General Staff had suddenly put into action a cipher system of such apparent simplicity, but real subtlety, that at sight of the first message, Captain Painvin, the great genius of the French Cipher Bureau, threw up his hands in dismay and almost in despair (Fig. 7.1).

Even a layman can readily appreciate the character of this cipher. The precise message that occasioned the dismay of Captain Painvin is not known, but here is one of exactly the same type, which could not be distinguished from it except by the most careful expert examination. The message as received read thus:

```
LU v AG
CHI 128
XXDFG DXGAF DDFGA FDXDA DAAAX AXFAG GXDGF AAGFA GGDAF GFXFA
AFXDX DFFGD AXDFG FDGGD GADAF AXXDG XXFAX XDDDG DDAAX DFFFG
XAAFG DGDGA DDGAD GGFAA FGXFX GGD.
```

This type of cipher – later known to experts among the Allies as ADFGVX, because the letter V was added for convenience in transmitting numerals – was the most famous of all the German ciphers. It was in constant use for field operations on the Western Front and for communications with the Black Sea area, from its first introduction on March 5, 1918, until the very close of the war, and carried a larger volume of messages than any other cipher in the history of the war.[1]

[1]ADFGVX is what is known as a *fractionating cipher system*. In a fractionating system, each letter is enciphered using two or more letters, making the ciphertext longer than the plaintext. The ciphertext is then transposed, breaking up the enciphered pairs of letters. The ADFGVX cipher system was used for German command messages at the division level and above. This is significant because traffic analysis of the intercepted messages could provide the Allies with indications of German division and corps locations and troop movements.

J.F. Dooley, *Codes, Ciphers and Spies*, DOI 10.1007/978-3-319-29415-5_7

Fig. 7.1 Lt. Georges Painvin decrypting a cipher message (From the David Kahn Collection at the National Cryptologic Museum. Used with permission)

Table 7.1 Sample ADFGX square

	A	D	F	G	X
A	G	Z	K	N	B
D	S	A	R	E	Y
F	L	C	P	H	X
G	U	O	W	D	Q
X	I	T	F	V	M

At first sight, the message given above looks very simple. There are only five letters used in the construction of it, A D F G X. Obviously, therefore, it is impossible for these five letters taken singly to represent the 26 letters of the alphabet. The first idea that would occur to any person with some knowledge of ciphers is that we have here a two-letter substitution; for if the number of letters in the alphabet is reduced to 25 by treating I and J as the same letter, we could obviously represent each of these 25 letters by a group of two. Table 7.1 will show clearly how this could be done.

In the present table, A in the clear text would be represented in the cipher by DD; N in the clear text would be represented by AG; the word "any", then, would be spelled DDAGDX. This would give a cipher that would look very much like the message just quoted.[2]

But Captain Painvin, who had been studying the ciphers of the Germans since the beginning of the war, was too familiar with their methods to believe for a moment that they would use a cipher as simple as this. The very simplicity of it warned him of its real subtlety and difficulty and was in part a cause of his dismay.

There were other reasons for his state of mind. In the first place, there was every reason to trust the insistently repeated rumors of the approaching great drive; and this of itself would mean an enormous increase in the number of messages which would be broadcast by the Germans, intercepted by the Allied radio stations, and poured in floods upon the Cipher Bureaus of the Allies. It had always been obvious that immediately preceding and during times of special military activity, there was always a notable increase in the number of messages.

Besides, it would seem that the number of different codes and ciphers already in use by the Germans was enough to bewilder and dismay those charged with the task of attending to deciphering them. There were, for example, tri-numeral field and trench codes, differing for every sector of the front and changing at irregular intervals from 2 to 5 weeks.[3] There were three-letter field and trench codes, equally numerous and short-lived.[4] There were half a dozen meteorological codes, used for aiding artillery practice in different sectors. There were aviation codes and various other minor codes and ciphers. For long distance communications, there were the "Wilhelm" cipher, used for communications between Berlin and German agents on the north coast of Africa, and several varieties of double transposition ciphers, for communication with the Black Sea district.[5] There was the naval code, for communications to submarines, warships, and Zeppelins,[6] and there was the so-

[2] The square that Manly describes is known as a Polybius square after the ancient Greek historian. This square is usually written with numbers rather than letters and uses a keyword to jumble the alphabet that is inscribed in the square (Kahn 1967, p. 83). The ADFGVX system enciphered each plaintext letter using two ciphertext letters (called *confusion* by Claude Shannon); it then used a transposition to separate the pairs of ciphertext letters moving the letters in each pair farther away from each other in the final ciphertext (called *diffusion* by Shannon) (Shannon 1949).

[3] This is likely the *Schlüsselheft*, a three-digit code put into service on March 5, 1918, just days before the ADFGX cipher. The first break into this code was made by American cryptanalyst Lieutenant Hugo A. Berthold that very night. His partial solution was sent to Captain Painvin at the French cipher bureau the next morning. The *Schlüsselheft* code was never replaced, but its super-encipherment table (the *Geheimklappe*) was replaced every 2 weeks or so and daily near the end of the war.

[4] These are likely the *Satzbuchen* three-letter codes. These codebooks were changed every 10 days to 2 weeks.

[5] In Manly's Article II, we saw what happens to a double transposition cipher when a cipher operator forgets to do the second transposition.

[6] There were several German naval codes that were acquired by the Allies during the war. It is likely, because of interagency secrecy, that in 1927 Manly did not know any of the stories behind them. The most famous is the acquisition of the *Signalbuch der Kaiserlichen Marine*, the main German naval code. On the night of August 25, 1914, the German light cruiser *Magdeburg* ran aground just off the island of Odensholm at the entrance to the Gulf of Finland. The *Magdeburg*

called Colonial code, used for communications with German agents in various countries. Truly a bewildering situation, for the keys to some of the ciphers were changed daily, so that every new day brought a new problem for the cryptographers, not to speak of the "Wilhelm" cipher, with its 22 mixed alphabets and its 30 keys varying in length from 10 to 18.[7]

In spite, however, of the fact that he had only recently recovered from a long, serious illness, Captain Painvin's dismay did not paralyze his energy or confuse his mind. He and all the rest of the cipher analysts of the Allies immediately began a study, feverish in its intensity, but sane and dogged, and determined and cool in method. A friendly rivalry arose in which, like newspapermen, every expert was eager to make a "scoop," by being the first to solve the problem, but each loyally communicated to all the rest every new method of attack as soon as it was invented or discovered. And as we shall see, Captain Painvin's revenge was not long delayed.

On April 6, the officers of G2 at American Headquarters received from Captain Painvin a decipherment of a message of April 1, together with a note explaining completely the complex method of enciphering and giving the keys by which all messages for that date could be read. The message was originally enciphered in the following manner:

First, the letters of the clear text were enciphered by the five-letter substitution table – Table 7.1 was the one used for April 1. Then the cipher letters thus obtained were transposed in such a way that no two of them that belonged together were left together in the message as received. The key for this transposition was 13-3-18-7-6-11-4-19-9-16-8-5-17-2-10-15-1-14-12-20. Finally, the text thus obtained was separated into groups of five letters. Each bigraph represents one letter in the clear text that is found by referring to the table above.

This was so wonderful that there were some who openly suggested that the cipher had not been solved, but the keys had be captured or stolen. It seemed absolutely incredible that a cipher, the method of which was to split every letter into two parts and scatter these parts in all directions, could possibly be solved by any exer-

was part of a German squadron trying to enter the Gulf and attack Russian shipping. Throughout the night of August 12–16, efforts were made to free the Magdeburg. None were successful. Ultimately, her captain decided to scuttle the ship. As the crew was abandoning ship, the bow charge detonated prematurely, and, simultaneously, Russian warships arrived and began shelling. Two radio officers of the *Magdeburg* jumped overboard—clutching copies of the *Signalbuch* and German maps of mine fields in the Baltic. The Russians boarded the *Magdeburg* and found—untouched—a forgotten third copy of the codebook. The other two copies were recovered from the water, still mostly readable. On October 13, 1914, the Russians handed over the untouched copy of the codebook to the English in London. The Germans never believed that the Allies had a copy of their main naval code and didn't change the *Signalbuch* until 1917 (Kahn 1991; pp. 15–30).

[7]The "Wilhelm" cipher was a polyalphabetic cipher system that used a table of 22 mixed alphabets for encryption. There were 30 keywords that ranged in size from 10 to 18 letters. Each keyword was a monoalphabetically enciphered German word that was changed periodically; there were 30 keywords that were rotated through. Each key letter told the operator which of the 22 alphabets to use for the current plaintext letter substitution (Friedman 2006; Childs 1919).

cise of human ingenuity. But a little later, Captain Painvin sent in the keys for the messages of another day; and shortly after, the keys for still a third.[8]

Comparison of the three sets of keys for these 3 days showed that no two of them were alike and indicated – as had already been suspected – that both keys were changed every day.

Captain Painvin's triumphs came too late to be of service in connection with the March drive, but on May 27 the great battle called the Second Battle of the Marne began in a major offensive by the Germans. Naturally, the flood of these ciphers swelled enormously, and on May 28, 138, messages in this cipher alone were intercepted. The keys for this day were solved by Captain Painvin on May 31 and the next day, June 1, saw the solution of the key for May 30. All previous records, however, were broken the next day. The solution of the keys for June 1 was accomplished within 24 h and this at a time when the solution was of the highest value to the Allied commanders.[9]

The following samples will give some idea of the character of the enormous volume of messages transmitted in these exciting days. This one belongs to the Argonne sector. It was sent at 3:37 AM June 1 and intercepted by the radio station at Clermont:

Decipherment – "ANGRIFF NACH ARTL VORBEREITUNG 6 VORM GEMEINSAM MIT 7 AK WEITERFUEHREN."

Translation – "After artillery preparation at 6 A.M. extend attack in cooperation with 7th Army Corps."

Another message in the same sector, sent at 8 AM on the same day and intercepted by the radio station at Neufchatel, reads as follows:

Decipherment – "WO STANDORT UNSER BEUGIES ABMARSCH UNBESTIMMT."

Translation – "Where is our position at Beaugies[10]? Departure undetermined."

This was sent at 7:40 PM The same day:

Decipherment – "NACH MITTEILUNG 50 1 D IST I D VOM BOIS DAETROTTE GEWORFEN. 50 I D MACHT GEGENANGRI."

---

[8] In fact, Painvin had not come up with a general solution to the ADFGVX cipher system. Instead, he identified three special cases that allowed the Allied cryptanalysts to break many days' messages. The special cases were (1) finding two or more messages with identical plaintext beginnings which was possible because of the stylized formats of many German Army messages; (2) finding two or more messages with identical plaintext endings, as in a series of messages from the same German Army division or corps; and (3) finding several messages with exactly the same number of letters. Despite the use of special cases, Painvin and the other Allied cryptanalysts were able to recover about half of the ADFGVX messages sent between March 1918 and the armistice on November 11. A general solution to the ADFGVX cipher was not found until William Friedman and his first three cryptanalysts in the Army Signal Intelligence Service discovered it in 1933 (Rowlett et al. 1934).

[9] What makes this 1-day achievement all the more startling is that on June 1, the Germans changed the cipher system from ADFGX to ADFGVX, adding a 6th letter to increase the size of the Polybius square to 6×6 to accommodate the ten decimal digits. Painvin figured this out quickly and produced the first key within 26 hours (Kahn 1967, pp. 344–345).

[10] This is Beaugies-sous-Bois in northern France.

Translation – "According to report of the 50th Infantry Division the 52nd Inf Division has been thrown from Bios Daetrotte. 50th Infantry Division is making counter attack.

This, sent at 1:10 PM June 1, was intercepted at Souilly:

Decipherment – "LEUTNANT TANGE IN BRAUSNE QUARTIER MACHEN. A FUNK A 7."

Translation – "Lieutenant Tange. Establish quarters in Brausne. Army Wireless Detachment 7."

With characteristic shrewdness, the German General Staff chose this day, June 1, for a transformation of this cipher by introducing a sixth letter, the letter V. Fortunately, Captain Painvin had been solving the cipher with the five mysterious letters since the beginning of April, and he was thoroughly prepared to deal with the equally mysterious group of six. The story of the search for rapid methods of solving this cipher is too long to tell and would be of interest only to the cryptographer.[11] Suffice it to say that America was able, through the efficient work of Lieutenant

[11] Painvin's method of solving the ADFGX cipher relied on finding messages with stereotyped beginnings that the Germans used often. Such messages would fractionate the same and then form similar patterns in the positions in the ciphertext that had corresponded to column headings in the transposition table.

In a technique similar to the Babbage-Kasiski method for polyalphabetic ciphers, Painvin used repeating sections of ciphertext to derive information about the length of the key being used. If the key was an even number of letters in length, his key observation was that, due to the way the message was enciphered, each column consisted entirely of letters taken either from the top of the Polybius square or from the left of the square, but not a mixture of the two. This meant that after substitution, but before transposition, the columns would alternately consist entirely of "top" and "side" letters.

With the ADFGX cipher, each "side" letter or "top" letter is associated with five plaintext letters. In the Polybius square below, the "side" letter "D" is associated with the plaintext letters "e t b r k," while the "top" letter "D" is associated with the plaintext letters "i t h s x."

	A	D	F	G	X
A	d	i	a	m	n
D	e	t	b	r	k
F	g	h	u	l	c
G	v	s	f	w	z
X	o	x	p	y	q

Since these two groups of five letters have different cumulative frequency distributions, then a frequency analysis of the "D" letter in columns consisting of "side" letters will have a distinctively different result from those of the "D" letter in columns consisting of "top" letters.

This observation allowed Painvin to tentatively identify which columns consisted of "side" letters and which columns consisted of "top" letters. He could then pair them up and perform a frequency analysis on the pairings to see if they were noise or real pairings that corresponded to plaintext letters.

Once he had the proper pairings, he could then use frequency analysis again to figure out the actual plaintext letters. The result was still transposed, but at that point, all he had to do was solve a simple complete rectangular transposition. Once he was able to reconstruct the complete rectangle for one message, he had the daily key, and he would then be able to solve all messages enciphered with the same transposition key (derived from https://en.wikipedia.org/wiki/ADFGVX_cipher; retrieved 10 August 2015).

Childs, to contribute processes of much value and that in October solutions were obtained in less than 2 hours. But the use of this cipher was not confined to the operations on the Western Front, and the high value that the Germans placed upon it is well indicated by the tasks to which they next subjected it.

With the collapse of German power in the East and the breakdown of overland lines of communication, there was an imperative need for the maintenance by Germany of some sort of connection with what remained of her scattered forces and stores. The ADFGVX cipher was, therefore, made use of to keep in touch with General Kress von Kressenstein in Tiflis, with General Mackensen[12] in Romania, with the Eichhorn Army group[13] in Southern Russia, and with the innumerable propaganda delegations scattered through the length and breadth of the Near East.

The Germans appear never to have suspected that the Allies had solved this cipher. Even as late as November 6, the Chief Signal Officer at Berlin sent to the Chief Signal Officers in Tiflis and Bucharest the following message, which sufficiently indicates the confidence reposed in the security of the system:

> Do not destroy November cipher key for wireless station circuit. The key for December 1st will be the same as for November 10th etc. until December 21st, which will be the same as November 30th.

The main difference between the manner of using this cipher on the Western and Eastern Fronts was that – owing perhaps to difficulties of transportation – the keys for the East were not changed daily, but remained in operation for a week or more at a time.

Although by no means all the keys were solved, more than half the messages transmitted were read; and those that remained undeciphered belonged to periods when military operations were insignificant and the number of messages was small.

The most interesting message deciphered from the intercepts of the East was probably the long message, sent in ten parts, in which General Mackensen appealed for help and gave his plans for withdrawal from Romania and his estimate of the plans and intentions of the Allied forces. Here is a translation of a decipherment of parts of it:

> "To the Highest Command. Review of the situation. Up to date it has had to be reckoned with that the enemy will attempt a crossing of the Danube with the forces assembling at Lompalanka and vicinity of Rustchuk, with the object of cutting the railroad communication between Orsova and Craiova, and to thrust forward on Bucharest. Since November 1st, 1918, however, it appears that the Serbian armies, together with three French divisions, are engaged in an advance toward Belgrade-Semendria, and the intended attack at Vidin and Lompalanka seems to have been abandoned.

---

[12] General August von Mackensen was the commander of a combined German-Austrian army that invaded Serbia in September 1915. He was also commander of German forces in Romania beginning in 1916. He was responsible for the capture of the Romanian capital, Bucharest, and he spent the rest of the war in the Balkans.

[13] General Emil von Eichhorn was commander of the 10th Army and conquered much of Lithuania in 1915. His command was later expanded to *Heeresgruppe* (Army Group) *Eichhorn*, and in 1916, he conquered Latvia and Estonia. He was later made the military governor of the Ukraine and was assassinated on July 30, 1918, by a Russian revolutionary socialist.

> "It is therefore extremely probable that the Serbian armies, reinforced by the French, intend to cross the Danube at Belgrade-Semendria and march into Southern Hungary, while the French Army marching up south of Svistov and Rustchuk retains the task of directing an offensive toward Bucharest. In conjunction with this operation it is not impossible that Roumanian forces from Moldavia will enter Transylvania through the Tolgyes, Gymes and Oitos passes, thereby threatening the lines of communication in the rear of the army of occupation, which have up to now, as a result of – – –
>
> (9th part Missing)
>
> "– – – is threatened with attack, and the further occupation of Wallachia, as laid down in order of Headquarters Staff 2 1A. N.R. 11161 OP, is useless, and in view of the stocks on hand of munitions, provisions and coal cannot be carried out. In case a general armistice cannot be expected in the immediate future, it is proposed that the army of occupation be withdrawn from Roumania at once and start the march to Upper Silesia through Hungary, together with the German units of the 1-Army. Approval is requested. (Signed) K.M.R.M.1 a GR-OP"

So much interest attaches to the ADFGVX cipher, both on account of its importance and on account of the romantic story of its attack and solution that little space is left for the very important "Wilhelm" cipher of 22 alphabets and 30 keys, which has already been referred to. These messages were all easily recognizable by the first two groups, consisting always of the letters "Füer GOD." They were obviously intended for some distant station, the location of which was for some time unknown. But their chief use was for propaganda purposes, and the nature of their contents soon indicated their destination as being some point in North Africa, apparently Tripoli. The attempt to stir up revolt in North Africa appears in the following deciphered message, intercepted on August 8, 1917:

> Arabian soldiers and others of the Turko Regiment 1-3, who deserted to our lines in France, confirm each other in complaining about their being badly treated by the French, and getting much worse rations than French troops. Therefore, they gladly desert to the Germans, and are pleasantly surprised at the good treatment they receive from the Germans as the Germans had been pictured to them by the French as cruel barbarians. In Germany they learn for the first time the truth regarding the harmonious cooperation of Germany and Turkey, the chief Mohammedan power. General Staff, Political Section 30062.

The impression produced by this message is confirmed by the following, which was intercepted on May 2, 1918, and deciphered immediately:

> On April 24th (or '26th'), between Ancre and Somme, a Moroccan Division, consisting of 2 regiments of Turks, a Zouave Regiment and one regiment Foreign Legion, which had been assigned to attack on the most difficult battle front, suffered the most terrible losses, as it was thrown against the German lines in close order. This manner or utilization is characteristic of the brutal exploitation of colonial peoples by the French. Colonial peoples are continually places in the most dangerous positions. The divisions were nearly annihilated. Please spread this by propaganda in North Africa, including Morocco. General Political (Section) 24500.

The location of the receiving station would doubtless have been discovered immediately but for the fact that no messages passed from it to Berlin, a fact that caused much speculation. It finally appeared that the receiving station had no sending apparatus until January 28, 1918, when an intercept refers to the shipment of such an apparatus by a submarine.

This was the longest-lived of all the German ciphers. It existed without change of alphabets or keys from 1916 to the end of the war. The solution of the system as a system was due to the patience and skills of Captain Brooke-Hunt of the British Cipher Section, but the American G2 had the pleasure and honor of working out a number of keys and the general system upon which they were constructed.

On October 30, 1918, at 11 PM, American General Headquarters received the first official intimation of the collapse of Turkey from a decipherment of the last of the "Füer GOD" ciphers ever sent.[14] The information was of particular interest as it came from the most authoritative German source, the Political Section of the General Staff in Berlin.

As has already been indicated, every branch of the German Army had its own code or cipher. For example, between the aviation centers, hangers, and the giant airplanes, there was a special aviation code used, very much on the order of the three-letter trench code, but distinguished by the fact that every group of three letters must begin with one of the three letters F L G, the abbreviation for the German word Flieger, "Flier." In one instance, as has already been related, the Allies were able to read large portions of a burned codebook of this type which had been captured when one of the giant airplanes was brought down.[15]

The meteorological codes carried messages that were long misinterpreted by the Allies. They consisted sometimes of five-letter groups and sometimes of four-letter groups like the following:

```
DDMM UUUU HHTT RRSS, etc.
```

These were originally supposed to be merely weather reports and as such were neglected. Later it was learned that the first group gives the day and month and the second, the hour to which the message applies. These two groups are then followed by several pairs of groups that give information for the direction of artillery fire. The third and fourth groups indicate the altitude and the ballistic weight of the air; the fifth and sixth, the direction of the wind; and the seventh and eighth, its velocity. This information might well have been utilized by our own forces. Perhaps it was.[16]

The wireless stations themselves had their own secret languages, chiefly cipher rather than code. Contrary to the usual practice on the Western Front, this branch of the service used substitution ciphers. One form was a two-letter substitution cipher,

---

[14]The Armistice of Mudros was signed on October 30, 1918, and went into effect the following day.

[15]There were three of these aviation codes, designated FLG, FLGWX, and FLGVZ. Except for the story that Manly relates of the burned codebook recovery, the Allies did not break these codes during the war (Friedman 1919, p. 88).

[16]These codes varied from German army to army. The Americans characterized one set of three-letter meteorological codes used by the German 5th Army as the "Fritz" codes. By July 1918, the Germans had established about 25 meteorological stations along the Western Front. Each of the stations sent weather data intended to help with artillery fire seven times a day. Each message had the same format, a starting word indicating which message this was, followed by 11 code groups that indicated things like ballistic weight of the air, wind direction and speed, and the expected ballistic flight time of the shell (Friedman 1919, p. 89).

in which the two alphabets were made from key words such as "Goetz von Berlichingen" and Die Braut von Messina," the first of these being inverted in forming the alphabet. They also used a substitution cipher made with mixed alphabets that were slid one space after enciphering each group of five letters.[17] But aside from the great ciphers already discussed, our code and cipher men were mostly concerned with the two types of field and trench codes. The tasks involved in this work were enormous, from the number of codes in use – a separate one being assigned to each army division – the ingenious methods used to encipher or to disguise the code groups, and the frequency with which changes were made either in the books themselves or in the encipherments.

The three-letter field and trench codes were used between divisions, as well as between divisional brigade and regimental headquarters, and sometimes even battalion headquarters, along the entire front. About ten different codes would be in use at the same time. Each of these codes contained between 2,000 and 2,500 code groups, composed of three letters each. At first each group began with one of the three letters K R U. Later two more were added and used in a way that greatly complicated the task of the enemy expert.[18]

The two forms of this code which were of special interest to our own experts were that used in the Verdun sector by the Fifth Army and that used by Detachment C in the St. Mihiel sector. For purposes of reference, the French designated the latter of these as the Albert code and the former as the Marcel. Successive editions of each of these codes came out with considerable frequency. They were all known by the same name, but each new form was given a special number.[19] In general, one of these books would remain in effect for a period of 4 or 5 weeks. Marcel code No. 3, however, was changed after it had been in used for only 8 days, October 4–12, 1918. This was, doubtless, due to the fact that a copy of the book was known to have been captured by our men on October 8. Obviously, if the codebook had not been captured, our men would hardly have been able to read any message in code of so short life. Usually only a beginning in attacking a new code was made during the first week of its use, and the messages were not read freely and completely until the book had been in use 2 or 3 weeks. The Germans discovered this and changed their books with greater frequency toward the close of the war.

Notwithstanding all the difficulties attending the solution of codes of this character, both Lieutenant Sellers, one of the four young officers first selected for this work, and Lieutenant Friedman – who had been the instructor of Lieutenant Childs and Lieutenant Sellers at Riverbank and who later was placed in charge of code

[17] Neither of these would have been hard for Allied cryptanalysts to break. The first is just a two-alphabet polyalphabetic using mixed alphabets created with the keywords. The second is a true polyalphabetic, but each group of five letters is enciphered using the same alphabet; so each group of five letters is a monoalphabetic cipher.

[18] The letters added were S and A, creating the KRUSA code family.

[19] The Albert and Marcel codes were three-letter codes of about 2,000 code words, used between divisions and between wireless stations. They were changed approximately every 2 weeks. The names derive from the names used by the Germans in the message headers (Friedman 1919, p. 8).

decipherment with the A.E.F., one of the ablest code analysts developed by the war – succeeded in working out solutions of both the Albert and the Marcel codes and the tri-numeral codes.

The tri-numeral code was used by the most advanced posts to communicate with each other and this battalion headquarters or even regimental headquarters. It was first introduced in March 1918, in connection with the great spring drive. As the name indicates, the code groups are composed of three figures. Naturally, there are 1,000 of them, ranging from 000 to 999.[20]

Originally designed for the uses just indicated, this code carried an enormous number of messages. A chart captured early in October 1918, speaks of it as the only means then available for encoding messages sent within the danger zone, that is, within three kilometers of the front line, whether sent by radio, ground telegraph, telephone, carrier pigeons, or dogs.

As the tri-numeral codes were in use in the front line, many of them were captured. For instance, between September 25 and 29, 1918, the American First Army captured 19 tri-numeral codebooks and 17 keys (*Geheimklappen*).

As the mention of keys indicates, these codes were regularly enciphered with an enciphering table that was changed frequently. Attempts were made further to render difficult the work of the Allied experts by sending practice messages and messages subjected to various kinds of distortions. Obviously, such messages, unless recognized by the code expert, made him waste a great deal of his time and involved him in enormous labor. Fortunately, the Germans, for the protection of their own men, labeled these messages by enciphering in the second or third groups from the beginning or end of the messages, either the letters Ü B, meaning Übung (exercise), or O S, meaning Ohne Sinn (nonsense).

It is interesting to recall that about a week before the Armistice, our reports from AEF Headquarters informed us that during the preceding week, there were various evidences or nervousness, haste, and decreased efficiency in the handling of the German field messages. Many were encoded without the usual precautions, and some were even sent in clear text without encoding.

The state of mind of the senders of such messages is easily imagined when we further recall that about the middle of October, when the German hopes for the termination of the war were at their highest, the messages in the tri-numeral code contained a large number of references to peace and to freedom.

References

Childs, J. Rives. 1919. *German Military Ciphers from February to November 1918*. 1016. Paris, France: United States Army Expeditionary Force. Friedman Collection. National Archives, College Park, MD.

[20] This is the same *Schlüsselheft* code introduced in March 1918. It was designed to be used into and out of a 3 km "danger zone" closest to the front line (Friedman 1919, p. 66).

Friedman, William F. 1919. *Field Codes Used by the German Army during the World War*. #209. Washington, DC: War Department. Friedman Collection. National Archives, College Park, MD.

Friedman, William F. 2006. *The Friedman Legacy: A Tribute to William and Elizebeth Friedman*. Sources in Cryptologic History #3. Ft. George Meade, MD: National Security Agency/Center for Cryptologic History.

Kahn, David. 1967. *The Codebreakers; The Story of Secret Writing*. New York: Macmillan.

Kahn, David. 1991. *Seizing the Enigma*. Boston: Houghton Mifflin Company.

Rowlett, Frank R., Solomon Kullback, and Abraham Sinkov. 1934. General Solution for the ADFGVX Cipher. Washington, DC: U.S. Army Signal Intelligence Service.

Shannon, Claude. 1949. "Communication Theory of Secrecy Systems." *Bell System Technical Journal* 28(4): 656–715.

Chapter 8
The AEF Fights

Abstract By late fall of 1917, American Army units were beginning to occupy positions along the Western Front. Over the course of the next year, they would fight in six major and many minor engagements with differing levels of success. Overall, the entry of American troops into the front lines relieved pressure on the Allies and put increasing pressure on the Germans. By August 1918, two million Americans were in France and over the next 100 days would make an increasingly important impact on the fighting and the ultimate outcome of the war.

8.1 Germany's Final Offensives

As soon as the Treaty of Brest-Litovsk with Russia was signed on March 3, 1918, the German Supreme Army Command began moving upwards of 50 divisions from the Eastern Front to the West. Despite successes in the East, the war was not going well for the Germans in the late winter of 1917–1918. The Army was exhausted and low on replacements as the number of 18-year-olds newly available for conscription was decreasing every year. There were food riots and the beginnings of starvation in the cities, a result of the extremely effective British blockade of German ports. Sailors in the High Seas Fleet were increasingly restive as the fleet had been bottled up in port since the Battle of Jutland in May–June 1916. The Majority Social Democratic Party was splintering into moderate and radical groups in the Reichstag, and the radicals were gaining influence. The Kaiser and the Supreme Army Command were worried.

Generals Hindenburg and Ludendorff, effectively in charge of the Army Supreme High Command and increasingly in charge of the entire government, decided that the addition of the Eastern Front divisions gave them the opportunity—possibly the last one—to push towards Paris and force a negotiated settlement to the war in Germany's favor. It was crucial to the Germans that they finish the war in the West before the massive American forces and materiel could arrive in France. Ludendorff planned a massive, multi-attack Spring Offensive (*Kaiserschlacht*) to begin in March that was designed to split the Allies and break the stalemate on the Western Front. After nearly 4 years, this offensive was to be the end of stagnant trench warfare and shift the conflict back to a war of movement.

In preparation for their Spring Offensives, the Germans changed their main division level and above crypto system on March 5, 1918. Gone were the large code-

J.F. Dooley, *Codes, Ciphers and Spies*, DOI 10.1007/978-3-319-29415-5_8

books; instead, the Germans introduced a small, frontline three-digit code, the *Schlüsselheft*, for use in the danger zone near the front line and moved back to a fractionating substitution and transposition cipher, ADFGX (updated on June 1, 1918, with the addition of another letter to ADFGVX), to use for more strategic communications.[1] At this crucial period, as the German forces were moving into their battle order for their big offensive, the Allies were completely blind to strategic German radio communications.

Code named "Michael," the first, huge German assault, began on March 21, 1918, along the Somme River. The objective of Michael was to split the British and French forces, occupy the communications centers at Arras and Amiens, and roll up the British toward the sea. It was then supposed that the French would sue for peace once the British Army had been defeated. The Germans used 69 divisions and over 6,700 guns along a 40 mile front, facing two British Armies, the Third and Fifth, comprising 33 divisions, of which 10 were out of the line in reserve. While the Allies knew that a large German offensive was in the offing, the size and ferocity of the German attack took them, especially the British, completely by surprise. The massive German advance crushed the 15 divisions of the British Fifth Army on the first day, advancing nearly 10 miles, and effectively eliminating the Fifth Army as a fighting force (Eisenhower 2001, p. 107). The French, who had promised to support the British right flank, panicked as they saw the Germans opening up a possible path in the direction of Paris and brought up their reserves to protect Paris, rather than help the British. The Germans continued pushing both the British and the French back and advanced over 40 miles in 4 days, all the way to Montdidier in the west and Noyon in the south. The British and French fell back and concentrated their defenses around the access to the Channel ports and the rail and communications junction at Amiens. By March 26, the British and French resistance was strengthened, and the Germans began to outrun their supply lines. The German advance was finally halted on April 5, leaving a large 50 mile wide and 40 mile deep bulge in the Allied front lines.

The problem with differing objectives of the British and French during the early days of the German offensive finally pushed the Allies to agree on a unified command, with French Marshal Ferdinand Foch as the Supreme Allied Commander. As part of the series of meetings in late March and early April, US General John J. Pershing volunteered to allow some American forces to fight alongside Allied troops; however, he also continued to insist on an independent American Army.

8.2 Cantigny

In response to a plea from the new Allied Supreme Commander, General Ferdinand Foch, Pershing sent the US 1st Division to the Montdidier sector the last week in April. This was the first time that an American division had been in place in a "hot" sector of the front. The 1st Division moved into place opposite Cantigny on April

[1] See Chap. 7 for a more complete description of the *Schlüsselheft* and ADFGVX and Painvin's solution of the cipher.

24, supporting the French. The area around Cantigny, just west of Montdidier, formed a small salient in the line that the Allied command wanted to straighten out in order to shorten their defensive lines. This was also the first opportunity for the Allies to see how the Americans could fight on their own. The Americans were ordered to take Cantigny in mid-May, with an attack date of May 29. The 28th Infantry Regiment of the US 1st Division, supported by two machine-gun companies, 12 light French tanks, and 386 artillery pieces, was used in the assault. The attack came off flawlessly. On the morning of May 28, after a short artillery barrage, the Americans left their trenches at 6:45AM. By 7:30, they had occupied Cantigny and some high ground just east of the village and were entrenching to hold their new positions. That's when the trouble started. Because of a French commitment to another offensive farther south, the French artillery that had been supporting the Americans was removed, leaving the Americans with only a few pieces of light artillery. Over the next 24 hours, the Germans launched at least six counterattacks and, in between the counterattacks, shelled the American positions relentlessly. The 28th Regiment held onto their gains and was relieved by the 16th Infantry Regiment on the night of May 31, and the German artillery assault continued for another 3 days before the Germans finally accepted the loss of Cantigny. Overall, the 1st Division had 1,067 casualties, including 750 dead. They captured 225 prisoners and inflicted about 1,400 German casualties (Coffman 1968, pp. 156–158). By the time a French division relieved the 1st Division and left the Montdidier sector on July 13, they had suffered 1033 officers and men killed and 4197 wounded (Hallas 2000, p. 83). While not a large engagement, or strategically important, Cantigny was the first proof that the Americans could fight independently and win.

8.3 Belleau Wood

The next major engagement of American forces was part of a larger operation to stop a major German offensive west of Verdun. On the morning of May 27, 1918, the Germans launched the third attack of their Spring Offensive (the second was against the British in Flanders and had been stopped in April), this time further south along the Aisne River. For a second time, the offensive was a complete surprise to the Allies. The Germans overran the formidable Chemin des Dames fortifications the first day and pushed the French back over the Aisne River. By May 29, the Germans had taken Soissons and were heading toward Chateau Thierry on the Marne, just 50 miles from Paris. General Foch rushed reinforcements to the area, including the American 2nd Division. The 2nd was a mixed division with the 3rd Brigade (the 9th and 23rd Infantry Regiments) of Army soldiers and the 4th Brigade (the 5th and 6th Regiments) of Marines. On May 31, the Americans began moving into the line just northwest of Chateau Thierry near a pleasant hunting forest called Belleau Wood.

By June 1, the American 2nd Division was in the front line north and west of Chateau Thierry, with just under 27,000 tired and hungry, but fit soldiers and Marines. The division was put under the French general Jean DeGoutte, commander

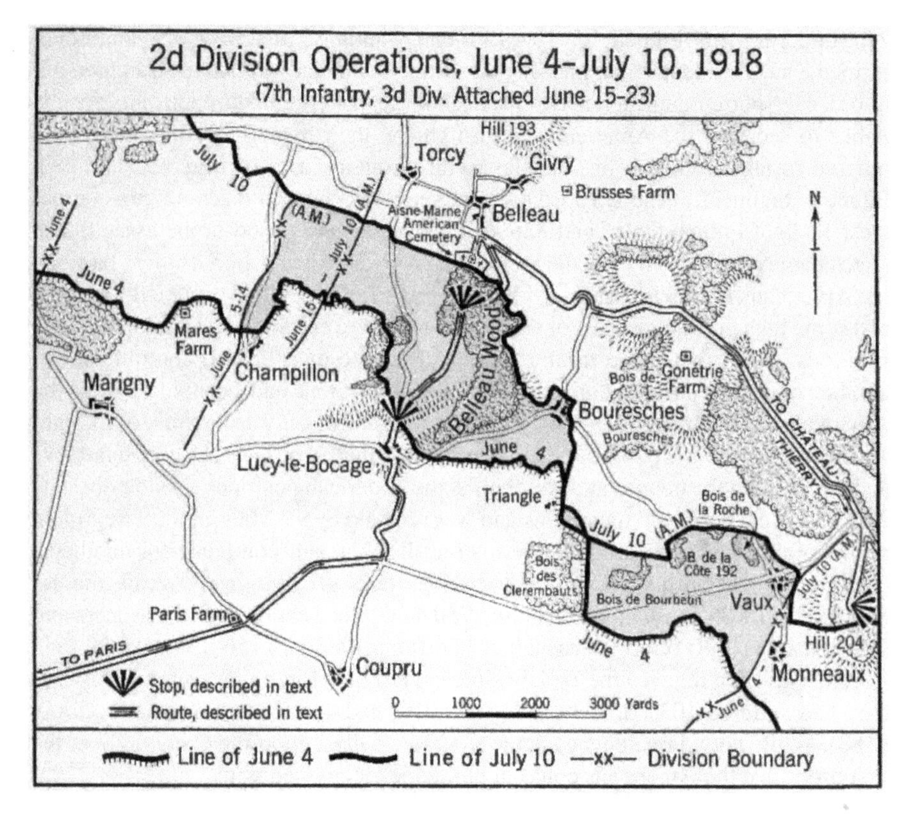

Fig. 8.1 The plan of attack through Belleau Wood, June 6, 1918 (American Battlefield Monuments Commission) (*Public Domain*. From ABMC Blue Book; used with permission. https://www.abmc.gov/sites/default/files/publications/AABEFINAL_Blue_Book.pdf)

of the XXI Corps. The Marines were on the left side of the line with the 5th Marine Regiment holding down the left flank near a strategic ridge called Hill 142 and the 6th Marines to their right; both Marine regiments were opposite Belleau Wood. The Army brigade was to the right of the Marines with the 23rd Infantry Regiment in the middle and the 9th taking the right flank near another strategic hill, Hill 204, just southeast of the Wood. The 167th French Division was to the left of the Americans and the French 10th Colonial Division was on their right (Fig. 8.1). Other French forces were actually in front of the Americans, slowly falling back through the American lines as the Germans continued to attack (Eisenhower 2001, pp. 141–142). As the French withdrew, the Germans occupied the territory in front of the Americans, including Belleau Wood, the village of Bouresches, and the strategic hills opposite—and looking down on—the American lines. At one point, a French officer stopped a Marine captain, Lloyd L. Williams, and suggested that it was time that the Americans also withdraw. Williams responded with the famous line "Retreat? Hell, we just got here!"[2] (Coffman 1968, p. 216). Despite the fact that the Americans were technically not engaged with the Germans yet, they received heavy

[2] Captain Williams would be killed in action at Belleau Wood on June 12, 1918.

shellfire during June 3 and 4, suffering more than 200 killed and wounded (Hallas 2000, p. 86).

By the afternoon of June 4, the French had finished withdrawing, and the Americans were in charge of their sector waiting for the German offensive to reach them. They didn't have to wait long. Late on the afternoon of June 4, the Germans began shelling the Americans and launched an assault on the Marine lines just west of Belleau Wood. The Marines, trained as expert riflemen, waited for the Germans as they approached. When the front line of the German assault got within 100 yards of the Marines, they opened up. The German front line disappeared. There was a short pause as the second line approached, and then it disappeared as well. At that point, the rest of the German assault melted away, and the Germans settled down to spend the next couple of days raining artillery shells down on the Americans (Hallas 2000, p. 86).

By June 5, the French were satisfied that the German offensive had stalled and it was time to counterattack. DeGoutte set as objectives for the Americans Hill 142, clearing the entire Belleau Wood and taking the villages of Bouresches and Vaux. The French assured the Americans that Belleau Wood was lightly defended and that it would be easy to clear of Germans. This was true on June 1 when the Americans first arrived, but by June 6, the Germans had had time to reinforce their positions and prepare an unpleasant welcome for the Marines. After a brief artillery bombardment, the 5th Marine Regiment stepped off at around 5:00am on June 6, with one battalion heading for Hill 142 and the other toward the west side of Belleau Wood. Twelve hours later, the 6th Marine Regiment did the same, heading for the south side of the Wood and the village of Bouresches. By the end of the day, the Marines had taken Hill 142 and Bouresches, but had barely made it into Belleau Wood. They had also taken 1,087 killed and wounded, the most casualties in a single day for the Marines until the landing at Tarawa 25 years later. Twenty days and nearly four thousand more casualties later, a Marine battalion commander would finally radio "Woods now US Marine Corps entirely" (Eisenhower 2001, p. 145).

While the Marines were sustaining nearly 5,000 casualties over the course of 20 days in Belleau Wood, the two Army regiments of the 4th Brigade were having their own problems attacking and capturing the strategic town of Vaux and the adjoining hills from the Germans. They spent most of the month of June gathering intelligence and sustaining casualties from German artillery fire. Over the same period in June, in reaching their objectives, the Army regiments suffered nearly 4000 casualties. The 23rd and 9th Infantry Regiments attacked the village of Vaux on July 1, 1918, and because of an excellent artillery barrage and even better intelligence, they met their objectives on the first day, eliminating the last salient in the Allied lines in that sector.

While a victory (largely for the Marines), the action at Belleau Wood was poorly planned and executed and resulted in many unnecessary American casualties. The 2nd Division was one of the best trained in the AEF, but many of the officers, particularly in the Marine 4th Brigade, insisted on using outdated tactics like advancing closely in ranks across open ground in the face of machine-gun fire rather than using a skirmishing tactic with smaller groups of soldiers spread out farther and leapfrogging each other across the battle ground. These outdated tactics caused hundreds of unnecessary casualties. The action at Vaux was much better executed because, first, it was easier terrain for the 3rd Brigade and, second, because they were given more time to get ready and to acquire better intelligence about the village and the German order of battle. From June 6 to 26, the 2nd Division suffered a total of 9777 casualties, including 1811 dead, a casualty rate of about 37 %.

8.4 Chateau Thierry and the Marne

While the 2nd Division was slogging through Belleau Wood in June 1918, the US 3rd Division was moving up east of Chateau Thierry in support of other French troops as the Allies tried to stem the German offensive. The 3rd Division moved up along the south bank of the Marne river in mid-June and was joined in early July by the US 28th and 42nd Divisions. By early July, the Allies were anticipating the fifth (and it turned out the last) German offensive that the Germans called their "Friedensturm" or *Peace Offensive*. Signals intelligence and information gathered from prisoners gave the French and Americans the place, the Marne River east of Chateau Thierry, and the date, July 15, and even the hour, so the French and Americans had prepared a defense in depth across the anticipated line of battle (Hallas 2000, pp. 101–102). East of Chateau Thierry, the south bank of the Marne River is lined by an east-west string of hills that create a natural barrier for a modern army. The only gap in this line of hills is the valley where the Surmelin River empties into the Marne from the south. That is where the US 3rd Division was positioned. Their job was to stop the Germans if they crossed the Marne River at the Surmelin valley and push them back (Fig. 8.2).

Two infantry regiments of the 3rd Division, the 30th and the 38th, were placed along the south bank of the Marne from the town of Mézy across the Surmelin valley. On their right was the 125th French Division. The commanders of the 30th and 38th Regiments, Colonels Edmund Butts and Ulysses Grant McAlexander, had positioned their forces so that their front line was thin and was designed to slow down the German assault across the river and then fall back onto a much more strongly defended second line of defense. For McAlexander's 38th Regiment, holding down the right flank of the 3rd Division meant one battalion of four companies along the river, with a 2nd Battalion astride a railroad embankment a few hundred yards further back. A third prepared line of defense was readied on the top of a hill, the Moulin Ridge, overlooking the valley and was where the 1st and 2nd Battalions would fall back if necessary. This defensive line looped around to the west looking into the valley and east to protect McAlexander's extreme right flank should the French 125th Division fall back—which they did, driven back later in the morning by the German 36th Division. McAlexander's 3rd Battalion was placed on another hill topped by the Bois d'Aigremont across the valley from the other battalions. Thus, when the men of the 38th Regiment fell back into their third line of defense, the regiment would be able to keep the Germans in a cross fire and stop their advance.

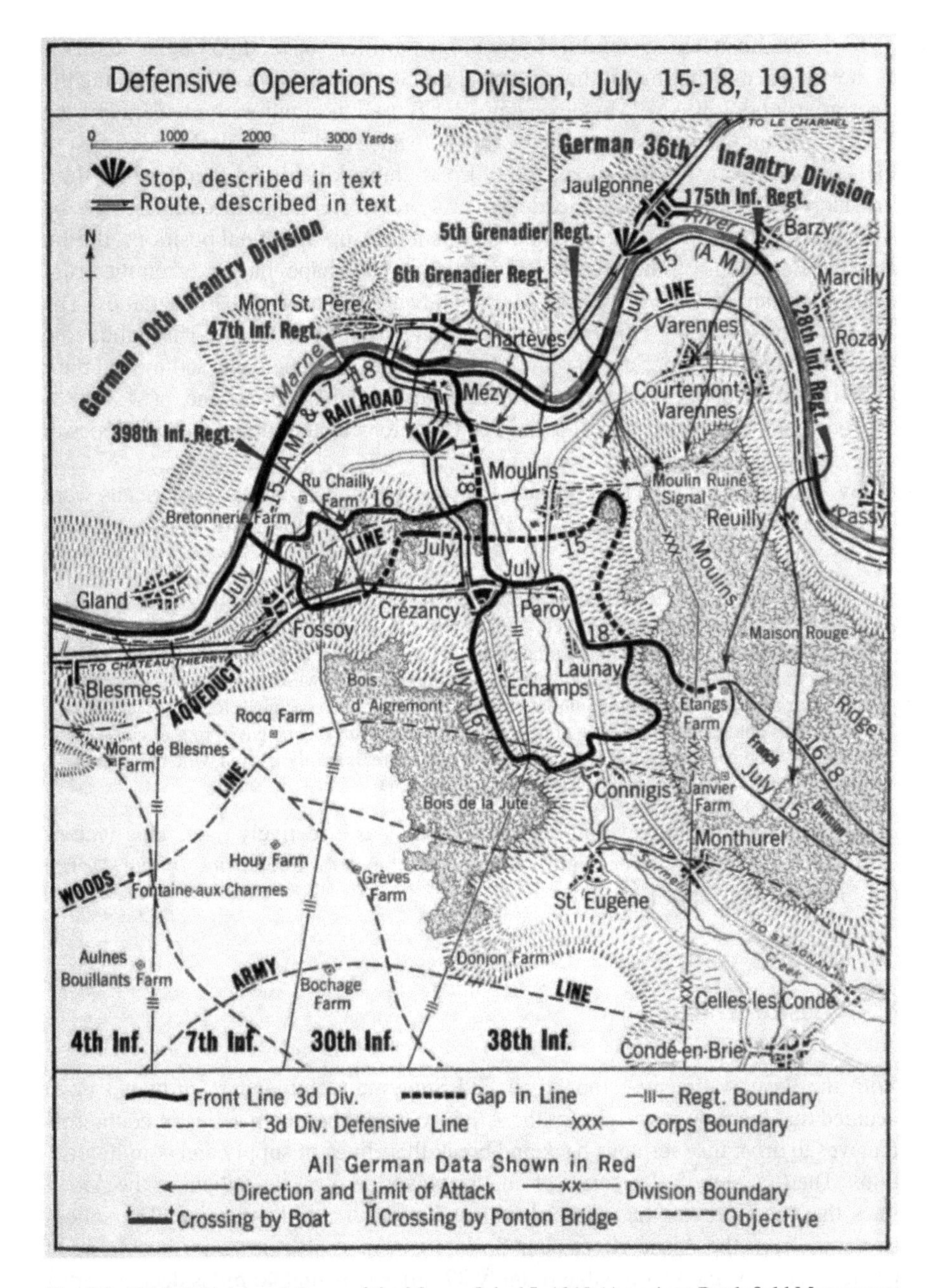

Fig. 8.2 US 3rd Division defense of the Marne, July 15, 1918 (American Battlefield Monuments Commission) (From ABMC Blue Book; used with permission)

Early on the morning of July 15, the French artillery opened up on the Germans as they began marshaling for their attack. Two German divisions began crossing the river at around 3:30AM. McAlexander's 1st Battalion along with elements of the 30th Regiment in Mézy slowed the German advance and slowly pulled back first to their second defensive line along the railroad embankment and then to their third line atop the adjoining hills. McAlexander's orders were that these lines must be held at all costs. By the time the 38th Regiment took up their final positions, the 1st and 2nd Battalions on the Moulin Ridge were in bad shape, having taken the brunt of the German advance across the Marne and suffered nearly 50% casualties. The 3rd Battalion adjacent to the Bois d'Aigremont was in better shape. Throughout the rest of the day, the Americans repulsed attack after German attack and moved their light artillery forward to shell German forces in the valley, and at 4:30 PM McAlexander even moved the 1st and 3rd Battalions further down the hill slopes to squeeze the German advance even more.

By the end of the day, the German advance was halted and the Americans were still astride the Surmelin valley; the Germans began withdrawing back across the Marne River that night.

> Lieutenant Kurt Hesse, a member of the 5th Grenadiers, which had been shot to pieces by the 3rd Division doughboys, recalled, "I have never seen so many dead; never have I seen such a frightful war sight. On the other bank, the Americans, in close combat, had completely annihilated two of our companies. Lying down in the wheat they had allowed our troops to approach and then annihilated them by fire at a range of 30 to 50 paces. This enemy was coldhearted; this was already recognized; but this day he gave proof of a bestial brutality. 'The Americans kill everyone' was the cry of fear on July 15, and which for a long time, caused our men to tremble." (Hallas 2000, p. 104)

By July 17, the Second Battle of the Marne was effectively over. This victory earned the 30th and 38th Infantry Regiments the sobriquet *Rock of the Marne* (Eisenhower 2001, pp. 155–161).

8.5 Aisne-Marne

With the Peace Offensive stopped, Allied Supreme Commander Ferdinand Foch decided that the time was right for the Allies to begin their own series of counteroffensives to drive the Germans back and break their lines of supply and communications. The first step in this series of attacks was to remove the salient in the Allied lines that the Germans had created with their earlier April offensive. The salient stretched from the Aisne River near Soissons in the north down to the Marne at Chateau Thierry and was a bulge roughly 37 miles long and 40 miles deep. For this offensive, Foch chose the American 1st and 2nd Divisions (detached from the US III Corps and operating under the French XX Corps) to attack east from just southwest of Soissons and the newly created American I and III Corps to attack north from the Marne. The American I Corps under Major General Hunter Liggett was made up of the 4th, 26th, 42nd, and 77th Divisions, while the III Corps under Major General

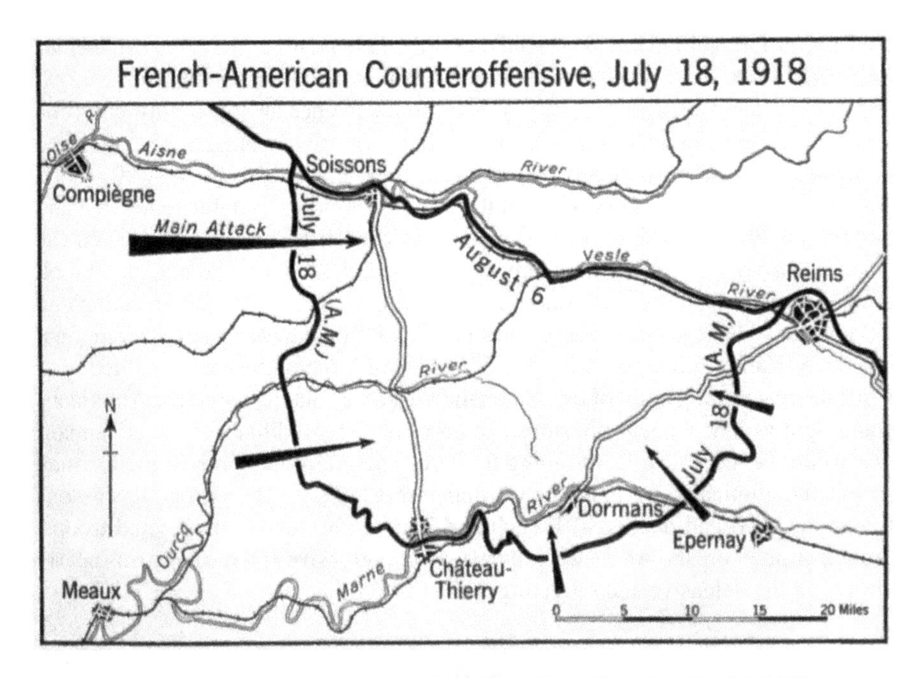

Fig. 8.3 French-American attack along the Aisne-Marne Front (American Battlefield Monuments Commission) (From ABMC Blue Book; used with permission)

Robert Lee Bullard comprised the 1st, 2nd, 3rd, 28th, and 32nd Divisions. A total of 300,000 Americans participated. The 1st and 2nd Divisions, rested from their work near Belleau Wood and Chateau Thierry, constituted the main eastern thrust of what is known as the Aisne-Marne offensive (Fig. 8.3).

By 1918, the Allies had begun to change their tactics. Early in the war, each side learned the advantage of sustained artillery bombardments to destroy the enemy front lines before infantry attacks. That was now changed. The 1918 strategy was short or nonexistent preliminary bombardments, and the use of creeping barrages to soften up the enemy as first the tanks and then the infantry moved forward. This also gave the attacker back the element of surprise. The offensive began on July 18 at 4:25AM when the French and American artillery let loose with a terrific barrage just before the infantry went over the top. In the words of a French lieutenant, "the whole forest exploded with gunfire. We were surrounded by guns spitting out fire, muzzles aflame" (Lloyd 2014, p. 4). The Americans were learning modern warfare, and the Aisne-Marne offensive was another success for them. The 1st and 2nd Divisions along with the French 1st Moroccan Colonial Division took their main objectives on the first day of fighting. After the first day, however, the German resistance stiffened and the advances slowed. On July 18, the 1st and 2nd Divisions advanced some 4 miles; it would take 5 more days for the American troops to go seven additional miles. Nevertheless, the Americans and Moroccans continued advancing until relieved, the 2nd Division on the night of July 19 and the 1st on July 25. The two American divisions once again suffered heavy casualties with the 2nd

Division suffering 4300 killed or wounded in the first 4 days of the offensive including nearly 1,000 dead and the 1st Division 6,900 casualties, including some 2,000 killed. The I and III Corps attacked from the south and had an equally difficult time driving the Germans from their defensive lines. The two prongs of the offensive joined up near their final objective, at the town of Fismes on the Vesle River on August 6. At that point, Foch stopped the advance in order to stabilize his line and prepare for the next offensives that would begin what has come to be called the *Hundred Days*. While the general advance stopped, the fighting did not, and it took until September 7 before elements of the US 28th Division succeeded in taking and holding the village of Fismette on the north bank of the Vesle River, their final objective (Hallas 2000, pp. 121–143). The Aisne-Marne offensive is credited with being the real turning point of the war on the Western Front. It proved the Americans could fight as independent divisions and corps and showed that the American soldier would be a decisive factor going forward. The offensive removed the German salient and eliminated for good the German threat to Paris. The victory at Soissons turned the tired and depressed French and British into newly invigorated troops. From that point on, the Allies were always on the offensive and the Germans nearly always on the defensive and retreating.

8.6 St. Mihiel

On August 10, 1918, just after the official end of the Aisne-Marne offensive, General John J. Pershing got what he had wanted since the American entry into the war—an independent American Army. On that day, the US First Army was officially organized with Pershing as commander. No longer would American troops—most of them anyway—be fighting under French or British generals; the Americans would have their own sector of the front and fight as an independent unit under American command (Eisenhower 2001, p. 175). It didn't take long for Supreme Allied Commander Foch to assign the Americans a sector and give them their first offensive objective.

On the eastern end of the Western Front, just southeast of Verdun, was a bulge in the line near the French town of St. Mihiel. The St. Mihiel salient had existed since September 1914 as a small thorn in the side of the Allied lines and an opportunity for the Germans to attack north and west should they so desire. The Germans never had decided to attack, and 4 years later the salient was considered a "quiet" sector of the Front. But it made the Western Front longer and it was a German encroachment into French territory. If the Allies were to launch offensives along the front with the objective of pushing into German territory, the St. Mihiel salient had to go. Pershing and the new First Army were eager to be the ones to eliminate that thorn.

For the next 5 weeks, Pershing organized the US First Army and moved his divisions into place around the St. Mihiel salient. He planned on attacking on the west and south sides with the hope of puncturing the German defenses and encircling the German troops in the salient. His longer-term grand plan was to use his new line on the Moselle River as a springboard for an attack on Metz and as a way to cut the main railroad lines from Germany into occupied France, thus cutting off German reinforcements and supplies for a large section of the Western Front.

But Pershing's grand plan was not to be. On August 30, just as the Americans were taking control of the St. Mihiel sector, Supreme Commander Foch met with Pershing and proposed instead a grand offensive north rather than east. The new offensive would work in conjunction with a British and French offensive and was designed to push the Germans out of France and eventually Belgium. All that was good strategically, but then Foch came to the second part of the plan. He proposed minimizing the St. Mihiel attack and breaking up the American forces, moving eight divisions under a French army west of the Aisne River, a further 6 or more divisions between the Aisne and the Meuse Rivers, also under French command, and the shell of the US First Army, about 6 divisions, east of the Meuse, all to attack north. Pershing absolutely refused to break up the newly created First Army; the Americans would fight as an independent army or not at all. After a contentious meeting that lasted several hours, the two generals parted with no agreement.

Three days later on September 2, Foch, Pershing, and General Petain, head of the French forces, all met at Foch's headquarters. Pershing finally proposed that the St. Mihiel offensive continue as planned and then the American forces would turn north and occupy the sector from the Aisne River to St. Mihiel. This would allow them to participate in a general offensive with the Americans driving north between the Aisne and Meuse Rivers through the Argonne Forest toward Sedan. This would make the Americans the southern part of a pincer movement that Foch wanted to use to force the Germans to withdraw out of northern France. Foch agreed and the impasse was over. But it would be at a tremendous cost to the rank and file American soldiers. The Americans would now have to make an all out attack to reduce the St. Mihiel salient and then swing their entire army—600,000 men; 2700 artillery guns, aircraft, and tanks; and all their ammunition, food, hospitals, and other equipment—90° to the north and move 60 miles across just three passable roads to begin a new offensive through arguably the most difficult terrain on the Western Front just 10 days later. All this with no rest (Eisenhower 2001, pp. 187–188).

With this plan in place, the Americans prepared for the St. Mihiel attack (Fig. 8.4). Pershing's plan was to attack in strength from the southeast side of the salient. In this area, he positioned the IV Corps consisting of the 1st, 42nd, and 89th Divisions on the left and the I Corps composed of the 2nd, 5th, 82nd, and 90th Divisions on their right. On the western edge of the salient, Pershing placed the 26th Division of the V Corps along with the French 15th Colonial Division and elements of the US 4th Division. On the tip of the salient was the French II Colonial Corps, made part of the US First Army for this attack. The French Colonials would mount a holding attack, taking the town of St. Mihiel and holding the German forces in place while the Americans used a pincer movement to come up behind them and cut them off. Overall, Pershing could deploy 550,000 American and 110,000 French troops in the attacks. The 1st and 26th Divisions had the most important role, converging on the village of Vigneulles in the center of the salient and closing the pincer. The Americans would then wheel east and drive to the Germans secondary line of

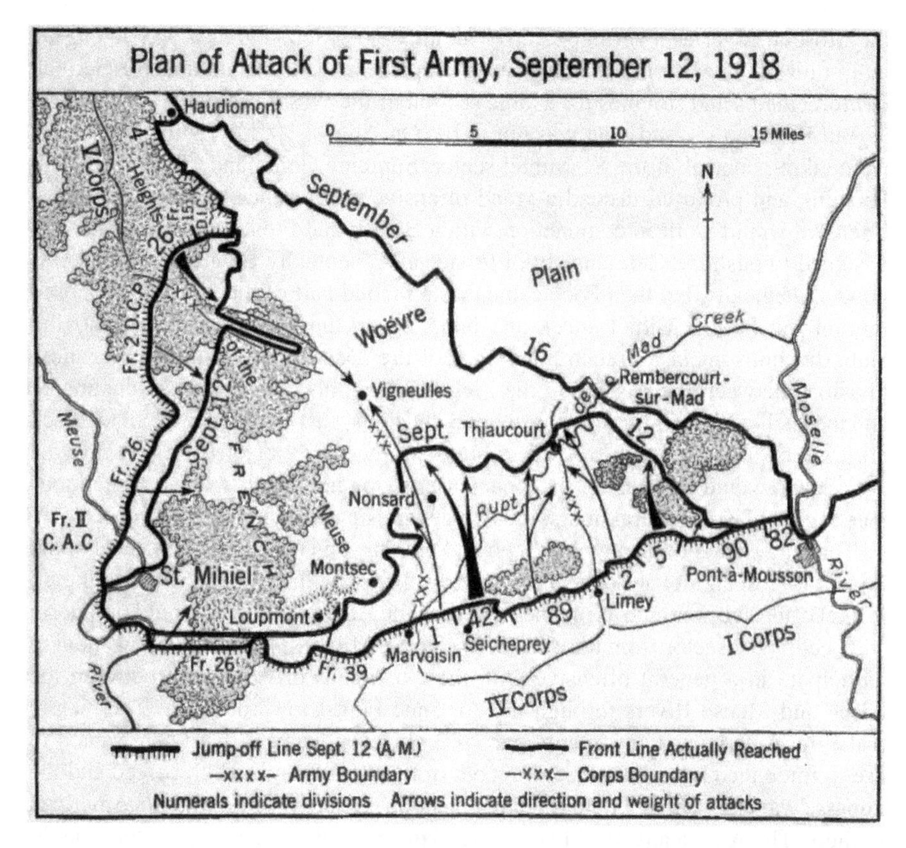

Fig. 8.4 American First Army reduction of the St. Mihiel salient (American Battlefield Monuments Commission) (From ABMC Blue Book; used with permission)

defense near the Moselle River. Pershing's plan was a true combined arms assault as it included the use of 1481 aircraft (all of French manufacture with about 40% of them crewed by Americans) and 419 tanks (all of French manufacture with American crews) in addition to the artillery support and infantry. Luckily for the Americans, within the salient they were facing just seven German divisions, all of them understrength, with four more divisions in reserve. This gave the First Army a significant advantage in numbers going into its first completely independent action.

The American assault began early on a rainy and blustery September 12, 1918, and was an immediate and surprisingly easy success. Unbeknownst to the Americans, in response to the British and French offensives further north, the Germans had ordered the evacuation of the salient on September 8th. German troops had started moving back the day before the Americans attacked, so the Americans caught the Germans with their artillery in disarray and many of their troops already on the road heading northeast toward the German secondary line of defense, called the Michel line. By early on the morning of September 13, the 1st and 26th Divisions had met at Vigneulles and begun their wheel to the northeast. That same evening, Pershing began removing troops from the line and sending them off toward the Meuse-

Argonne area. By September 15, the battle was over; the Americans had liberated St. Mihiel, pushed the Germans back to their Michel defensive line, and captured 16,000 prisoners and 443 artillery guns (Lloyd 2014, p. 130). By the standards of the Western Front, casualties were light for the 2-day battle; across all American divisions, 7,000 soldiers were killed or wounded in the attack. Overall, while not very important strategically, the battle for the St. Mihiel salient was important for the Allied cause for two reasons. First, it removed the salient as a possible starting point for German attacks on the Americans once they turned north to join in the next set of Allied attacks. Second, and probably more importantly, it was a major morale boost for the Americans, who proved they could plan and carry out a major offensive operation independently. It was also a major morale boost for the British and French, who now saw the large numbers of Americans (nearly two million American combat troops were in France by September 1918) as effective combatants and equals in the fight against the Germans.

8.7 Meuse-Argonne

Even while the battle for the St. Mihiel salient was still underway, Pershing began moving some of the 600,000 troops of the First Army 60 miles north to the area between the Argonne Forest on the west and the Meuse River on the right. This was the sector from which the Americans would attack in the next phase of Foch's multi-offensive plan to break the German's Hindenburg defensive line and push them out of France and Belgium. Pulling the First Army out of a battle and moving them and their equipment 60 miles north on poor and rain-sodden roads; setting up headquarters, communications, hospitals, and other infrastructure; and putting the troops in a new front line ready to attack in just 10 days was a logistical nightmare. The relative inexperience of the Americans showed in the traffic jams, missing equipment, lost troops, and overall confusion that reigned from September 16 until the offensive began on September 26. However, the move was managed, and the Americans, some of them arriving just hours before the jump off time, were ready at the scheduled hour.

The terrain over which the Americans were going to attack was some of the most inhospitable on the Western Front. In a sector approximately 20 miles wide, on their left, the Americans would have to fight through the Argonne Forest, dense and criss-crossed with deep ravines and heavily fortified and replete with machine-gun nests. In their center, they would have to fight through three different German lines of defense, each about 5 miles apart and each overlooked by high ridges from which the Germans could see the advances and rain artillery shells down on the attackers. On their right, they would come up against the Meuse River and just on the other side a series of heights from which the Germans could also shell the attacking Americans. The Germans manning each of these defensive lines while fewer in number than the Americans would prove to be well armed and determined to fight. One American soldier who fought in the Meuse-Argonne campaign said "Every goddamn German there who didn't have a machine gun had a cannon" (Hallas 2000, p. 239).

Pershing brought nearly all the American troops he had to the Meuse-Argonne. Attacking almost due north from his starting point, he planned to use nine oversized divisions of the I, III, and V Corps in line for his offensive. On the left was Hunter Liggett's I Corps with the 77th, 28th, and 35th Divisions in the initial assault; he would be responsible for taking the Argonne Forest. In the center, V Corps under General George Cameron consisted of the 37th, 79th, and 91st Divisions, all very inexperienced divisions. The 79th Division would be responsible for taking the German high ground and observation posts on the Butte de Montfaucon, a hill right in the middle of the sector. On the left, General Robert Bullard's III Corps was made up of the 4th, 33rd, and 80th Divisions. Assigned to move up the west bank of the Meuse, Bullard's Corps had the easiest job of the three. Each Corps had a division in reserve and the Army had six more divisions in general reserve. Because of the recent St. Mihiel campaign, Pershing was forced to rest his most experienced troops, and the majority of the attacking force the first day of the campaign were new or relatively inexperienced soldiers. Pershing also had 2,780 artillery pieces, 380 tanks, and 840 airplanes at his disposal. Over the 47 days of combat in the Meuse-Argonne, Pershing would have 22 oversized American divisions participate in the action, approximately 1.2 million combat troops, two full field armies. To the American's left, just on the other side of the Argonne Forest, were the 31 divisions of the French Fourth and Fifth Armies (Fig. 8.5).

Opposing the Americans at the start of the offensive were just five German divisions. However, the Germans had been fortifying the area between the Meuse and

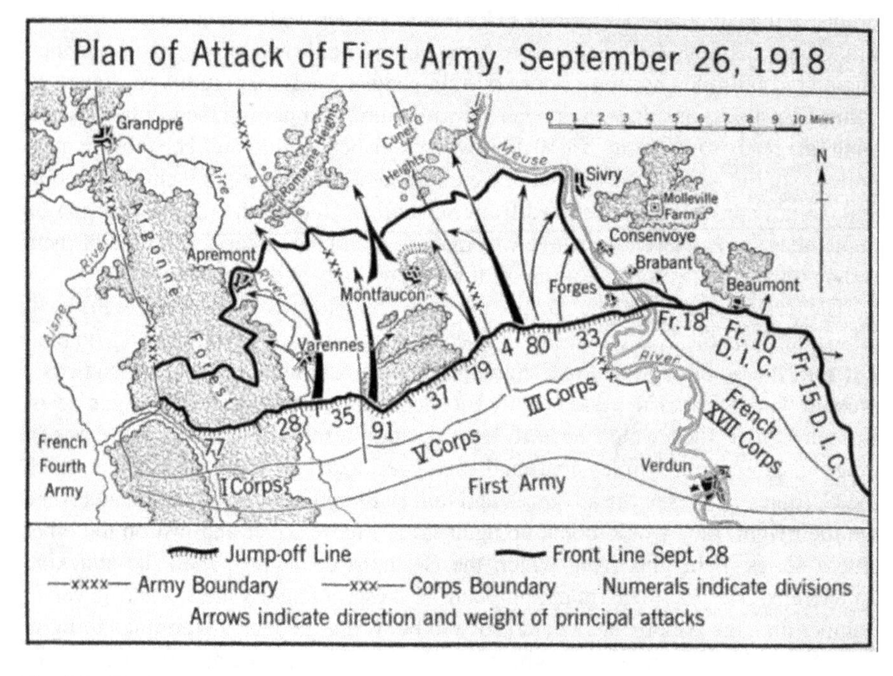

Fig. 8.5 The Meuse-Argonne offensive (American Battlefield Monuments Commission) (From ABMC Blue Book; used with permission)

the Argonne Forest since 1914 and had excellent rail and road access into the region. Within 24 hours, the Germans could move in four more divisions. In another 24 hours, two more divisions could arrive, and within 72 hours of an American attack, the Germans could bring in nine more. So within 3 days, the Germans could have a total of 20 divisions in the area, all ensconced behind a series of very formidable defensive lines. Because of the Germans ability to quickly reinforce their defensive lines, Pershing decided he had to attack as aggressively as possible and overrun at least the first German defensive line on the first day of his attacks—an average distance of six or more miles from the Army's starting point.

Naturally, things did not go according to plan for the Americans. At 5:00 AM on September 26, the First Army began its assault. Bullard's III Corps on the right, with the best terrain, made the most progress and managed to make their first-day objectives by mid-afternoon, 6 miles from their jump off point, and actually past the Butte de Montfaucon on the east. Liggett's I Corps made similar gains, with the 77th Division managing a mile and a half gain through the Argonne Forest and his other two divisions making gains of nearly 5 miles. Cameron's V Corps was another story, however. With three inexperienced divisions, two of which did not even have their regular artillery support because the field artillery was still being trained, Cameron also had the most difficult task of the day, moving across broken, but open terrain with an initial objective of taking the 300 foot high Butte de Montfaucon while under steady and heavy artillery fire. Cameron had also chosen the 79th Division to make the attack on the Butte de Montfaucon. This turned out to be a tactical mistake. The 79th had been cannibalized repeatedly to provide replacements for other divisions since its arrival in France, including nearly all of its experienced officers. Consequently, more than 60 % of the infantry had had less than 4 months training and the artillerymen even less. In addition, the division commander Major General Joseph Kuhn, while an experienced and competent officer, had no experience leading men in combat. Kuhn also decided to attack with his two brigades in column one behind the other, despite the fact that he had a nearly 2 mile front to cover. Consequently, his brigades and regiments soon lost contact with one another in the early morning fog, and their line of advance quickly became jagged and dangerously uneven. Once the fog lifted, the 79th came under intense German artillery and machine-gun attack and was quickly bogged down. Eventually, two regiments of the 79th made it 5 miles, but still fell a mile and a half short of the Butte de Montfaucon, and were unable to take their objective. Thus, at the end of September 26, the First Army had not quite made all the objectives set for it by Pershing. His gamble of a quick advance before the Germans could reinforce their defensive lines had failed.

Things got somewhat better on September 27, with the 79th Division finally taking the Butte de Montfaucon around noon that day and the other Corps continuing to advance. But, with German reinforcements beginning to show up, the First Army also failed to take its 2nd day objective of the second German defensive line at the Romagne Ridge. On September 29, the relatively inexperienced 35th (Kansas and Missouri National Guard) Division, attacking on the left flank of the Argonne, was shattered by a determined German counterattack by the 1st Guards Division, the best German troops in the line. Many of the 35th officers and noncommissioned officers were killed or wounded and the regiments so intermingled that communications were nearly impossible. Forced to retreat, the remnants of the division were relieved the next day and would not fight as a division in the Argonne again. None of the other American divisions made any progress on September 28 or 29 in the face of increasing German resistance, and by September 30, ten more German divisions had entered the Meuse-Argonne valley, and Pershing's idea of a quick victory was gone. At this point, he halted the advances and started relieving his original attack divisions with his reserves.

By October 1, Pershing's more experienced divisions from the St. Mihiel engagement, including the 1st, 42nd, 28th, 3rd, and 5th, had arrived and were ready to be put in the line to continue the advance. Liggett's I Corps now had the 1st and 28th Divisions and included the 77th Division that was deeply engaged in the Argonne Forest and was kept in the line for the next attack as well. The next stage of the advance was designed to clear the Argonne and to capture the German second and strongest defensive line, the Kriemhilde Stellung, which snaked across the Romagne heights. Capturing the Kriemhilde Stellung was the key to the Meuse-Argonne offensive. Once past this defensive line, the Americans would have a clear path toward Sedan.

The attack went off at 5:30AM on October 4. The 1st Division, now on the right side of Liggett's I Corps and just east of the Argonne in the Aire River valley, made good progress the first day against very heavy German resistance. Each of the 1st Division's two brigades was up against an entire German division, so progress was slow. On October 4, the Division reached their objectives, but they failed to make as much progress on the following days. In the Argonne Forest, Liggett's 77th Division continued to make slow progress through the forest; their artillery was not particularly useful, and the fighting broke down into a series of small unit actions. Every regiment, battalion, company, and platoon was essentially fighting by itself, and the orders were never to give up ground already won.

In the center, Cameron's V Corps was replaced with the 3rd and 32nd Divisions, both experienced now. With only two divisions, Cameron was given a more limited objective, but one that was still difficult owing to the new German divisions in line. His Corps made good progress on October 4 and 5. Bullard's III Corps on the right now had the 4th, 33rd, and 80th Divisions in the line and was to take out a bulge in the Kriemhilde Stellung line and also take the town of Cunel just north of the German defensive line. Bullard's Corps also made slow progress and did not capture its objectives in the early days of the new offensive. The army history of the campaign lays out the problems that the First Army encountered, "…fighting all along the front from that time on was of the most desperate character. Each foot of ground was stubbornly contested, the hostile troops taking advantage of every available spot from which to pour enfilading and cross fire into the advancing Americans" (American Battle Monuments Commission 1993, pp. 175–176).

After the disappointing results of October 4 and 5, Pershing decided to eliminate the German artillery on the east bank of the Meuse River that was continually shelling his troops in the Meuse-Argonne valley. To do this, he assigned the French XVII Corps with three French divisions and two American divisions, the 29th and 33rd, to cross the Meuse River on October 8 and attack the Heights of the Meuse on the east side along a 5 mile front to eliminate the German threat from that side. This new attack was also a slow, but steady slog up the heights and against determined German resistance, but the combined French-American force cleared the heights by late in the day on October 9.

By late in the day on October 6, the 1st Division had made enough forward progress up the Aire River valley east of the Argonne that they had effectively flanked the Germans in the Argonne Forest. Liggett took advantage of this turn of events and brought up his reserve division, the 82nd, and launched a flank attack into the Argonne on the morning of October 7 to force the Germans to retreat out of the forest and to capture their artillery. The 82nd was successful in pushing the Germans back, and this allowed the 77th Division also to make forward progress from the south. By October 9, the Germans were withdrawing from the Argonne, and the Americans had completely cleared the Forest by October 10.

With the clearing of the Argonne Forest, the reduction of the artillery on the Butte de Montfaucon, and the clearing of the Heights of the Meuse, the Americans began to make significant progress. By October 9, attacks from Bullard's 3rd and 80th Divisions had pierced the Kriemhilde Stellung line in several places, and Cameron's 32nd Division had broken through at the Romagne Heights. By October 10, the advance became general across the front, but still met extremely fierce and stubborn German resistance only breaking the German Kriemhilde line in isolated places. Despite days of terrible, rainy weather and horrific traffic jams, the First Army's engineers had managed to get enough roads repaired and built to eliminate much of the congestion that hampered the Army during the early part of the campaign. This allowed ammunition, supplies, and the field artillery to be brought up in a much more timely fashion and allowed for frontline divisions to be replaced more easily. At this point, Pershing once again halted the advance in order to rest the Army and give time for replacement divisions to move up. Days later, on October 14, it would all start again.

Despite the advances the Americans had made up to October 10, they had still not broken through the entire Kriemhilde Stellung line. The Germans had been working on the Kriemhilde since 1914, and it was a particularly strong defensive line. Not a line, really, but rather an elongated maze of entrenchments that would force the Americans to cross multiple lines of interconnected bunkers and trenches to penetrate it. Organized as a series of strong points, it would have to be reduced one position at a time. "The battle, therefore, could not be decided by any single master stroke. Whether Pershing liked it or not, reducing Kriemhilde would be a matter of attrition: bodies for territory" (Eisenhower 2001, p. 251).

The First Army's next attack stepped off early in the morning of October 14 as a general attack across all three Corps' front lines. The Germans had also been reinforcing their lines, and the 7 oversized American divisions (about equivalent to 14 German divisions) faced 17 German divisions in the line. The fighting continued to be brutal in terrible fall weather and the Americans gained ground through the Kriemhilde Stellung line a foot at a time. By October 21 when Liggett, the new First Army commander, called a halt to the offensive, the Americans were through the Kriemhilde and facing the third and weakest German line of defense.

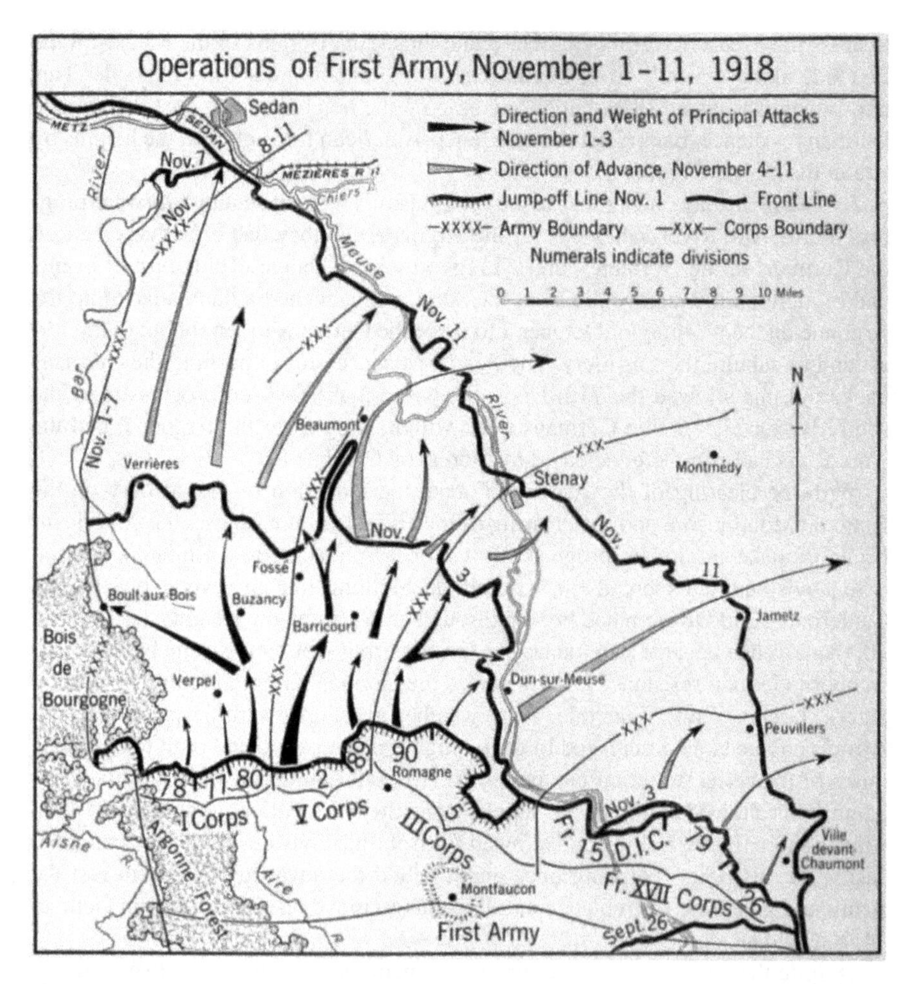

Fig. 8.6 Meuse-Argonne attack, November 1, 1918 (American Battlefield Monuments Commission) (From ABMC Blue Book; used with permission)

First Army was given the opportunity to rest, reorganize, and replace front-line divisions from October 21st till the 28th. It was obvious at this point that the way was relatively clear for the First Army's advance to its ultimate objective, the city of Sedan and the railroad junction there. Cutting the Sedan-Mézières railroad would remove most of the German's ability to resupply and move their armies on the Western Front and would force them to withdraw into Germany or sue for peace. At the same time the Americans were slogging up the Meuse-Argonne valley, the British and French (along with the American II Corps and other American divisions still assigned to Allied corps) had been hammering the German lines at three different places along the front from Flanders to the Somme, making good progress and pushing the Germans further and further back (Fig. 8.6).

By the end of October, the First Army was up over one million men strong, including 135,000 French soldiers, with an infantry strength of over 500,000. They had also been resupplied, the logistical problems encountered earlier in the campaign largely overcome. The First Army made their final assault of the war beginning on November 1, with a general assault involving all three Corps. The V Corps in the middle took the brunt of the action, and all three Corps punched through the third German defensive line on the first day of the battle. By the second day, the Americans had overrun the last of the German artillery positions, and from then on, the Germans were in full retreat. By November 5, the Americans had crossed the Meuse and were on the heights overlooking the Sedan-Mézières railroad, and by November 9, the Americans had cut the railroad and were within a few miles of Sedan. At this point, the Americans paused to allow the French Fourth Army to take the city, revenge for the French defeat there in 1870. At 5:10AM on November 11, the armistice was signed, to go into effect at 11:00 AM that day. Fighting continued right up till the appointed minute, with American troops crossing the Meuse just south of Sedan at 11:00 AM itself (Eisenhower 2001, pp. 271–272). The men of the AEF could barely believe that the fighting was over so suddenly. Thomas Johnson, a newspaper reporter embedded with the AEF, tries to describe it:

> For most of them, dirty and dog-tired in body and spirit, it was something unnatural, almost incredible. They stood up in trenches and cold wet foxholes, stretched themselves, looked about in wonderment, while, so close often that a stone would hit them, other figures stood up too, and stretched themselves. They were gray-clad and had been enemies, whom our men had tried to kill, lest they themselves be killed. (Eisenhower 2001, p. 284)

The Meuse-Argonne campaign had lasted 47 days of nearly continuous combat. It cost the American First Army 26,277 killed and 95,786 wounded; this was more than half of the number of Americans killed in the war. Several thousand more troops were casualties because of the advent of the Spanish Flu, which became an issue for troops on both sides of the front. In the planning for the Meuse-Argonne campaign, Pershing was overly optimistic about the fighting abilities of the still inexperienced American troops and underestimated the strength of the German defenses. Logistics was a massive problem with roads clogged, supplies missing, and troops trying to move in and out of the front lines at the same time, with artillery unable to move up when necessary to keep pace with the infantry. Overall, the AEF improved their effectiveness as a fighting force in 1918, at the price of a large number of casualties. The Americans were enthusiastic and brave, and their mere presence greatly improved the morale of the other Allied troops. But even by the summer of 1918, they were still not trained well enough and were inexperienced in the field. However, by war's end they had been bloodied and turned into excellent troops. In 19 months of war, the United States suffered 50,280 battle deaths and 205,690 wounded in battle for a grand total of 260,496 battle casualties. An additional 57,460 died of disease, just over half of those from the Spanish Flu which began infecting troops in training camps in the United States in September 1918 and spread (both from the United States and from Eastern and Southern Europe) to US troops stationed in France. There were also 7920 deaths from accidents, for a total of 115,660 American deaths in the war (Ayres 1919, pp. 123–126).

References

American Battle Monuments Commission. 1993. *American Armies and Battlefields in Europe (World War I): A History, Guide, and Reference Book*. Vol. 24. 24 vols. 23. Washington, DC: Center of Military History U.S. Army: U.S. G.P.O. Supt. of Docs. http://www.history.army.mil/catalog/browse/pubnum.html#23.

Ayres, Leonard P. 1919. *The War With Germany: A Statistical Summary*. Washington, DC: Government Printing Office. https://archive.org/details/warwithgermanyst00ayreuoft.

Coffman, Edward M. 1968. *The War To End All Wars: The American Military Experience in World War I*. New York: Oxford University Press.

Eisenhower, John S. D. 2001. *Yanks: The Epic Story of the American Army in World War I*. New York: The Free Press.

Hallas, James H. 2000. *Doughboy War: The American Expeditionary Force in World War I*. Boulder, CO: Lynne Rienner Publishers.

Lloyd, Nick. 2014. *Hundred Days: The Campaign That Ended World War I*. New York: Basic Books

Part II
MI-8 and the Home Front

Chapter 9
MI-8 and Civilian Messages

John Matthews Manly

Abstract This Manly article, Article VIII, is more about the work of MI-8 and G2-A6. This article covers the civilian messages that were intercepted by various government agencies (mostly the postal censorship bureau) and forwarded on to MI-8 for evaluation and solution.

In both organization and functions, MI-8, the Code and Cipher Section of the Military Intelligence Division of the General Staff at Washington, differed widely from G2-A6, the Code and Cipher Section of the A.E.F. In France, the Code and Cipher Section was concerned solely with the solution of enemy messages in code and cipher. It had nothing to do with the preparation of code and cipher systems for our own use or with the transmission of our own communications, or even with the discovery and solution of other modes of secret communications used by the enemy. Its administrative problems were confined solely to those that concerned its own work.[1] MI-8 was originally organized for much the same purposes, but almost immediately, the need for an extension of its activities arose, and it was forced to take on numerous varied activities. The addition of each of these came solely as a response to an imperative need that was not otherwise provided for. The final outcome of this growth was that this section, organized in June 1917, with a personnel consisting of one officer and one clerk, numbered on Armistice Day, November 11, 1918, 165 persons (Fig. 9.1).

Before entering upon the details of experience with code and cipher messages, it seems desirable to give the reader a brief survey of the functions actually undertaken by MI-8 during the war.[2] First of all, it seems desirable to correct a widespread error.

---

[1] G2-A6 was the 6th subsection (Radio Intelligence) of the Information Division (the A) of the Intelligence Section (the 2) of the AEF General Staff.

The Code Compilation Section of the Signal Corps under Captain Howard Barnes did compilation of field and trench codes separately. The Signal Corps also ran the Radio Section that handled interception of enemy messages and goniometric analysis (Center of Military History 1991, pp. 22–26).

[2] By the end of the war, MI-8 in Washington consisted of six different sections:

Communications Section.
Code and Cipher Compilation Section.
Code and Cipher Solution Section.

J.F. Dooley, *Codes, Ciphers and Spies*, DOI 10.1007/978-3-319-29415-5_9

Fig. 9.1 Group photo of the MI-8 cryptanalyst staff in early 1918 (John Manly is second from the *left* in the *front row*, Charles Mendelsohn is at the *far right*) (Item 604.4 from the Friedman Collection at the George Marshall Foundation Library. Used with permission)

It is often supposed that a principal part of the work of MI-8 was sleuthing or "gumshoe work." Nothing could be further from the truth. Its only connection with spies or supposed spies was to examine suspicious documents sent in by other agencies of the government under the supposition that they contained code or cipher or other forms of secret communication and determine whether this was true or not. The activity of MI-8 began with the reception of the suspected documents and ended with the report as to their suspicious or innocent character. And it may be remembered that although it was instrumental in ascertaining through its examinations of secret messages the guilt of a number of enemy agents, it was also instrumental in relieving of suspicion an enormously larger number of persons who form one reason or another had awakened the suspicions of government agents or of patriotic but sometimes hysterical neighbors.

In June 1917, no enemy messages in code or cipher were sent in for decipherment, and consequently, the work of the one officer and his clerk consisted entirely in encoding and decoding messages sent and received by the Military Intelligence Division (M.I.D.). The volume of our own communications increased so rapidly and

---

Shorthand Solution Section (based in New York).
Secret Inks Section (two laboratories, one in New York adjacent to the postal censorship office and one in Washington).
Training Section, which took over the training function from William Friedman at the Riverbank Laboratories in April 1918 (Yardley 1931, p. 47). Yardley erroneously leaves out Training.

became so great that it was necessary to establish our own telegraph and cable office and ultimately to provide it with two officers, seven telegraph operators, and four code clerks, who maintained continuous service day and night.

Before this was accomplished, it became obvious that a new codebook should be prepared to replace the old army codebook and that cipher systems should be prepared for communications between military intelligence officers at the various camps in America. The personnel required for this was one officer and four civilian assistants (Yardley 1931, pp. 39–41).

In connection with this work of communications, it became necessary to establish special instruction in the use of code and cipher and teach our own military attaches and other intelligence officers how to encipher messages in our own systems so that they would be safe against enemy attack and how to decipher accurately messages received by them. A similar school of instruction had to be organized later to supply code and cipher clerks for the A.E.F.[3]

It was at first supposed that the decipherment of enemy messages would be limited to code and cipher attack, but it soon became clear that other forms of secret communication must be reckoned with. First of all came the need for a secret ink laboratory, the establishment of which will be related in the present article.

Soon it appeared also that among the letters sent in to MI-8 under the suspicion that they were written in cipher, there were many which turned out to be more or less innocent communications in strange forms of writing, such as Korean, Bengali, Malay, and various languages of the Near East. This made it necessary to include among the persons selected for code and cipher work experts in a large number of strange languages.

Among the curious sorts of writing sent in for examination were also a number that turned out to be comparatively unknown systems of shorthand. To deal with these, we were obliged to obtain the aid of a shorthand expert, who organized a special section for dealing promptly and accurately with all known systems.[4]

Out of the two activities last mentioned, it developed naturally that when interpreters and stenographers were needed for use in France, the task of finding the proper candidates and testing their qualifications devolved upon MI-8. It then appeared that interpreters and stenographers, although competently trained in German, were unfamiliar with German trench slang and with the technical terms of German army

---

[3] William F. Friedman and his team trained the first 80 or so officers designated for work in the AEF in France at the Riverbank Laboratories in Geneva, IL. This civilian research laboratory was owned by Col. George Fabyan who donated the services of his Cipher Department, headed by Friedman, in the spring of 1917. Riverbank trained a group of four officers in the late fall of 1917, a second group of about 76 officers in January–February 1918, and a final group of about six in March–April 1918. At that point, MI-8 took over the training of all code and cipher clerks destined for the AEF.

[4] This was Franklin W. Allen of New York. Allen was a partner in the firm of Hulse and Allen that supplied law reporters. Allen donated his services for free and established MI-8's Shorthand Section in New York City. Allen's group of six shorthand experts could recognize over 50 different systems; he also had connections to other experts nationwide that he could call on (Kahn 2004, pp. 31–32).

organization and operations. We had, then, the additional task of compiling treatises on both of these subjects for the instruction of the interpreters and stenographers who were to be sent to the front.

It should be sufficiently clear that these varied functions and the personnel required by them imposed also upon MI-8 a very considerable administrative problem. The burden of this was from time to time lightened by more or less cheering incidents. One day there appeared among the candidates to be examined for interpreters a husky brute from Pittsburgh. He looked to me like a man who belonged in the infantry or the tank corps; but through some strong influences, he had got himself transferred to what he regarded as a less dangerous branch of the service. As he spoke German well, we had no means of rejecting him. I therefore congratulated him on the fine report our examiner had given on his German and told him he was greatly to be envied, and he would probably be wanted right up on the front line to interview prisoners as soon as captured and that he would have a chance to see some lively fighting. He did not return the next day for further instructions. We later learned that he got another transfer.

When I reported for duty in MI-8 on October 1, 1917, the stream of suspected messages was a mere trickle. MI-8 was a central bureau for the examination of any and everything seized by any agency, official, or unofficial, in the United States under the suspicion of being an enemy message, but few of these agencies had begun to function and even the little that they sent in proved for the most part to be entirely innocent. One or two examples of approximately this date will illustrate this.

One day a Department of Justice agent came in with a carrier pigeon over which he was very much excited. He had seized it in a loft belonging to a person supposed to be a German sympathizer, and he pointed to very curious perforations in some of the feathers. These bore a striking resemblance to shorthand of some unknown form of writing. He was confident that he had made a great discovery. Although we had no pigeon fancier in MI-8, we consented to take the case for examination. The officer to whom it was assigned made a careful study of the markings but was able to discover no system or meaning. The next day, in resuming his study of the feathers, he observed that one of them, which on the previous day had been entirely free of markings, was now perforated in the same manner and to almost the same degree as the others. A microscopic examination of the feathers soon confirmed his inference that the perforations were not the work of human hands but of certain minute insects that the microscope revealed in great numbers, thus vanishing one of the first spy cases that were brought to our attention.

Shortly before Christmas 1917, the naval officer stationed at Camaguey, Cuba, stated that one evening about 9 PM, while listening in on the wireless set at his camp, he had heard a strange set working and copied the following message:

```
JA COD DIST HAS HANT TGM JZTT 2700
MIL HAAR OCH HGRT 5600 MIL SETR JA
COD NATT DWQ OCH LYCKLIG RESA OM VI
EJ HZRES MORGON SGG
```

He was highly excited about it and said:

> The operator sending the message was not a Cuban or an American, for their styles of sending are well known to me. I believe him to have been a German, for he used the German ch, which is four dashes instead of the two distinct characters used in this part of the world. As the wave length was two hundred meters, I should say that this station is located somewhere on the island of Cuba, not more than eighty or ninety miles from here.

After carefully examining the message, we replied:

> This is not a code message, but gossip between the wireless operators on the Swedish ship *Nordic* and another unidentified ship whose wireless call is 'DWQ.' The following is the translation;
>
> Yes good just have had 'TGM' (JZTT) 2700
> miles here and heard 5600 miles (SETR)
> yes good night 'DWQ' and happy journey
> if we do not communicate tomorrow 'SCG.'

Later a similar report came from an officer in Hawaii who was puzzled by the unfamiliar groups of letters and figures. In this case also, the supposed enemy cipher message turned out to be nothing but the ordinary chatter between operating stations. One of the operators was boasting of the number of these stations he had been able to hear and gave the three-letter groups which were the station calls of New Brunswick, New Jersey; Nauen, Germany; Rome; Balboa; Pearl Harbor, Hawaii; and Cavite. The message closed with "Mani 73," the conventional salutation "many best regards."

Equally innocent were some wireless messages in colloquial Japanese. It is not strange that the symbols necessary to transmit the syllables used in spelling ordinary Japanese were unintelligible and suggested to the uninitiated mysterious secret language.

In the midst of these and other similar experiences, it is not strange that Captain Yardley and I were somewhat excited when a customs officer on the Mexican border sent in to MI-8 a blank piece of paper that had been found in the heel of the shoe of a woman suspect who was attempting to cross into the United States. The very fact that this apparently innocent piece of paper was carefully concealed suggested that it might contain a message written in invisible ink. Although we had often read of such messages, neither of us had ever had any experience with them. We knew only that heat and other simple reagents could develop some forms of invisible ink. We were very anxious, however, to make no mistake in the handling of this, our first experience of this character, and we consequently called upon the National Research Council. Although the men we interviewed were among the leading chemists of the country, they had never specialized in developing invisible inks and were able to tell us little more than we already knew. Thrown back upon our own resources, we resorted to heroic measures. We borrowed a flat iron and, in the solemn depths of the basement of the Army War College, heated it to what we supposed was the proper temperature and applied it to the edge of the paper. In our ignorance and excitement we got the iron too hot, but it would be hard to describe our excitement when we saw traces of writing appear through the scorched surface.

The message was not only written in invisible ink but was in modern Greek. Translated, it reads as follows:

> I beg to have sent ... to Galveston
> in order that the Swedish ambassador
> may give the 119,000 dols which were
> asked for in my letter of 5/8. It is
> necessary not to have trouble with the
> I. W. W.

Having developed the writing and obtained the translation, we reported the results to the officers of the customs and, as usual, left the further disposition of the case in their hands.[5]

So far as MI-8 was concerned, the chief result of this incident was to impress upon us the absolute need of a laboratory for the development of invisible inks. The matter was brought to the attention of Colonel Van Deman, the chief of the division, who immediately took the necessary steps to find two trained chemists and send them to England and France to learn the methods that had been worked out by the experts of those countries, who had been working intensely upon this subject since the beginning of the war. Meanwhile, it was necessary to make immediate provision for carrying on the work without delay, and in response to a request from Colonel Van Deman, the British Intelligence Office sent to us Captain Collins,[6] one of the most experienced officers of the British Secret Ink Bureau.

This experience in developing invisible ink by heat reminds me of an amusing incident which occurred long afterward when we had a well-equipped laboratory and an experienced personnel for handling all sorts of secret writings. The postal censorship of Cuba, through our liaison officer, sent in a letter containing an apparently blank piece of paper addressed to a man in the island of Jamaica and requested that we ascertain whether or not the blank paper contained invisible writing. The writing was easily developed, and to our astonishment and amusement, we read the following message:

> Ah Grandson,
> Now you must make me a
> sacrifice of $50. to be placed at the
> 4 corners of the nearest grave in the
> cemetery where I will so as to send me
> back to my resting place. Then after
> twenty days the gentleman will get you
> and tell you where to meet him to get
> your fortune.
> So good bye
> Your Grandfather

[5]These last two stories are almost identical to the first two stories in Yardley's chapter on Secret Inks in *The American Black Chamber* (Yardley 1931, pp. 55–61).

[6]This is Dr. Stanley Collins, arguably the best British invisible ink expert. Collins was chief of the British postal censorship chemistry department. He came to the United States for a couple of months, beginning in May 1918 to instruct the Americans in invisible ink techniques and to help them set up their secret ink laboratories. The lab in New York was set up in November 1917 and was headed by Dr. Emmett K. Carver, while the lab in Washington was created in July 1918 and was headed by Dr. (later Colonel) A. J. McGrail (Macrakis 2014, pp. 160–163; Yardley 1931, pp. 61–76; Kahn 2004, pp. 32–33).

We do not know the whole history of the case, but it is obvious that some friend thought the person for whom the letter was intended had too much money and, wishing to relieve him of $50, conceived the brilliant idea of causing him to receive a letter purporting to come from his dead grandfather which could be read by the application of a degree of heat sufficient to suggest grandfather's present dwelling place. No doubt some sort of séance had previously been held and full instructions given for reading the letter when it arrived. The erratic spelling is perhaps attributable to the conditions under which Grandfather is supposed to have written.

It was only natural that a very close watch should be kept upon the Mexican border. After the postal censorship of foreign mails was established, not only all persons but also all written messages crossing the international boundary were scrutinized with great care. It was felt that a neutral country bordering on our own would certainly be used as a base for the operation of German agents, just as this country had been before it entered into the war. It was known that Germany had many able representatives in that country and it was suspected that they were attempting to secure from the United States information of military value for transmission to Germany. Such information would naturally concern the enlistment and training of troops, the manufacture and assembly of arms and ammunition, the development of airplanes and the training of aviators, the movement of ships, the food supply, and the general state of public opinion and feeling.

In September 1918, a large number of messages purporting to have passed between German agents at various points in Mexico and in Texas and other southern states, together with not a few purporting to come from an agent in New York, were brought to us for examination. Some of them were in English, some in Spanish, and some in cipher. They were regarded as of the greatest importance because they gave the names and addresses of a number of German agents and contained reports on the movement of ships and the operations of submarines, the progress of enlistment, and the conditions of interned prisoners. They referred to certain maps as indicating bridges and other points at which railway communications might be broken by carefully laying bombs. They reported on the number of airplanes at certain aviation fields and the skill of the aviators, and they contained a good deal of propaganda information for distribution to German sympathizers in Mexico. The dates of these documents ranged from the beginning of January to the middle of September.

Certainly, there could be no question of the importance of these messages or of their value to the agencies charged with the defense of the United States against the plots of enemy agents if they were indeed genuine. Of their genuineness, there seemed at first no question, for all of them were either signed by, or addressed to, Hermann Rueckheim, who was well known to be one of the most active agents of the German information service in Mexico.[7] Rueckheim had his office in Nuevo

[7] No record of a Hermann Rueckheim has been found. Manly may have changed the name.

Laredo, just across the line from Laredo, Texas, and there was abundant evidence that his operations reached into the United States.

Here is a message in a simple transposition cipher—a type in common use among German agents—ordering him to arrange for procuring promptly the telegraphic news of the world for distribution among the German sympathizers in Mexico.

```
EENZOTEESE KWUETEMWOM TYMIELPENE
EODTXEFISN NRIUAPETTT DKENSHHEWF
ITSGAOLRIU ERELRNEGEE NENARONEAR
SFPRADMBND TFEENEIESI NERDGRTNIR
```

When deciphered with the key—Hindenburg—the German text of the message is found to be:

> Empfehlen mit zeitung Laredo, Texas, arrangement fuer direkten Dienst New York treffen um diesen per telephon oder sonstwie an Sie Weitergeben.

This may be translated:

> We recommend that arrangements be made with a Laredo, Texas paper for direct service with New York so that the news may be transmitted to you by telephone or some other method.

Readers who are interested in this method of using a transposition cipher will find it in the following diagram:

H	I	N	D	E	N	B	U	R	G
5	6	7	2	3	8	1	10	9	4
E	E	N	Z	O	T	E	E	S	E
K	W	U	E	T	E	M	W	O	M
T	Y	M	I	E	L	P	E	N	E
E	O	D	T	X	E	F	I	S	N
N	R	I	U	A	P	E	T	T	T
D	K	E	N	S	H	H	E	W	F
I	T	S	G	A	O	L	R	I	U
E	R	E	L	R	N	E	G	E	E
N	E	N	A	R	O	N	E	A	R
S	F	P	R	A	D	M	B	N	D
T	F	E	E	N	E	I	E	S	I
N	E	R	D	G	R	T	N	I	R

As will be seen, the cipher message is written in columns under the letters of the keyword—Hindenburg—and then the columns are read downward in the order of the alphabetical priority of the letters of the key.[8]

The messages spoken of a moment ago were so numerous that we must content ourselves with samples, and we will take first some of the English messages dated in August.

> I drop this before I answer your last,
> dated the 1th, of the present, and urge to
> communicate you that one of our Agents will
> depart to the United States of America, so
> we appreciate you all give the best attention,
> he is got all the necessary instruction to
> inform us how beyond the BORDER, and soon
> he will communicate with the other agents in
> Texas, Louisiana, Oklahoma, and New Mexico,
> and wish you let me know when he cross the
> boarder, we are sure he will furnish us all
> kind of information on this trip, such as the
> number of the troops, plane of some important
> City, in Texas, Louisiana, Okla.
>
> For the moment I give my best regards, & be on
> the spy is our motto.
>
> * * *
>
> By these lines I let you know that we had
> reach the triumph, the information of Mr. G. at
> New York, was receive here on time, and we had
> enough time to inform our submarines, so
> they put in a fine job on the oceans.
>
> * * *

[8] To decode this message, one writes it as a rectangle with ten columns (the number of letters in Hindenburg) and writes down column numbers in alphabetical order as in the table in the text.

One then pulls the columns off in order by their alphabetical order

```
01 empfehlenmit
02 zeitunglared
03 otexasarrang
04 ementfuerdir
05 ektendienstn
06 ewyorktreffe
07 numdiesenper
08 telephonoder
09 sonstwieansi
10 eweitergeben
```

and then finally breaks them up into words by reading across the rows from top to bottom:

> Empfehlen mit zeitung Laredo, Texas,
> arrangement fuer direkten Dienst New
> York treffen um diesen per telephon
> oder sonstwie an Sie Weitergeben.

By one of our agent, I have receive a
knowledge, and it was sent direct to my office,
and it says, this. The German Military POWER
hasn't suffer any damages at all during the last
battles had, and our Empire has arrange a great
number of men for the final triumph, he is
getting tired of this, so he is going finish the
job and says this, the one who drop the last drop
of blood, of the last men are the one lost

* * *

I was receive of your telegram on which you
notify me about the captured of one of (spies)
wish you investigate if you are Mr. M.P.R if it is
him, we should work with eagerness and save him,
try to investigate the matter with out delaying.

And remember that all our agents in U.S. are under
different names so be carefully who is this man,
and send me his wright name at once.

* * *

The Spanish messages, which are much more numerous and are signed with both Mexican and German names, are, from the linguistic point of view, equally as interesting as the English. We may perhaps save time and space by giving as an illustration of this a message composed in Spanish but written in cipher. It is a simple substitution cipher of the kind Edgar Allan Poe dealt with in *The Gold Bug* and is deciphered with all its errors as follows:

Sirvase, Vd. Dnmunicar a nuestro jfe
pue ge podidn ver un bien numero de
airoplanos hsta hoy he contado mas de cien
auque debo adbertirle que son de luy mala
construccion y sus pilotos son inenespertos
pues adiario se ven vasios acsidentes por lo
que ago constar para el conosimiento de
nuestro gobierno.

Corrected decipherment:

Sirvase Ud. Comunicar a nuestro jefe
que he podido ver un buen numero de
aeroplanos. Hasta hoy he contado mas de cien,
aunque debe advertirle que son de muy malo
construction y sus pilotes son inexpertos,
pues a diario se ven varios accidentes, por
lo que se hago constar para el conocimiento
de nuestro gobierno.

Translation:

Please advise the chief that I have been
able to see quite a number of aeroplanes.
Up to date I have counted over a hundred,
though I must inform you that they are badly

constructed and their pilots are inexperienced,
for accidents occur daily. The above is for
the information of our government.

It is obvious that if genuine these messages were indeed of the highest importance, but although we were assured that there could be no possible doubt of their authenticity, we were unable to believe in them. Our reasons for skepticism were numerous and to us at least convincing.

In the first place, the same errors in Spanish occur in papers purporting to come from different persons in widely separated localities and are found even in documents signed with the name of a Mexican gentleman known to be a man of thorough education and the finest culture.

In the second place, the faulty idioms in English are the same in all of the English papers, whether purporting to come from persons of Mexican birth or those of German birth, and there are no Germanisms in the English of the messages signed either by the head of the German Bureau or his German correspondent in New York.

In the third place, the same kinds of error in handling the cipher appear in messages signed by all the agents who used it. Moreover, the type of cipher was one that we could not believe would be used for official German communications even if the agents were half educated Mexicans or Americans, whereas it was in very common use among inexperienced persons almost everywhere.

Finally, the German name of the supposed chief of operations is misspelled even in documents purporting to come from him and signed with his name. In view of these facts, we could not resist the conviction that someone had perpetrated a big hoax on the agent who had been pluming himself upon obtaining so important and valuable a collection of documents.

Such experiences as this might well have dampened the ardor of a cipher expert, or even that of the agent who had been induced to believe in the authenticity of the documents. But it is the business of a Cipher Bureau never to allow its interests or energies to flag, for although a thousand suspicious documents may turn out upon examination to be entirely innocent or insignificant, the very next one might be of the greatest importance.

References

Center of Military History. 1991. *Reports of the Commander-in-Chief, AEF, Staff Sections and Services*. Vol. 13. 18 vols. United States Army in the World War 1917–1919. Washington, DC: Center of Military History U.S. Army: U.S. Government Printing Office. http://www.history.army.mil/catalog/browse/pubnum.html#23.

Kahn, David. 2004. *The Reader of Gentlemen's Mail: Herbert O. Yardley and the Birth of American Codebreaking*. New Haven: Yale University Press.

Macrakis, Kristie. 2014. *Prisoners, Lovers, & Spies: The Story of Invisible Ink from Herodotus to Al-Qaeda*. New Haven, CT: Yale University Press.

Yardley, Herbert O. 1931. *The American Black Chamber*. Indianapolis: Bobbs-Merrill.

Chapter 10
Civilian Correspondence: Foreign Letters and Hoaxes

John Matthews Manly

Abstract This article, number IX, is similar to and possibly a continuation of Manly's Article VIII. It deals with mostly civilian letters in uncommon foreign languages and with hoaxes that people attempted to perpetrate on their friends. Other letters included chess moves, music annotations, contest entries, surveyor's notes, and, the largest group of all, enciphered love letters. Obviously, during war the postal censorship office would intercept any encrypted letter and forward it on to MI-8. The interesting point here is that most of these letters are entirely innocent, either jokes or encrypted for lovers' privacy.

The problems which confronted the code and cipher experts on the battle front in France were entirely different from those which confronted the experts on this side of the water, who constituted MI-8, the Code and Cipher Section of the Military Intelligence Division. In France, practically all the messages presented for decipherment were official German messages that had been intercepted by wireless intercept stations of the Allies. These messages were, to be sure, in many different codes and cipher systems, but in the main it was easy to tell which system had been used for any particular message. In fact, the most important messages could not only be classified from their general appearance but were also distinctly labeled by the Germans themselves. It is easy to see why this labeling was necessary. As the messages were all transmitted by wireless, the station to which a message was directed obtained it from the air just as all intercept stations did. It could just as easily obtain messages intended for other stations and written in forms of code and cipher, which were unknown to it and, therefore, unintelligible, and it would certainly have taken a very large number of such messages and wasted its time in vain attempts to read them if their destination and nature had not been clearly marked. Messages were, therefore, plainly marked with the letters designating the sending and receiving stations, thus "From GSK to DCA" of "From GGI to DAX." These were stations in the Argonne sector. Messages from the political section in Berlin to the propaganda station in North Africa were marked only with the designation of the receiving station, "Füer (for) GOD," and these messages were always enciphered in the same system but with different keys depending on the date on which they were sent. A message from Berlin (LP) might be sent to Constantinople (OSM) for repetition to another station in the East. In this case, it would begin with the letters "OSM v (from) LP Fuer (for) USK," but if many forms of cipher were used between these

J.F. Dooley, *Codes, Ciphers and Spies*, DOI 10.1007/978-3-319-29415-5_10

stations, the message would also contain at the beginning a word which indicated the method by which it was enciphered. Thus, it might be called "Richi," "Alachi," "Omochi," or "Itochi," each of which would indicate a different method of encipherment.[1]

Not only was it possible in most cases to tell the source and destination of a message, it was even possible to make a rough estimate of the relative importance of the different classes of messages even before they were read, for certain systems were used for certain classes of messages and others for others.

Any particular message might of itself contain information of the highest military value, but even if it did not, it was almost certain to contribute in some measure to our knowledge of the situation of operations or intentions of the Germans in some part of the world. Even the practice messages and fake messages sent out during periods of idleness were still the work of the enemy and deserved attention as such.

The problems of G2-A6 were, therefore, fairly simple. I do not mean that they were easy. On the contrary, they were amazingly difficult. I mean that they were perfectly definite and that the work was all genuine code and cipher work. It was the duty of the Signal Corps to intercept as far as possible all messages that were flying through the air, to make accurate copies of them, and to forward these copies to the code and cipher experts. It was the duty of these experts, so far as was possible, to read every message so received and to report to headquarters both the decipherment and accurate translation of every message.

No such simple state of affairs existed on this side of the ocean. MI-8 was established to serve as a central bureau for the examination of everything that had anywhere in the United States come under the suspicion of carrying a secret message. The material submitted as suspicious was very great in amount and was unbelievably varied in character. It consisted in part of real messages, some in invisible ink, some in cipher, some in all sorts of codes, and some certainly not intended for any kind of secret communication, but every such piece of material, no matter what its source or nature, had to be carefully examined to determine its true character and whether or not it justified the suspicion that it had aroused.

That the volume of material sent in as suspicious was enormous should cause no surprise. One has only to recall the intense excitement prevalent at that time and the hysterical character that it often assumed to understand this. Actions and things that under ordinary circumstances in times of peace would attract no attention were closely observed and interpreted by excited imaginations. Much that was written in newspapers and magazines and talked of in ordinary conversations led people to believe that America was full of accredited German spies, many of whom, according to common belief, had received the most thorough and subtle training. These spies were supposed to be on the alert to procure and forward to German officials the fullest and most detailed information concerning every operation and event in America that could have even the remotest bearing upon the preparations for the war and the conduct of it. A scrap of paper on which a man had made aimless and

[1] There were all different German ciphers. See Childs (1919). For example, Alachi was a double transposition cipher.

meaningless scribbles while waiting for a call in a telephone booth would readily be taken for the work of a spy and, if it were sufficiently unintelligible, would be regarded as code or cipher. Half-heard conversations awakened suspicion, especially if they contained the names of any places connected with the seat of war or with preparations for the war. Rough sketches and drawings of all sorts were interpreted as possible maps or designs for bombs and mines. The general tendency to interpret as dangerous everything that was not understood was greatly stimulated by a considerable number of articles in magazines describing ingenious methods of conveying information secretly. Every man, woman, and child became a detective, anxious to preserve the country, discovering and reporting some thread of the network of plots which was supposed to be infinite in its ramifications.

Besides the suspicions natural and excusable under such conditions of nervous strain, there can be no doubt that some malicious persons took advantage of the general state of feeling to cast unjust suspicions upon persons to whom, for any reason, they bore a grudge. I recall one case in which I was asked to give an opinion concerning a man who had been reported to the Department of Justice as being a master spy for the Germans. It was asserted that he possessed a very ingenious and effective method of sending secret messages. Had the accusation been true, there is no doubt that the man would have been sentenced and severely punished—probably hanged. Fortunately, I knew the accused man very well and was thoroughly familiar with the apparatus that he was accused of using in sending secret messages, and I was able to testify not only that I had confidence in him but that it would be an absolute impossibility to use the apparatus in any manner for the conveyance of secret information. In this case, the source of the accusation was unquestionably business jealousy and personal enmity. There can be no doubt that numbers of persons were unjustly brought under suspicion by similar practices.

Of course, the hysteria which made such things possible is greatly to be regretted, but there is no evil that may not produce some good, and it has to be remembered that but for this highly excited condition, many really suspicious and dangerous actions and objects would never have been noticed and reported.

One of the most important functions of MI-8, therefore, was to separate the innocent messages and papers from those which were not and thus relieve of suspicion persons who had been wrongly suspected, and MI-8 can justly pride itself on this part of its work in counteracting the natural hysteria of the time and also upon the fact that there was never any leak of information through any of its officers or clerks which allowed the fact that anyone was under suspicion to become public knowledge.

The number and variety of things sent in for our examination can hardly be imagined. That some of them were even the cause of suspicion will now seem ludicrous, but it must be remembered that they fell into the hands of people who did not understand them and that everything that was not understood was rightly or wrongly the object of suspicion.

First of all, perhaps, we may mention the letters in uncommon languages. Most of those were written by, or to, soldiers in the camps and, being unreadable by the officers into whose hands they fell, were naturally sent to us for examination. Some of them were seized by the postal censorship as they were entering or leaving this country in the foreign mails. Naturally, letters in the common Western European languages were recognized at once and forwarded without question unless they contained some peculiar feature, but letters in any of the languages of the Far East or the South Seas were likely to be regarded as cipher and a decipherment requested. Among the letters thus sent in for examination as cipher, such languages as the following occurred: Korean, Chinese, Arabic, mixed Chinese and Arabic, Tagalog and other Malay dialects, Mexican Indian, Basque, Lettish (Latvian), Finnish, Polish, Croatian, and ancient Egyptian hieroglyphics.

Some of these were quite simple. Others raised puzzling questions even after they were read. For instance, from one of the camps, there came a bundle of strange letters written in Lettish, accompanied by a bundle of pamphlets in the same language. Even when translated the letters were very queer, but a careful study of the letters and the pamphlets convinced us that they were the work of a religious enthusiast who had gone half crazy over a system of Oriental mystic theosophy. The pamphlets were spotted here and there with dots of ink that might have been used to mark letters conveying a secret message, but we were convinced that they were accidental and innocent of any such intention.

Almost any pamphlet that seemed unintelligible to the finder was apt to be regarded as mysterious German propaganda and submitted for a serious examination. Thus, on more than one occasion, we had to pass on the ritual books of various well-known fraternal orders such as the Masons and the Odd Fellows.

Another letter in Lettish was sent in for examination because so many of the letters had strokes through them, but the strokes were there simply because the letter was written on an American typewriter which did not possess the keys for some of the special forms of letters used in this language. The letter was from the editor of a Lettish newspaper who was trying to secure subscribers.

A characteristic instance was that of a Polish letter intended for a soldier in one of the camps but held up for examination because it could not be read. It turned out to be a letter from the soldier's old father and was full of the most ardent patriotism and devoid of any proper cause for suspicion. The old man was proud and pleased that his son was going to the defense of the country in which he had lived so many years and which he had come to love as his own.

Of an entirely different character were some brief messages in Chinese characters that at first could not be read even by the Chinese experts. With the aid of a few suggestions as to the possible ways in which a cipher message might be constructed in Chinese, these messages were finally read and found to relate to certain arrangements for a Chinese funeral.

Some telegrams in an unknown tongue addressed to an officer in one of the camps turned out to be in Latin, though not very good Latin, and related to the marriage of a friend. In reporting on this case, our expert recommended as an appropriate punishment that the officer should be sentenced to parse the Latin.

Illiterate letters in almost any language were liable to awaken the suspicion of any officer who had only a knowledge of the standard form of that language. Thus, numerous letters in even such familiar languages as French, German, Spanish, Italian, Hungarian, Bohemian, and Polish written by persons who could not spell correctly or construct good sentences in those languages were momentarily the cause of awakening the suspicion that they contained some sort of code or cipher.

The letter containing Egyptian hieroglyphics certainly looked formidable and suspicious, but when it was submitted to Professor Breasted, the well-known Egyptologist,[2] he reported that it was a series of exercises from a standard instruction book in the language of ancient Egypt.

With this may be placed some sheets of paper that awakened suspicion because of the presence of a great many figures and unintelligible groups of letters. After careful examination, we were able to report that the papers represented a correspondence course lesson in accounting. We wrote in our comments:

> The references to Mont are to Robert Montgomery's Auditing Theory and Practice. The second edition is probably the one used, as the pages indicated by the numbers are those on which chapters begin and end in that edition.
>
> The word Racine refers to Samuel F. Racine's Guide to the Study of Accounting – a companion to Montgomery's book. The references are apparently to the sections corresponding to Montgomery's chapters.
>
> The supposed messages appear to be lesson assignments and the words "energy", "force", and "engage", are, perhaps, a slogan suggested by the instructor.

Among the other educational exhibits that were sent in under the suspicion of being code and cipher were two geological reports. Fortunately, the first one bore the name of a well-known geologist and contained references to a number of authorities on geology. This made it easy to determine the nature of the document and check up on the data it contained.

A surveyor's notes of the survey of a piece of land also seemed mysterious enough to one observer to cause it to be sent in for examination.

Two of the most frequent types of material sent in as suspicious were collections of postcards and postage stamps. This was particularly true shortly after the appearance in one of the popular magazines of some articles on code and cipher, explaining, among other things, how secret messages could be conveyed by the use of such collections. To test these cases, we were, of course, obliged to secure the services of experts on postage stamps and postcards, and—much to our disappointment—they were unable to find among our cases a single one in which this means of secret communication had been used. So far as the protection of the country was concerned, this was lost labor, but it must have been a great satisfaction to the owners of the collections to be relieved from suspicion—if, indeed, they ever knew they had incurred it.

Closely connected with these subjects were the numerous postcards coming from all parts of the world which contained many groups of three or four capital letters occasionally followed by place names and usually bearing the signatures of

[2] This is James Henry Breasted (1865–1935), at the time director of the Oriental Institute and the Haskell Oriental Museum at the University of Chicago.

foreigners. These turned out to be the initials of clubs and associations of collectors. The number of them was a perfect revelation to most of us, but was, of course, perfectly well known to the experts.

Innumerable letters were mailed during the war and are doubtless mailed today, which contain no written message and are sent only for the purpose of supplying the recipient with a new foreign stamp. These came to us by the score, under the suspicion that they contained messages in invisible ink.

A scheme for advertising a certain brand of soap brought many letters from South America addressed to a firm in New York City, which aroused the suspicion of the postal censor because a part of the scheme was a contest in forming Spanish words from the letters in the name of this soap. The largest number of words was to receive a prize. This naturally resulted in long lists of words, many of which were rare and unusual. As the words in the lists made no sense, they were naturally taken for code.

A very peculiar case was furnished by letters from a woman in California to persons in the Republic of Salvador. Under the flaps of the envelopes were certain letters. For example, in envelope number 5233, the right side of the flap bore the letters SSA, SN, SH, and SJ. On the left side were the letters SA, SM, STA, ST, AG, J, M, and J. The other letters bore the same initials. These were certainly mysterious and might well have carried a secret message. Fortunately, one member of our staff was very familiar with the religious practices of Latin countries and recognized these letters as the exact initials of saints in parts of the Litany of the Saints.[3] It is customary in these countries to write such initials under the flap of the envelope of personal letters.

But it was not only letters that were sent in for examination. Many other scraps of paper excited suspicion. These were so numerous that only specimens can be given. Some of them were clearly the scribbles of demented or half-demented persons. In one instance, the papers bearing the scribbles were the labels from fruit and tomato cans. In another instance, the suspected paper was the enlistment blank of a boy who had obviously become very excited in filling it out and covered it with meaningless scribbles.

On two occasions, a large number of sheets of soft white paper marked with raised dots, which obviously bore some sort of meaning, were brought in for interpretation. The two clearly did not belong to the same system of marking, but both turned out to be writings for the blind: one in Braille and the other in New York Point.[4] Although the character of the writings was recognized at once, it was, of

[3] The Litany of the Saints is a prayer in the Roman Catholic Church, along with other denominations that are close in liturgy to the Catholics (like the Anglican Church). It consists of several prayers, all of which are supplications, to the three parts of the Triune God and to a list of the saints. It is most typically sung at Easter. See http://en.wikipedia.org/wiki/Litany_of_the_Saints (accessed August 13, 2014).

[4] New York Point is a raised-letter system for the blind like Braille. It was invented by William Bell Wait, a teacher at the New York Institute for the Education of the Blind. Because it was more cumbersome to use than English and American Braille, it eventually lost support. See http://en.wikipedia.org/wiki/New_York_Point (accessed August 13, 2014).

course, necessary to read them for such writing might readily have been the vehicle for secret information. The sheets in New York Point were easily read. Those in Braille offered more difficulty because they were not in the standard system but a private adaptation of it. They had to be deciphered just as if they had been written in an unknown cipher. The contents of all were, however, entirely harmless.

One of the cases that excited most interest and for a time offered considerable difficulty was a typewritten sheet containing entirely unknown sequences of letters. An officer of MI-8 made a very careful examination of it and discovered that by far the most frequent of the letters were those that are placed near the center of the universal keyboard. He finally decided, therefore, that the sequences had no meaning and that the frequencies of the letters were due to their relative positions on the typewriter. We traced the paper back to its source and there learned its true origin and meaning. It came from the office of a large business house in Washington in which there was a certain employee—whom we may call Mr. X—who smelled German plots and German spies everywhere. Two or three of his fellow employees, therefore, decided to have some fun with him. One of them, Mr. Y, would, therefore, several times a day, go stealthily to one of the windows and, carefully looking around from time to time as if to make sure that he was not observed, would make mysterious signals as if to some other person in a window opposite. This greatly excited Mr. X; for whose benefit, of course, the farce was planned. He was already on the point of reporting Mr. Y to the Department of Justice as a German spy when his companions decided to clinch his decision by supplying him with better evidence. They, therefore, prepared the typewritten sheet that was later submitted to us and laid it on the desk of Mr. Y among some other papers. One of the typists then confided to Mr. X that she had seen this suspicious document on Mr. Y's desk and suggested that they should try to get it without his knowledge. It was arranged that she should pass by Mr. Y's desk with a bundle of papers in her hand and should accidentally knock from his desk, as she passed, the pile of papers containing the suspicious sheet. In gathering them up, it was thought she could easily obtain possession of it and carry it off among those she had in her hand. This maneuver was successfully executed, and that night Mr. X triumphantly told to an agent of the Department of Justice the tale of the mysterious signals and delivered to him the captured sheet of paper. Fortunately, we did not have many hoaxes of this sort to deal with. They were not easy to detect, and in any case, the detection of them was a great waste of time. MID was too busy for such cases, or for the puzzles and personal ciphers that many persons wished to submit out of curiosity.

Another hoax came in about this time from one of the soldier camps in the form of a letter in German addressed to Sir Lucifer von Satan in Hell. It read in part:

> My dear Satan,
>
> I received your letter of June 28 in which you graciously give up your throne in my favor. I have accepted this with a full sense of its importance and I write to assure you that Hell will lose none of its fearfulness by the change of management. I readily recognize that you have done precisely right in abdicating before the crash came, for the day has come when German power and German influence will encircle the Earth, for only through German efficiency can the world advance.....

> You will from today on release from the torcher chamber the following four arch traitors, namely: Judas Iscariot, General Burnadotte, the Duke of Augustenburg, and Benedict Arnold. Judas Iscariot is to go to Russia and become an apostle for the faith of Kerensky.....

The letter breaks off abruptly and is not signed. No doubt the intention was to sign it with the name of the Kaiser. It is obviously a burlesque piece of writing, the work of a practical joker, and, though not in good taste, is at least entirely innocent of any serious meaning.

Music came in for its share of suspicion. This took two forms. Because some of our censors had read in articles on the subject that cipher could easily be written with the notes of the musical scale, every handwritten musical score that passed under the eyes of these censors in the international mails was quite properly suspected and submitted for examination. As it is practically impossible to write good music and at the same time make the music symbols convey a secret message, any score that upon examination by a competent musician justified itself as music was allowed to pass unchallenged. Of course, it could not be safely assumed that every piece of bad music was intended to convey a secret message, but we felt that the writer of bad music ought to receive some punishment even if he was not plotting against the government.

More common than musical scores, perhaps, were papers containing syllables that turned out to be the tonic sol-fa system,[5] a method of writing vocal music formerly much in vogue in England.

Less difficult than the formation of a secret message with musical notes is the writing of a message in secret ink on a sheet of written or printed music.

Long and tedious as has been this list of suspicious papers, it is not yet complete.

On one occasion, we received some drawings with a request to ascertain whether or not they were of military importance, perhaps diagrams of fortifications or munitions plants. They turned out to be a tailor's designs for the lapels of coats and bore in writing the designations of the new styles that they represented. Equally amusing and belonging to the same field was a series of letters and numbers suspected of being code but finally determined to be a list of styles and pricing of underwear.

Advertisements in papers and magazines did not escape the attention of the lynx-eyed patriots. One of these was the advertisement of an Orthodox Jewish bookshop, and the cause of suspicion was the presence of the word "tallyism" in the following sentence: "Our tallyism are the best in the world." It was submitted by a university professor, whose suspicions were removed when he learned that the word is a term for the Jewish prayer scarfs commonly called tallith.

None of these prayer scarfs were ever brought in for our examination, but other suspected objects of a religious nature were. Chief among these were the amulets

---

[5]The tonic sol-fa system is a way of teaching sight singing invented by Sarah Ann Glover of Norwich, England. It is a solfège that uses pronounceable syllables to teach pitch and sight-reading. The English syllable equivalents are do, re, mi, fa, so, la, and ti. See http://en.wikipedia.org/wiki/Tonic_sol-fa (accessed August 13, 2014).

that are sometimes worn by devout Catholics.[6] They are, of course, marked with characters that may well seem very mysterious to anyone unfamiliar with them.

Another was a cipher advertisement printed in some unknown paper, which was found in the possession of a person who had fallen under the suspicions of the Department of Justice. The cipher text read as follows:

OMIKRON
STDIS PHO CPKZST STGLRZSB, IPST
LRRSK KSGT OLHU8LT. FSTKLTM7GSHS
CDHSHULTZS PHO 7MHZTL8LHO R1OPHE
OST HSA BMTU DHOSTST K7GDIIS
DHZSTSKKDSRSH LC CSDKZSH

The cipher contains four errors but was easily deciphered.[7] Translated from the German, it reads as follows:

Letters and plan received, for all
very thankful. The promised mine
(or bomb) parts and the contraband
cargo of the New York and other ships
are of the greatest interest.
Sigma

The meaning of the message is clear. Whether it was a joke or a serious message was not a matter for MI-8 to determine, but for the Department of Justice. In this, as in many other cases, we never learned what disposition was made of the evidence from the messages that we had deciphered.

Since it is possible to use almost any sort of material for the transmission of secret messages, pictures sometimes are used in this way. So far as I can recall, only one such was submitted to us for examination. This was a large picture that appeared in a daily paper of one of the principal western cities. It came to us accompanied by a letter explaining in great detail that the writer thought the real purpose of the picture was to indicate the location of munitions plants and soldiers' camps in the United States. A careful check on the locations of these plants and camps convinced us that this was not the case and that the only criminal feature of the picture was its bad drawing.

One issue of a Yiddish newspaper fell under serious suspicion because in one column of it there was a serious mixture of two fonts of type. It was, of course, a difficult task to prove that the mixture could not convey any meaning and was due only to an accident in the printing office.

The cipher articles to which we have referred discussed in so interesting a fashion the possibilities of using chess moves for cipher messages that we received an enormous number of clippings of chess problems and reports of chess games which

---

[6] The best-known amulets worn by Catholics are crucifixes, medals (e.g., St. Christopher medals), and scapulars.

[7] The cryptogram appears to be a mixed alphabet monoalphabetic substitution cipher. Its single letter frequency chart is very close to that of regular German plain text. The decryption is left as an exercise for the interested reader.

the senders thought might not be genuine. The volume of this material was so great that we were obliged to enlist the services of two chess experts to advise us on these cases. Other experts had to advise us on a similar case connected with the game of "Go." As is the case with music, it is, of course, possible to convey hidden meanings through what are apparently plays or problems in these games, but the experts assured us that it is almost as difficult to combine good chess play with a secret message as it is to produce good music under the same conditions.

But the largest class of innocent documents that passed through our hands was that of enciphered love letters. No doubt there are at all times thousands of people in the world who use cipher for the secrets of private correspondence, but until the war brought these letters to our attention, no member of the code and cipher section had the remotest idea of the extent and volume of this type of clandestine correspondence. Some of the more interesting and amusing of these letters will be discussed in the next article.

Reference

Childs, J. Rives. 1919. *The History and Principles of German Military Ciphers, 1914–1918*. Paris: United States Army Expeditionary Force. Friedman Collection. National Archives, College Park, MD.

Chapter 11
Civilian Correspondence: Families and Love Letters

John Matthews Manly

Abstract This Manly article, number XI (the number appears to be out of sequence), is on the interception and decipherment of civilian letters. At 1785 words, this is by far the shortest of the Manly articles. An interesting note is Manly's observation that there is apparently a large amount of innocent correspondence by civilians that is normally sent in code or cipher. The war just allowed that correspondence to be made visible because the postal censorship office was intercepting it. In this article, Manly exhibits a dry sense of humor at the expense of the lovers writing letters to each other in cipher.

One of the largest classes of cipher letters that MI-8 had to handle was furnished by the private correspondence of civilians. Of course, the enormous amount of this kind of correspondence carried on within the boundaries of our country entirely escaped our attention for, as is well known, there was no censorship of domestic mail, but letters in any kind of cipher that passed into or out of the country in the international mails were subject to detention for examination. The volume of cipher correspondence, even with this limitation, proved to be unbelievably great. At first we thought that most of these people had resorted to the use of cipher because of the war, but it became increasingly clear that this was not the case. The war, indeed, rather interfered with than promoted the use of cipher for private correspondence, for there was practically no chance in the world that a letter that was plainly written in cipher could escape examination and reading.

Obviously, at all times, thousands of persons in this and other countries make use of cipher for the purpose of concealing their thoughts from other persons who might accidentally or otherwise come into possession of their letters. A letter written by one member of a family to another often contains a sentence or two in cipher in order to communicate some idea that other members of the family who might read the letter are not to be allowed to share. A family letter sent in from the Cuban censorship for examination furnishes a good illustration of this. The letter consists of several pages in Spanish about family affairs, but on the last page, just before the end comes a passage which, when deciphered and translated, runs thus:

> Hernandez is as silent as ever and not even on account of these days has he deigned to write two lines. I of course do not waste my time writing him but his children do and his answer was – silence. It seems he has no soul. I do not understand how he can live in that way, like a stone. I am even ashamed when anybody asks me about him, for sometime one does not even know how to find excuses for him. There, as I have heard, he passes the day alone in a

J.F. Dooley, *Codes, Ciphers and Spies*, DOI 10.1007/978-3-319-29415-5_11

> field in which he has with some chickens and some calves, and even forgets how to speak. Well, may God give him good health there, and may he get along the best he can since he runs away from anything like love and affection.

An extract from another letter affords a pathetic or perhaps even tragic glimpse of a family situation. The writer, a workman in Pennsylvania, sends a letter to his brother in Spain, in which the following sentences (in Spanish) appear:

> To see that our mother does not know of me it is necessary that they know that I am not writing and I refused to write, so I shall continue to write to you and shall sign the letters with another name and instead of placing at the end of the letter, "your brother," I shall place "your friend," just as if it were a friend who was writing you. I write this to you in numbers so that in case this letter is lost they will not read it so easily.

Young people often invent cipher systems and use them in their writing for the sheer delight in the mystery of the thing. But the most common use of cipher letters is for clandestine lovemaking. It would be difficult to estimate the number of thousands of such letters that passed through the hands of the postal censorship during the brief period that it was in operation during the war.

As a rule, it is easy to recognize these private and amateur ciphers at a glance. Almost invariably, they make use not of letters or numbers but of odd characters. Some of them, perhaps, especially invented for the purposes of the cipher. There seems to be a general idea among persons of little experience in such matters that the difficulty of reading a cipher is in proportion to the unfamiliarity of the signs composing it. A cipher written with the ordinary letters of the alphabet, by assigning to each a special new value, strikes them as simple and comparatively easy to read, whereas one composed of question marks, brackets, asterisks, crosses, and other odd but common signs seems more difficult, and one composed entirely of freshly invented signs seems entirely indecipherable. As a matter of fact, the ciphers written with peculiar symbols are nearly always simple and easily decipherable. Almost invariably they belong to the type known as single alphabet ciphers, that is, a cipher in which each letter is represented every time it occurs by the same sign and each sign represents always the same letter. Poe's famous *Gold Bug* cipher is of this type, and all ciphers of this type can be solved and read by the very elementary methods that he explains in the story of the *Gold Bug*.

Here is a letter partly in a cipher that the writer undoubtedly thought would be impossible to read. Its only difficulty, however, arises from its comparative brevity and the number of errors that it contains. Contrary to common opinion, it is easier to solve the cipher of a long message than of a short one; although, of course, after the solution is reached, it may take longer to write out a decipherment.

The message enciphered by the writer reads:

> Lula
>
> Distroy the letter in which
> I mentioned Bertha in before Otto
> finds it. You understand.
> Homer Johnson or Dago Jack Jossinea.
>
> Do not fail Lola Lindsey.

Although ciphers in letters or figures are not as common as those in the odd invented characters, they are, nevertheless, sufficiently numerous. The writer of the following was apparently trying "to start something," and the action of the censor in holding up the letter for examination may have killed an incipient romance. The letter begins thus:

```
20 1 14 17 15, 4 9 7 1 14 5,
21 9, 19 23 9 5 20 5; 19 23 5,
12 5 5 21 3 20 9 2 1, etc.[1]
```

The decipherment and translation reveal the following:

Ramon,
Tell me if you want me to write you.
I ask you because you know that I wrote you from Guantanamo and Caimanera and you did not write and did not even do so from courtesy.
Tell me how your affairs are going; from those concerning me to what you eat. You know I was your secretary.
Adios.
Regards from your friend,
Rosausa.

The order displayed in some of the letters would suggest that cipher affords a sort of protective coloring to the writer and encourages him to express himself more unreservedly than he would in ordinary writing, but perhaps it is best not to illustrate this phase of the subject.

The following message, written in numbers at the top of a letter from a girl in New York City to her friend in Mexico City, doubtless relates to a subject of common interest to them that they wish to keep from the knowledge of their parents:

```
5, 14-3, 22, 1, 14, 21, 16-1-14, 16, 23, 9,
16, 20, -21, 5, 14, F, 16-22, 14, -5, 14, 7,
13, 16, 19, 1, 4, 16, -18, 22, 5, -5, 20, 13,
5, 24, 9, 3, 1, 14, 16, -1F, 5, 19, 16-14, 16, -
13, 5, -F, 22, 20, 21, 1, -21, 1, 14, 21, 16-
3, 16, 13, 16-11, 16, 20, -4, 5, -13, 5, 24,
9, 3, 16, -
```

Deciphered and translated this reads:

In regard to suitors, I have a lover
who is a Mexican but he is not as much
to my taste as those of Mexico.

Sometimes the most incongruous and apparently incompatible subjects are mixed in with the most ardent expressions of love. This is perhaps a tribute to the confidential and familiar relations of the lovers.

The most picturesque example of this that I recall is a love letter from a rat catcher, but perhaps the rats were musk rats:

[1] Manly does not give the reader access to the complete ciphertext of any of the cryptograms in this article, nor does he talk about either the type of cipher (except for the monoalphabetic substitution cipher here) or any methods of cryptanalysis.

Dearest my own,

Received your loving letter O.K. Was more than glad to hear from my love. Oh, hun, so he was to stay with you from Sat. till Sun. night. Well, then, you wasent lonesom or homesick and even asked him to come next Sat and stay till Sunday night. He says he is going and he took you a lot of things for you and babe. And you a new Easter hat. Gee it must have been a nice day tomorrow to want to come so quick again.

Well, love, I dident have very good luck, only 32 rats; that only means 35 dollars.

Hun, shall I use 8 of that for you know what? Love, I am a lot better, am going to fish for a while, tho then I am going on the road with Dad.

Hun, when the time come I will help you if I can.

Cheer up, for God's sake dont go with him again.

For if you do, it will kill me.

It come near it before.

Oh, love, I wish you were here to trap and fish with me. I think the reason you saw me was those nights you spoke of I praied to see you, that or those nights in my sleep ....

It is one of the sordid incidents of a state of war that private correspondence of so intimate a nature that it seeks the cover supposed to be afforded by cipher should come to any other eyes than those of the persons for whom it was intended, and some readers will feel that it is an unwarrantable profanation of sacred feelings that extracts from such correspondence, even when disguised by alterations of names, should be exposed to general knowledge, but the public needs to be warned that in time of war, nothing is more certain than the fact that a cipher letter that falls into the hands of a censor is sure to be suspected and deciphered. At such times, ordinary correspondence is passed with scarcely a glance, even if examined at all, but a message written in cipher is a challenge to suspicion.

Chapter 12
Civilian Correspondence: Prisoners and Spies

John Matthews Manly

Abstract A continuation of Manly's article IX, article X, begins with a discussion of letters to and from the prisoner of war camps, both German and Allied prisoners. It also mentions Franklin W. Allen, who was head of the MI-8 Shorthand Section during the war. The article also discusses attempts by German sympathizers in America to extract information from soldiers in an effort to get logistics information to share with the Germans. The article wraps up with a lengthy discussion of the German spy "Patricia." Herbert Yardley, in Chapter 4 of *The American Black Chamber*, also has a brief discussion of a German spy named "Patricia," although the details are different than in this article and are much more focused on secret inks.

One of the steadiest sources of supply for suspected messages was, of course, the group of prisons and prison camps in various parts of the country.[1] Those in charge of these places were naturally always on guard to prevent improper communications between the prisoners and the outside world. The prisoners and their friends, on the other hand, were bent upon communicating sometimes for entirely innocent purposes and sometimes with a view to furnishing money and means of escape. Various forms of secret communication were tried. The simpler minded occasionally wrote a few sentences in some elementary sort of cipher, hoping that prison officials would not read their letters carefully enough to notice the cipher. Some might use an invisible ink. These were naturally of the simpler sorts, such as could be procured in the prison without exciting attention, such as milk, the juice of lemons or onions, and the like.

In spite of the simplicity of these means, there is very little doubt that a considerable number of these messages got through without discovery, but some of the officials were very alert and submitted for examination any letter that seemed peculiar in any way.

Unlike the European nations which were engaged in the war, all of whom had captured and placed in prison camps thousands of prisoners of war, the United States had few prisoners of war and had been in active participation in the war for

[1] During America's participation in World War I, the War Department kept three prison camps for enemy aliens, German prisoners of war, and for officers and sailors of merchant ships interned when war was declared. The three prison camps were at Fort Oglethorpe, Georgia; Fort McPherson, Georgia; and Fort Douglas, Utah. The Office of the Adjutant General of the Army was in charge of running the prisons. At their peak the three prison camps held about 3,000 prisoners. The last prisoner was released in June 1920. See Commandant of War Prisons 1918 and http://net.lib.byu.edu/~rdh7/wwi/comment/yockel.htm (accessed on August 13, 2014).

J.F. Dooley, *Codes, Ciphers and Spies*, DOI 10.1007/978-3-319-29415-5_12

so short a time that we can cite few, if any, examples of our own men who attempted to send news home from the German prison camps to which they were confined. We know, however, that under these conditions the prisoners developed very ingenious methods of communication. One of their first problems was, of course, to inform the people at home of the method that was used.

Here is a letter from an English prisoner of war in Turkey that will repay close examination:

My dearest Dorothy,
My dear, what a treat this really is to get one
letter, or more, from you about once a week regularly with
family news. I think your solution of the kiddies problem excellent.
Half your troubles should disappear now. Excuse pencil as the ink
pot has just upset. Milk with half water usually is a good remedy
in a case of bad indigestion. If you boil it to make a dish
of Chamomile tea, you should just dip the herb in, tied up in a
sheet of muslin. Till flavored sufficiently. On Thursday I bathed till
half past six. Its topping being immersed in cold water this weather
then returning for pre-dinner cocktails (?) till time to remove
ourselves to our respective homes and pass the evening as best we can. Am
writing this week to Portal. The news of his engagement should appear
soon in the papers. Must finish later. Suppose you've heard
this little bit. "There was a young lady of fashion –
Whose swain overcame her with passion – She woke up
at seven – And remark'd "Thank heaven! – There's one
thing Lord – can't ration"!!"
I got your letter of 27th Feb, and one you forwarded
of Mum's on 6th Mar. just a few days ago. All the back
ones seem to be coming through now but I have no recent
news of you. The Lower Camp are great on collecting
animal pets these days. So far the collection is limited
to an eagle owl which "clicks" at you, a huge hawk, about
4 magpies which nearly swallow one's fingers & try to
commit suicide down the well, a young wolf, and several
other domesticated animals! The wolf is a fine little
beast of the name 'Snaps', He'll have to be respected
when he gets older!!
Best love as ever

The prisoner has to depend upon the keenness of the person who receives his letter to find out that he has written a message by using the first and last words and middle words of alternating lines to express what he has to say. Undoubtedly the attention of any intelligent person would have been arrested by the odd and almost silly character of the first half of the letter. The writer expected this, and the occurrence of such words as "treat," "ink," "writing," and "finish" to induce his correspondent to study out the system. It will be seen that the middle, first, and end words taken in proper order read:

Treat this letter with solution
half ink with half water in a
dish. Just dip the sheet till
immersed then remove and writing
should appear. Finish.

Clearly he has given a description for developing a longer message written in secret ink across the lines of this letter. Both his use of secret ink and the method he adopts for explaining to his correspondent what he has done indicate that he was very familiar with methods of secret writing. This use of the first, last, and middle words is one of the most ingenious that came to light during the war. In our example we have underlined the significant words. Of course, they were not so marked in his original letter.

The directions he gives for developing his secret writing—the use of a thin solution of ink—indicate that in writing he did not use any liquid at all but wrote with a dry pen. This made lines on the paper in which enough of the ink would settle when the paper was immersed in the ink solution to make the secret writing legible. His reason for using a lead pencil in the visible writing of his letter was not because he had upset the ink pot but another and a double reason. First, he does not want the ink of the visible writing to run and interfere with the development of the secret writing, and second he wishes to introduce the word "ink" into his letter at that point for use in his explanation of the method of development.

We had in this country, so far as I know, no spy messages written with dry pen, but the members of the congressional committee who visited MI-8 in the autumn of 1918 will remember that they were taken into the secret ink laboratory and given a series of interesting demonstrations of what could be done. They wrote with half a dozen different kinds of invisible inks, and the writings were developed in their presence. They then were asked to write with water freshly drawn from the faucet of the city water supply. This writing was then made dark and clearly legible by the application of certain reagents. And as a final demonstration, they were asked to write with pens that had not been dipped into any liquid but were perfectly dry. The writing with these was developed as easily and clearly as any of the others.

The paper used in these experiments was a sized but unglazed writing paper,[2] and this undoubtedly was the kind of paper used by the English prisoner of war in Turkey. If he had been a prisoner in Germany, France, England, or America, his plan would have failed, for he would not have been allowed to use such paper. He would have been furnished, as all prisoners were in these countries, with a paper bearing a hard, smooth finish that would have revealed immediately, not only to the microscope but even to the naked eye, any scratch made upon it with a pen.

Before this precaution was adopted, the Germans on one occasion worked a very ingenious trick for obtaining certain information that they wished to procure from England. They knew that at that time the method of testing for secret inks was to apply different kinds of developers in stripes across the suspected letter to see whether writing would appear under any of them. This method left the corners untouched. They, therefore, intercepted a letter from one of their English prisoners which contained writing in secret ink; then, imitating his hand, they wrote in the corners of his letter the questions to which they wished to obtain answers, feeling

[2] Sized paper is treated with a substance that reduces the papers' ability to absorb liquid. This allows the ink to remain on the surface and dry there rather than spreading out through the paper fibers.

sure that these questions would escape the test stripes but would be developed when they reached their destination. They then watched for the reply to this letter and by seizing it and developing the secret writing in it obtained the reports that the correspondent made to their questions.

Perhaps the most interesting letter from a prisoner who was planning his escape from one of our detention camps that came under our attention was one written in shorthand. Of course, the writer could not expect that the prison officials would pass such a letter, and he consequently attempted to get it mailed surreptitiously. He gave it to a friend who was being transferred to another camp and this friend threw it from the train, and the letter fell into the hands of the prison authorities. As it could not be read, it was taken to be deciphered and sent to MI-8. That the writer regarded it as important is indicated by the fact that it is headed "Only to be translated by a good friend. Private affairs." Although the letter was clearly written not in cipher but in shorthand, it was a system that was not known to anyone connected with the Military Intelligence Division. We, however, soon determined that the system was Gabelsberger,[3] and within 24 hours we succeeded in finding a trustworthy Gabelsberger expert who furnished an accurate decipherment of it in German.

The letter contains elaborate and interesting plans for the escape of the writer from Fort Oglethorpe, where he was interned. Of course, the plans were never carried out, for the letter did not reach its destination, but the device of enclosing yellow treasury notes in a glass bottle of preserved fruits is ingenious enough to make one regret that it was not tried. Some paragraphs of the letter are sufficiently interesting to be given here in translation:

> My dear Darling:
>
> The soldiers are leaving here and I am taking this opportunity to send you a few confidential lines. I would like to enter into secret correspondence with you whereby you write to me and I write to you on the reverse of the envelopes of our letters with lemon juice in invisible writing. This writing can be made visible by heating with a flat-iron and I do not believe that the censor will take the trouble to examine the envelopes for secret writing. It seems that the mail is now again forwarded quicker, as I received one of your letters in four days and in that case you could perhaps send me some old woolen underwear and a bottle of preserved fruit, perhaps apples or pears in transparent glass and into these fruits you could put ten dollars by taking ten or twenty dollar bills, which you would roll with the yellow side outward and insert them into a long thin testing tube, and this will not be noticed from the outside. Whole fruit will be less suspicious in transparent glass, so that I believe they will not take the top off and see if money is hidden in the jar. It would also be well to use an original jar with a label, if possible, because then it will appear as though coming direct from the factory. In case peace does not come soon, I shall try to get us together by telling you where to go. At present in Chattanooga, the nearest large city here, there is strike and disturbance. It would therefore not be worth while to come here; the people are too excited.
>
> Now after the military goes away about 1500 I.W.W.'s and other undesirables will come into camp. Then life will be still worse. Therefore, I want to get out. I shall write you much more when we can secretly correspond.
>
> ....

[3] The Gabelsberger shorthand system was developed around 1817 by Franz Xavier Gabelsberger (1789–1849). It was widely used in Germany, Austria, and in the Scandinavian countries.

> Write me also the name of a hotel in St. Louis where upon your arrival you can say you are awaiting your husband from the South where he is traveling. Upon receipt of my telegram telling you to depart, leave secretly for the place which we shall agree upon in writing entirely irrespective of what may be the contents of my telegram. Harry and Ally can then say upon inquiry by the police that you have gone to Chattanooga in order to visit me; that you had received a letter from me which had been put into the letter-box by one of the seamen when they moved and further they know nothing. I just want you to prepare all this so that I may perhaps use it, because upon conclusion of peace it is possible that the authorities will be mean enough to send me out to the country without letting me come back to New York, and if they want to be very mean they can send me back to Honduras whence I last emigrated and then I shall have no money and I cannot see you first and that surely would not do, my darling.

This letter was the first of many in various unfamiliar systems of shorthand, but they gave us very little trouble; for as a consequence of our efforts to secure a reading of this first letter, we came into contact with Mr. F. W. Allen of Hulse and Allen, New York City, a firm engaged in supplying lawyers with reports of legislative actions and court decisions in all parts of the United States. Having a wide acquaintance with shorthand writers, a genius for organization, and a patriotic interest in the work, Mr. Allen undertook to organize all that part of our work that related to shorthand.

He very quickly assembled a body of experts who could write and read all the shorthand systems in use in Europe and America and supplied us instantaneously with transcriptions of the large number of letters in shorthand that were sent to MI-8 for examination.

A little later when the demand came from General Pershing for stenographers to take down the interviews with German prisoners, Mr. Allen relieved us entirely of the burden of interviewing and selecting the shorthand writers who were sent to France in response to this demand.

Scientifically, the most difficult part of the task undertaken by Mr. Allen was the rapid identification of the system of shorthand used in a particular letter. This difficulty was solved by the development of a system of scientific tests similar to those in use for determining the system of code and cipher. As soon as a shorthand letter came to Mr. Allen, it was subjected to this series of tests to determine what system was used. If among the large number of shorthand writers in his office there was one who could read this system, the document was read at once. If it was written in one of the rarer and less-used systems, Mr. Allen at once located an expert who understood that system and sent the document to him for transcription. This he was able to do with great accuracy and rapidity because he had prepared for this special purpose a card index of writers of all known systems and had ascertained which of them could be trusted with documents such as these.

Among the systems regularly handled by Mr. Allen and his force, besides the well-known American systems, were Gabelsberger, Schrey, Stolze-Schrey, Marti, Brockway, Duploye, Sloan-Duployan, and Grillana. Dozens of others turned up occasionally and had to be provided for.[4] A thorough knowledge of English, Spanish,

---

[4] There are more than 50 shorthand systems used worldwide. The most commonly used system in the United States is the Gregg system, invented by John Robert Gregg in 1888. In the United

French, Italian, Portuguese, Dutch, Swedish, Danish, and Norwegian was, of course, necessary to transcribe these various systems successfully.

An almost equally steady stream of ciphers and supposed ciphers came from the training camps. Many of the doughboys wished to communicate to their friends on the outside bits of information that were regarded by the authorities as confidential. Some of them were only preparing for such communications after they should get abroad and while they were in this country were merely practicing their ciphers on their friends. A few examples of this may be interesting:

> Dear Dotty:
> Just a line to let you know that I can write our code pretty well now.

The following shows a touching confidence in the security of a cipher that was actually of the most elementary simplicity:

> Can you make out the letter I wrote you this morning?
> There is another fellow in here who is writing to his wife in this code and he and I made up this code, and he and I, his wife, and you, dear, are the only ones that can make out this code.

Sometimes the writer seems more familiar with his cipher and writes freely and facetiously. This letter, which came from Fort Screven, Georgia,[5] is a sample of many similar ones:

```
NI SIHDTO EAUACO MSHE MTEETO
DRNFT N OTFTY TP MSHE HSE N
HCOT DSH KMSP EA RTOT BOALUAC
RADNH TJ TOU MNEEMT ERN IK N
SL BNITO ERSI BOAK RSNO N RAIT
UAC SOTER T HSLT RAD SOT UOC
FALLNI DNER LU RTMLTE XU I'NIYH
N PA?E IEEP NE IAD NE NHDSOI
TIACKR PADI R TOT RADNH ERT
DTSERTO CI' ERT OT SIU R AD
SOTRSJN IK MAEH A? OSNISI? FAMP
IAOER DNIP PADIRTOT ERSE FAMP
DNIT KNJEHLT SI NPTS ERSE UAC
SOTRSJNIK SFAMT HI'TM CI'ER TOT.
```

Kingdom and most of the Commonwealth countries, the most common system is Pitman, invented in 1837 by Isaac Pitman. Both Gregg and Pitman are phonetic shorthand systems. A relatively new system, the Teeline shorthand system, invented in 1968 by James Hill is rapidly becoming the most popular system in the United Kingdom and Commonwealth countries. Teeline is a spelling-based shorthand system, as opposed to a phonetic one. See http://en.wikipedia.org/wiki/List_of_shorthand_systems (accessed on August 14, 2014).

[5] Fort Screven (Manly has misspelled it) on Tybee Island, Georgia, was built in the 1890s and was originally designed to protect the Port of Savannah. At its height, it was composed of seven batteries; the fort was used through World War II. It was declared surplus after the war and decommissioned.

The word divisions are not always genuine and the cipher contains errors, but it is not difficult to read.[6] The English original of it would run as follows:

> In answer to your last letter
> which I received last Saturday, I
> sure was glad to hear from you.
> How is every little thing? I am
> finer than frog hair. I hope you
> are the same. How are you coming
> with my helmet? By jinks, I don't
> need it now. It is warm enough down
> here. How is the weather up there,
> anyhow? Are having lots of rain and
> cold north wind down here. That cold
> wind gives me an idea that you are
> having a cold spell up there.

There can be no doubt that in general the reasons for the doughboys wishing to communicate with their friends were entirely innocent, and the precautions taken by the military authorities against the inclusion in their letters home of any information in regard to their locations or their ideas of the plans of the army or reports on any casualties seem excessive, in view of the fact that the information would hardly reach its destination on this side until too late to be of any military value even if it were immediately cabled back to Germany. This was the feeling of many of the soldiers who consequently, without any disloyalty, tried as best they could to evade the regulations by ingenious arrangements for conveying information. Very rarely did they resort to cipher for that would have been observed and blotted out at once, but everybody is familiar with one or more of their devices for letting the home folks know where they were.

---

[6] This looks like a transposition cipher. It contains 323 characters. The factors of 323 are 17 and 19 so either of those rectangles would be a reasonable completely filled rectangle. Neither the 17 × 19 nor the 19 × 17 rectangle gives a good transcription of the ciphertext, however. Likely if this is a transposition cipher, the rectangle is not completely filled. And while the frequency counts of the ciphertext are close to a normal German language letter frequency, the ciphertext contains no Ws or Zs, both of which would appear in the German translation of the English text that Manly gives. This argues against a simple transposition cipher. That German text is likely to be

> "In Antwort auf Ihre letzte Buchstabe
> die ich erhielt am vergangenen Samstag, ich
> sicher war froh, von Ihnen zu hören.
> Wie ist jede kleine Sache? ich bin
> feiner als Frosch Haar. Ich hoffe, Sie
> gleich sind. Wie kommst du
> mit meinem Helm? Von Jinks, ich nicht
> brauchen es jetzt. Es ist warm genug unten
> Hier. Wie ist das Wetter da oben,
> haupt? Sind mit viel regen und
> kalter Nordwind hier unten. das kalt
> Wind gibt mir eine Idee, die Sie
> mit einer Kältewelle dort oben." (translation from https://translate.google.com)

It is certain, however, that at times some German sympathizers made very constant attempts to form relations with the soldiers in the training camps here. The purpose seems to have been twofold. In the first place, they believed that they could in this way secure information as to the preparations that were being made for sending the soldiers abroad that would be of service to Germany. The other and more sinister purpose was to persuade soldiers, if possible, that when they got on the other side it would be to their advantage to furnish to German agents working behind our lines as much information as possible about the situation and conditions on the American front.

I shall present only two cases of this sort. The first is a type familiar to all. The writer has obviously made friends with a number of men in the Army and has arranged with each of them a set of code phrases by which they can convey to her information about their movements. Some of these code phrases are given in the letter:

> Dear Bill:
>
> Have been hoping to hear from you and will be sorely disappointed if you fail to write.
>
> Have been getting letters from Henry frequently; is now on a destroyer, stays out about five days .....
>
> Ivan wrote me a thrilling account of his trip over and mailed it in a U.S. Port. Wish I could write about it but guess it is best to keep mum.
>
> How in the world will I ever know if anything should happen to you? Not expecting it however, but wish you would give in my name to be notified if you should get sick or anything happening. So anxious to hear how you are pleased, and what port you are coming in, try to mail me a letter ashore so you can tell more and be sure to observe the code when signing. If your port is New York, say at the close of your letter, "wish I could see you", (of course, I know you are not really wishing that, but it won't mean anything more to a censor) and if you come into Hampton Roads or Newport News write, "would be glad to see you". Please add this to your other code and be sure to observe it always. To let me know you have rec'd this particular letter tell me you rec'd my letter telling you that Henry was on a destroyer, his base is on French coast according to our code sign. Ned is in officers' field artillery training school, Camp Zachary Taylor, I expected to have joined him before this.....
>
> Is Big Buddy at Newport? I wrote to him on my way from Norfolk but heard nothing from him.

This letter does not sound as if this "godmother" were arranging codes and seeking information for the benefit of the Germans, but she has apparently only recently entered into correspondence with her "sons," and there is no telling what indiscreet use she might make of the information obtained in regard to the movements of ships. The information she wanted was exactly the sort that the German submarines wanted for attacks on troop and supply ships. And if we lost none, it was not the fault of this woman.

Undoubtedly there were many women who were very active in their efforts to become acquainted with soldiers for the sake of obtaining information of military value or for the sake of persuading them to transmit information to the enemy after they got into France. To this type belonged a woman who was operating in California and the Southwest during 1917 and 1918. Her story is a curious illustration of the fact that coincidences often occur in real life that a writer of fiction would not dare to make use of.

In the summer of 1917, one of the agents of the Department of Justice reported to his chief that on a recent railway journey through the Southwest he had fallen into conversation with a lady who betrayed an undue interest in everything connected with the war and the preparations that America was making. As the agent bore a German name and was obviously of German extraction, the lady talked much of Germany and the Germans. For a time the conversation seemed to him to have no other motive than that of passing away the time pleasantly, and he joined heartily in her enthusiastic accounts of German scenes and events which revived pleasant memories. But the journey was a long one and the talk gradually became more intimate and confidential. She then revealed herself as an ardent pro-German and relying upon his German name and tastes obviously thought that he shared her views and sympathies. She finally proposed pointedly and definitely that they should work together in securing information for the aid and benefit of Germany.

In spite of the fascination of her beautiful eyes and her auburn hair, he declined to join her in this undertaking; but even had he been so lacking in chivalry as to take advantage of her confidence, he could take no active steps to prevent her from carrying out her plans and contented himself with reporting the case to his superiors. It should be noted that the name under which this lady traveled was "Mrs. Smith Hopkinson."

In August of the following year, one of the postal censors on the Mexican border seized and sent to MI-8 for examination a letter addressed to a woman in Mexico City. The principal reason for seizing the letter was that the addressee was strongly suspected if not actually known to be an intermediary for the German Information Bureau. The contents of the letter, and especially the drawings which it contained, confirmed in the censor's mind the suspicions awakened by the name of the addressee. He felt confident that the drawings conveyed some sort of secret message. One of them certainly looked like a topographical drawing. The other looked as if it might be cipher. The complete text of the letter was as follows:

> Fairmont Hotel
>
> Theo dear,
>
> The book is here, it is just lovely and reminds us of Hopkinson Smith's "Under a White Umbrella."[7] Laura is reading the latest war book "Men in War" by Andreas Latzko,[8] an Austrian Socialist. We wonder the book is in circulation here, as it is obviously bad for the moral of the country. Last evening I succeeded in seducing Laura away from it for an hour to read poetry – now we read it something like this,
>
> A thing of beauty is a joy forever
> Of man's first disobedience and the fruit
>
> the long strokes are rests, of course. Do you like it. If you do I'll send you more. I do hope the censor is an artist or he might cut it out.

---

[7] Francis Hopkinson Smith (1838–1915) was an American novelist. Charles Scribner's Sons published *A White Umbrella in Mexico* in 1889.

[8] Andreas Latzko (1876–1943) was an Austrian pacifist. His novel *Men in War* was written and published in German in 1917 after Latzko's experiences in the Austro-Hungarian Army on the Italian Front. It was an immediate success and was translated into 19 languages. It was banned in all the combatant countries during the war.

Last Sunday we donned our waterproofs and sallied forth for a wade in the most luscious rain I ever encountered. We did a good six miles. At the Muirwoods Inn we fell in with Hatty, our anthropologist, you know, and as I was hatless she insisted on taking my cephalic index. You know how it goes. You used to be "bugs" on it, too.

I shall be pleased to send the fashion sheets, and the face creams. My mother and sisters join in good wishes and remembrances to all of you and in hopes that you may visit us again in the very near future.

Fondly,

Patricia.

The wording of the letter was certainly peculiar enough to suggest that it carried a secret message, but we were unable to extract any meaning from the topographical drawing or the supposed cipher. We decided, therefore, to examine the letter for invisible ink.

By this time our chemists had become marvelously skillful in reading messages written in secret inks. Not only were they familiar with most, if not all, of the inks in use by the enemy and with the reagents necessary to render them visible, but they had also developed a number of methods which were not only available for a single kind of ink but were effective as general tests and developers.[9] The latest of them was of such a marvelous efficiency that it would render visible almost any kind of writing, even that made with clear water drawn from the city water supply.[10] This had been tested on many pieces of writing prepared by ourselves as tests, and we were naturally very eager to see what results it would give with a letter which we had good reason to believe contained some unknown sort of secret writing. This test had the additional advantage of not disfiguring the paper in any way or leaving on it any trace of its having been examined.

The test was a triumphant success. Secret writing came out on both sheets of the letter, fully justifying the censor in his suspicions and us in making the test for invisible ink. The secret message read as follows:

(First page)

I wrote you about the incarceration of the trio etc.

Let me know as soon as you can about the boys going to France.

If of no use in France they are preparing to flee.

(Second page)

I'm wondering if this ink is good. Let me know if these boys
would be of any use to you in France.

Preparations are being made for training and drilling
in use of big guns in U.S. Officers returning from France for that purpose.

---

[9] One of the three "universal" reagents was, first, the iodine vapor test. A second was a mixture of potassium iodide dissolved in hydrochloric acid and combined with aluminum chloride, sodium, and iodine and mixed with distilled water. These two tests discovered the disturbance of the fibers on the paper made by the pen that was used to apply the invisible ink. A third test was effective for organic-based inks (milk, lemon juice, etc.); this was the use of an ultraviolet lamp. The lamp would cause the message to fluoresce—but just once—so the investigators needed to photograph the message before it disappeared (Macrakis 2014, pp. 150–151).

[10] This is probably the iodine vapor test.

From other sources of information, we knew that reports had been regularly going from San Francisco and other points in the West and Southwest to the German Bureau of Information. This letter was obviously only one member of a long series of letters, but unfortunately it was the first and only one that had been suspected and discovered. And the discovery had, of course, been made only after the letter had taken the long journey from the postal station on the border to Washington, which naturally involved a delay of 3 or 4 days before the test was made. This delay proved fatal.

A report on the letter and a transcript of the secret message were immediately placed in the hands of the officer in charge of the case, who promptly wired the result to San Francisco. An attempt was made to find and arrest "Patricia," but she was no longer to be found at the Fairmont Hotel. Probably she had been warned of her danger by the failure of her letter to reach her correspondent in Mexico promptly and had taken flight.

The agent, who was working on the case, reported that "Patricia" was probably an assumed name, for he could not learn that any woman of that name had recently been a guest at the Fairmont Hotel. But shortly after the discovery of the letter, a guest who had attracted attention by her beautiful auburn hair, her general intelligence and charm, and the wide range of her curiosity had left the hotel very suddenly. In view of this fact, the reference to Hopkinson Smith's *A White Umbrella in Mexico*, near the beginning of the letter, takes on a certain significance. It seems likely that the signature "Patricia" was used for the first time in this letter and that the reference to the book and its author was an ingenious device for informing the addressee of the name of the writer. It is now impossible to prove that the redoubtable "Patricia" and the "Mrs. Smith Hopkinson," who in the previous summer entertained the agent of the Department of Justice, was one and the same person, but the evidence points strongly to that conclusion.

The object of these women in their negotiations with the soldiers was evidently to find "boys" who would be of use in France—that is to say, who could be induced to supply information about our camps and preparations to German agents behind our front lines, of whom there is said to have been a plentiful supply. Such arrangements, if completed, would, of course, be of the highest military value; for any sort of definite information was at a premium. Many of our officers who were at the front report that during periods when there was confusion in the designations and locations of our forces, German soldiers often stripped the uniforms from our dead or wounded and for a time posed as American soldiers belonging to this or that battalion for the sake of the military information which could be secured even in the few hours during which such a pretense could be successfully maintained. Of course, these men disappeared before contact was made with the regiment to which they claimed to belong.

The case of Patricia emphasizes strongly certain difficulties in our handling of such cases. Had such a letter been seized in England by the British censorship, the letter would, of course, have been examined without an hour's delay and would then have been sent on to its destination. All mail entering the country would then have been carefully watched for a reply. When found, it would have been read and forwarded to its destination, and the correspondence would have been carefully watched, with a view to ascertaining the names and activities of all persons who were concerned in the matter.

We were thoroughly familiar with this procedure, but for various reasons it was not often possible to follow it. In the first place, the United States had not yet reached the point at which the censorship of domestic mails was permitted. Only letters entering or leaving the United States passed under the eyes of the postal censors. Furthermore, in order to expedite foreign correspondence, censorships were established at many American ports and border cities. Most of these were very distant from the Code and Cipher Bureau and the central laboratory for secret inks. To meet this difficulty the postal censors at New York and some of the other important stations were given a certain amount of training in the decipherment of code and cipher, and a laboratory for secret ink was established in New York. For effectiveness in checking and discovering activities carried on with the aid of secret messages, tests of both kinds ought to have been made immediately at the censorship stations, but the establishment of Code and Cipher Sections and laboratories at all these places would not only have been very expensive—not a matter of very great consequence in the conduct of the war—but it would have been a long and difficult task to supply properly trained persons to carry on this work at so many places. It may be said, however, that had the war continued, our arrangements would have equaled those of our Allies in effectiveness, even though our problems were enormously greater and more complex than theirs.

There were a large number of instances in which suspected persons who were seized upon their entering the country were found to be in possession of articles of wearing apparel soaked in the well-known reagents used for invisible writing. Among the articles examined by our laboratory and found to contain such substances were socks, neckties, boot laces, white shirts, and buttons covered with dark cloth and various soaps, lotions, perfumes, toothpastes, and patent medicines.[11]

Some of these cases occurred so near the end of the war that they were not carefully followed up. In other instances, the only result of the discovery was the internment of the guilty person, as his activities were stopped before he was able to accomplish any overt acts of hostility. These cases cannot be given any sensational development, but it must be remembered that every one of these persons came with plans and purposes which might have resulted in enormous disasters: That they were stopped before they could get into action can cause regret only to those who value a dramatic story more highly than the safety of the country.

[11] In Article IV it was noted that when arrested in April 1918, Madame de Victorica had a secret ink embedded in a pair of white silk scarves and several handkerchiefs. The spy George Vaux Bacon had secret ink embedded in his socks when the British captured him.

References

Commandant of War Prisons. 1918, August 14. *Annual Report of the Commandant of War Prisons*. Barracks 2, Ft. Oglethorpe, GA: Adjutant General's Office. RG 407. National Archives, College Park, MD.

Macrakis, Kristie. 2014. *Prisoners, Lovers, & Spies: The Story of Invisible Ink from Herodotus to Al-Qaeda*. New Haven, CT: Yale University Press.

Yardley, Herbert O. 1931. *The American Black Chamber*. Indianapolis, IN: Bobbs-Merrill.

Part III
German Spies in America, 1914–1918

Chapter 13
Spies Among Us: The New York Cell, 1914–1915

Abstract Even before the European war started in the summer of 1914, the Germans were worried about the Americans supplying arms, ammunition, and food to the Allies. The German Ambassador to the United States, Count Johann von Bernstorff, was instructed to set up a spy network in the United States, centered on the East Coast ports. This chapter describes the beginning of that network, starting in the biggest and busiest American port—New York.

On June 28, 1914, while Europe was rocked by the assassination of the Archduke Franz Ferdinand of Austria, heir to the throne of the Austro-Hungarian Empire, and his wife, Sophie, the Duchess of Hohenberg, by Serbian nationalists, America was oblivious. With the July 5th assurance by Germany that it would fully support Austria-Hungary in a war with Serbia, the Austro-Hungarians proceeded to bully Serbia, and America didn't care. Three weeks later, when Austria-Hungary delivered the July Ultimatum to Serbia on July 23rd, a list of demands designed to be rejected, America—and most of Europe—was on vacation. The 20th of July saw the German government advise the North German Lloyd and Holland-America shipping lines to begin moving their vessels off the seas and into port. On the same day orders went out to the German High Seas Fleet to begin to concentrate.

To everyone's surprise, the Serbs agreed to all but one of the Austrian demands; however, this was not good enough for a country that was already aching for war with Serbia. Austria broke off diplomatic relations on July 24th, and Serbia began mobilizing its armed forces the same day. Austria began its own mobilization on July 25th, and the Russians began their "Period Preparatory to War" as well. Not to be outdone, on July 26th the German General Staff warned neutral Belgium to allow the passage of German troops through Belgium in the event of war with France. Austria declared war on Serbia at 11:00 am on July 28th, and an Austrian gunboat fired the first shots of the war into Belgrade at just about 1:00 am on July 29th. From this point on, there seemed to be no going back. On the same day the Russians ordered a partial mobilization against Austria-Hungary; the British called back their fleet and canceled all leaves in the Navy while at the same time making their fourth offer of mediation of the crisis to the Germans, Russians, Austro-Hungarians, and Serbs. The 30th of July brought about the full Russian mobilization and led to German mobilization beginning on July 31st. The modified Schlieffen Plan was put in action, and once begun, the German General Staff claimed it could not be stopped.

J.F. Dooley, *Codes, Ciphers and Spies*, DOI 10.1007/978-3-319-29415-5_13

On August 1st German troops occupied Luxembourg and began massing at the Belgian border. This provoked the French to mobilize immediately; although with typical French indecisiveness, they also withdrew their troops back 6 miles from the German border. Later on August 1st, Imperial Germany declared war on Russia. On August 2nd the Germans sent an ultimatum to Belgium, demanding free passage on their way to France. Belgium refused. Germany declared war on France on August 3rd and on Belgium on August 4th. In accordance with the Schlieffen Plan, designed to crush the French and capture Paris in 6 weeks, German troops began to invade Belgium that day. Great Britain, supporting its ally France, and living up to its treaty obligation to protect neutral Belgium, declared war on Germany on the evening of August 4th. The next day, a British ship sailed out into the North Sea and cut all of Germany's undersea telegraphic cables, effectively cutting her off from direct communication with her embassies and consulates outside of Europe. World War I had begun (Keegan 1999, pp. 48–70).

In the meantime, the biggest events of note in the United States in July 1914 were the deaths of four anarchists who were trying to kill John D. Rockefeller, when their homemade bomb exploded prematurely, and the major league debut of Babe Ruth with the Boston Red Sox.

13.1 Bernstorff Builds a Spy Network

While the Americans may not have been paying much attention to the events in Europe and while they certainly wanted to have no part of a European war, others were not so disinterested. Johann Heinrich von Bernstorff had been the German Empire's Ambassador to the United States since 1908. Urbane, sophisticated, tall, slim, good-looking, intelligent, wary, and personable, von Bernstorff made an excellent protector of Germany's interests in the United States (Fig. 13.1). Among others in the diplomatic community, von Bernstorff was shocked by the assassination of the Austrian Archduke and worried about the consequences. But he was confident that the United States would not get involved in a European conflict. So he was quite

Fig. 13.1 Count Johann von Bernstorff (*Public Domain*. From the Library of Congress Prints and Photographs Division under the digital ID ggbain.03430; This work is from the George Grantham Bain collection at the Library of Congress. According to the library, there are no known copyright restrictions on the use of this work)

surprised when he received a telegram from the German Foreign Office on July 7th demanding that he return to Berlin immediately for consultations. He was even more surprised two weeks later when he got to Berlin and was told to go not to the Foreign Office but to the German Military General Staff headquarters. There he was escorted into the office of Major (later Colonel) Walter Nicolai, the head of *Abteilung IIIB*, the Military Intelligence Section of the Imperial German Army General Staff. Major Nicolai was one of those German staff officers who *was* interested in the United States and its commitment to neutrality.

In the event of a war, Nicolai said, the British were more than likely to cut off German communications (they did) and blockade German ports to prevent the delivery of food, munitions, and fuel (they did), all of which Germany would need desperately in a prolonged conflict. A British blockade, moreover, would cause the Americans to sell more supplies to the Allies than to Germany because it would be easier and safer to deliver goods to British and French ports (they did). Also, with the surge in German immigration into the United States in the last quarter of the nineteenth century and the first decade of the twentieth, there were several hundred thousand German reservists now living in the United States. The Imperial German Army would like to have many of those reservists—particularly the officers—return to Germany once a war started.

Since its creation in 1889, Abteilung IIIB had created intelligence networks in most of the European countries and especially in the Triple Entente countries of Britain, France, and Russia. But there were no German intelligence or counterintelligence agents in the United States, and Nicolai wanted Ambassador von Bernstorff to fix that. In addition to his diplomatic duties, primarily concerned with exercising his excellent diplomatic skills and his winning personality to keep the United States neutral, von Bernstorff was charged with creating an intelligence network in the United States. If he was unable to convince the Americans not to supply the Allies with munitions and foodstuffs or to supply both the Central Powers and the Allies, von Bernstorff was to make sure that American munitions in particular did not reach the Allies, *by any means necessary*. And so on August 2nd, as German troops were getting ready to invade Belgium and declarations of war were flying over the telegraph lines of Europe, von Bernstorff boarded the Dutch liner *Noordam* carrying $150 million in German treasury notes and instructions to create a spy and sabotage network in America (Landau 1937, p. 7; Witcover 1989, p. 39).

Von Bernstorff's first job—his job as German Ambassador—was to keep the United States neutral in the European conflict. His second job was to create a network that would ensure that the enormous potential of the American industrial economy was not used to aid the Allies. He was not trained for a job like this. He was a diplomat, the son of a diplomat, married to an American and much more interested in the social aspects of his job than in espionage. He was a fast learner though, but this second job would not be easy.

In 1914 the United States was divided in opinion about the European war. With a large German immigrant and first-generation German-American population, there was definite sentiment on Germany's side. Many Americans still did not trust the British, particularly after their lukewarm support for the Union and construction of commerce raiders for the Confederacy during the United States Civil War barely a generation ago. A large immigrant Irish population also was not enamored of the British because of their treatment of the population of Ireland. Despite this, most Americans—there were just about 100 million Americans in 1914—tended to support the Allies. Propaganda and newspaper reports of alleged German atrocities in Belgium were prevalent, particularly in the large cities along the East Coast, and far outweighed pro-German propaganda.

The Americans were also distracted from events in Europe by happenings along their southern border. The Mexicans had been engaged in their own Civil War since 1910. When the long established autocrat Porfirio Diaz imprisoned his main rival and stole the Mexican presidency in 1910, a revolution broke out. One year later, the rival, Francisco Madero, at the head of a coalition of revolutionary forces, forced Diaz to resign and go into exile. Elected president in the fall of 1911, Madero talked about land and social reforms and was popular with the people but failed to accomplish nearly all of his early promises. Overthrown in a coup d'état orchestrated by his Army Commander Victoriano Huerta in February 1913, Madero was assassinated a week later by army officers loyal to Huerta. The United States, under new President Woodrow Wilson, refused to recognize Huerta's presidency and a new outbreak of Civil War in Mexico began. In April 1914 in response to the arrest of nine American sailors in Tampico, Mexico, the United States Navy landed and occupied the city of Veracruz to prevent arms from reaching Huerta's army. The Mexicans resisted and the resulting skirmishes resulted in about 170 Mexican and 50 American deaths. In July 1914, after several more military defeats, Huerta gave up the presidency, fled to Spain, and Venustiano Carranza succeeded him as Mexican president. Carranza was opposed by a number of the former revolutionary leaders, including Pancho Villa. Carranza secured his presidency in 1915 by defeating Villa's forces at the Battle of Celaya, although the fighting would continue sporadically for another 5 years. The Wilson administration finally recognized Carranza's government in early 1916. Villa, feeling he had been betrayed by the Americans, proceeded to attack and loot American holdings in Northern Mexico. On March 9, 1916, Villa's forces attacked the town of Columbus, New Mexico, killing a number of residents. This sparked outrage in the United States and led, later that year, to the so-called Punitive Expedition in which the United States sent 10,000 US Army and National Guard troops under the command of General John J. Pershing into Northern Mexico to find and eliminate Villa's forces. These troops scattered Villa's forces early in their operations but never did capture Villa. Active operations were finished by late June 1916, but the expedition stayed in Northern Mexico until February 1917, shortly before the American entry into the European war. As we will see later, the Germans attempted more than once to use the unrest in Mexico to also cause unrest in the United States (Tuchman 1958, pp. 38–60).

When he arrived back in Washington in mid-August 1914, von Bernstorff immediately started setting up his intelligence network. He gave the bulk of the funds from Berlin to Dr. Heinrich Albert, the embassy's commercial attaché. Albert was to be the paymaster of the spy network and the head of pro-German propaganda. Von Bernstorff entrusted the job of finding a way to get German army reservists

Fig. 13.2 Captain Franz von Papen (Bundesarchiv bild 183-S00017. *Public Domain*. ggbain.03430 This file is licensed under the Creative Commons Attribution-Share Alike 3.0 Germany license)

Fig. 13.3 Captain Karl Boy-Ed (*Public Domain*. From the Library of Congress Prints and Photographs Division. hec-17235)

back to Germany to the military attaché, Captain Franz von Papen (Fig. 13.2). The job of slowing the transport of American munitions to the Allies went to the naval attaché, Captain Karl Boy-Ed (Fig. 13.3). All three of these officials set up their offices in New York City rather than Washington. This was to keep the German Embassy at a distance, because of the large German-American population in New York and because most of the munitions and transport from the United States to Europe was shipped out of New York. Von Bernstorff himself kept as far away from this hidden operation as possible so he would have plausible deniability should anything go wrong.

13.2 von Papen Tries – and Fails

Franz von Papen was born into a rich, Roman Catholic Westphalian family. He joined the Army and became a cavalry officer, joining the German General Staff in 1913. In December 1913 he was named the military attaché for the German

Embassies in Washington and Mexico City. This is not as impressive as it sounds because at the time the Germans viewed the American post as a minor one, not nearly as important as a posting to one of the major European capitals. Von Papen was not trained for intelligence work nor does it seem was he very good at it. Once he got his orders from von Bernstorff, von Papen set up shop in the fall of 1914 in an office suite at 60 Wall Street and started recruiting mostly stevedores and interned German sailors for espionage and sabotage work. His first job was to get as many of the stranded German reserve officers on ships bound for Europe so they could make their way back to Germany and join in the fight. While the German Army wanted all the reservists it could get, the officers were the most critical need because of the extraordinarily high casualty rate among the Regular Army officers in the first months of the war. German officers were taught to lead from the front, which was an effective strategy in the days of cavalry charges, muskets, and single-shot rifles but was not as effective in the face of modern machine guns and artillery. So the German Army rapidly depleted its supply of experienced junior officers in particular and was desperately training new officers as quickly as possible. The arrival of already-trained reserve officers would help bolster their forces on the Western Front.

German military personnel could not transit directly to Germany because the Allies were stopping all shipping and searching for German citizens who would then be interned, so von Papen's first idea was to get the German officers' fraudulent passports (Millman 2006, p. 12). Von Papen's hired hands would approach sailors from neutral countries like Spain, the Netherlands, Sweden, and the United States and offer to purchase their passports for around 10 dollars. Passports in those days typically were relatively easy to change and did not normally include photographs so changing the information on an official passport was a relatively easy task. Von Papen's pipeline of German reserve officers heading back to Germany with fake passports began to flow.

However, early in December 1914, the United States made the passport application process more involved and began requiring photographs for the first time. This made von Papen's work more complicated, so von Papen found and recruited Hans von Wedell, a German-American attorney with contacts in the New York underworld that included some expert forgers. Von Wedell set up his office around the corner from von Papen's. He paid a number of seamen and stevedores to apply for American passports, then took the originals, and started cranking out fake American passports for the reservists. Soon von Papen was sending hundreds of German reserve officers to ships heading for neutral European ports from whence they would transit to Germany. However, neither von Papen nor von Wedell was the most discrete conspirator, and the New York Police Department and the Secret Service were soon watching von Wedell's passport office. Von Wedell recruited one Carl Ruroede as his replacement to run the passport scheme late in 1914, and on December 25, 1914, he disappeared, heading first to Cuba and then back to New York to catch a Norwegian liner—using a fake passport. Von Wedell escaped on board the liner *Bergensfjord* but was arrested by the British on January 11, 1915, when the ship was stopped as it entered the English Channel. He was transferred to a British naval cutter. Unfortunately for von Wedell, a German U-boat torpedoed the cutter that was carrying him to Britain for extradition and he drowned. Ruroede didn't have much luck either as the Justice Department's Bureau of Investigation was on to him almost immediately. A Bureau agent posing as a longshoreman with German sympathies offered to get Ruroede some American passports. When the agent showed up with several phony passports and got paid by Ruroede on January 2, 1915, Ruroede was arrested, and the entire passport scheme was shut down (Witcover 1989, pp. 62–63).

Von Papen's next scheme wasn't his own idea but was brought to him by a German intelligence agent, Horst von der Goltz. Von der Goltz had been working in Mexico at the outbreak of the war but made his way to New York and got in touch with von Papen in September 1914. Von der Goltz's idea was to blow up the Welland Canal in Canada that linked Lakes Ontario and Erie. The Canal was one of the main conduits for shipping raw materials for the munitions industry in the United States. Disrupting traffic on the canal would greatly hamper the passage of supplies and troops bound for the Allies from the heart of Canada and the American Midwest. The Canal has a total of eight locks that lower ships just about 100 m (330 ft) from Lake Erie into Lake Ontario. Von der Goltz's plan was to blow up one of the locks, which would completely block the canal. Von Papen liked von der Goltz's idea and offered to fund it completely.

Von der Goltz proceeded to prowl the dockside and bars along the Hudson River, chatting up interned German sailors and sounding out their interest in the plot. He acquired 100 lb of dynamite from the DuPont Powder Company via a Captain Hans Tauscher, the New York representative of the Krupp arms manufacturing company. Von der Goltz hid the dynamite at the 123 West 15th Street brownstone of Martha Held, a former German opera singer who used her house as a gathering place for German diplomats, businessmen, and interned sailors. Dividing the dynamite into two 50-pound suitcases, von der Goltz and his accomplices entrained for Buffalo and checked into a hotel there, having been followed by Secret Service agents. The next day they started walking the Canal, getting the lay of the land and trying to learn how well it was guarded. This is when things started to fall apart. Von der Goltz's hirelings did not like the look of the heavily guarded locks and began to get cold feet. Von der Goltz ordered his men to stay at the hotel while he went further north to investigate more of the locks. When he returned to Buffalo 2 days later, everyone was gone. Stymied, von der Goltz returned to New York and made his way back to Germany in October 1914. German military intelligence gave him a new assignment and sent him back to the United States. On his way back to the United States via England, he was arrested, extradited to the United States, and spent most of the rest of the war in jail (Witcover 1989, pp. 58–61).

Despite these failures, von Papen's superiors in Berlin were adamant about limiting the ability of the United States to send arms and ammunition to the Allies. In November 1914, von Papen received a telegram from the German Foreign Office that said, in part, "It is indispensable to recruit agents to organize explosions on ships sailing to enemy countries, in order to cause delays in the loading, the departure, and the unloading of these ships." Von Papen continued to recruit dockworkers to give him information on ships loading and sailing from US East Coast ports (Millman 2006, p. 14). His success was very limited though. Von Papen wasn't very imaginative about what could be done to hamper shipping, and he depended mostly on others to come up with the innovative ideas.

The last of von Papen's early attempts at sabotage was the result of a telegram received by von Bernstorff from the German Foreign Office on January 3, 1915.

> Secret: The General Staff is anxious that vigorous measures should be taken to destroy the Canadian Pacific in several places for the purpose of causing a lengthy interruption of traffic. Captain Boehm who is well known in America and who will shortly return to that country is furnished with expert information on that subject. Acquaint Military Attaché with the above and furnish the sums required for the enterprise. [signed] Zimmermann (Witcover 1989, p. 68; Pohlmann 2005, pp. 51–52)

One possible target to disrupt the movement of supplies and troops across Canada and to ships heading for Europe was the Vanceboro Bridge in northern Maine. The bridge was on the border between Maine and the Canadian province of New Brunswick. Trains heading east to Canadian ports regularly crossed the bridge, and putting it out of commission would seriously disrupt that traffic and slow down resupply of the Allies. For this job von Papen recruited one of the German reserve officers waiting in New York to get a ship back to Germany. Werner Horn had served in the German army for 10 years before retiring into the reserves and heading off to Guatemala in 1909 to be a coffee plantation foreman. At the outbreak of the war, Horn headed to New York to try and take ship home to Germany, but he was prevented, like most of the others, by the British blockade. Von Papen was aware of Horn's predicament and enlisted him in the Vanceboro scheme. On January 19, 1915, von Papen wrote Werner Horn a check for $700, and by the end of January, Horn had a suitcase full of about 80 lb of dynamite and a train ticket to Maine. On the evening of Friday January 29, 1915, Horn left New York's Grand Central Station bound for Boston. Early the next morning, after arriving in Boston, he took a train to Vanceboro, Maine, arriving about 7 PM that evening. He took some time—and was noticed by some local citizens—to take a look at the international rail bridge before checking into the Vanceboro Exchange Hotel. Horn wandered around Vanceboro on Sunday the 31st and Monday the 1st of February, playing the tourist and pretending to be an interested Danish farmer looking for some land to buy. He checked out of his hotel early on Monday evening, telling the proprietor that he was going to catch the 8 o'clock train to Boston. But he didn't. He spent the evening hanging around the train station and finally wandered off down the road shortly after midnight on February 2nd. It couldn't have been pleasant for someone who'd just spend over 5 years in Central America as the temperature that night was close to 30 below zero, and there was a strong wind blowing.

Horn crept up to the bridge and began to cross to the Canadian side. He was forced to dodge two trains but finally made it across and attached his suitcase to a girder on the bridge. The fuse that von Papen had provided him was for 50 min, enough for him to get well away from the bridge and Vanceboro, but the passing trains had rattled Horn and, not wanting to have anyone hurt in the explosion, he shortened the fuse to just 3 min (he said later), lit it with his cigar, and took off back to the American side. At 1:10 am on the morning of February 2, 1915, the dynamite exploded. Unfortunately for Horn and von Papen, his bomb did very little damage to the Vanceboro Bridge itself; it was back in use within a week. But the explosion did manage to break most of the windows in the two hamlets on each side of the border and wake nearly everybody up. Some people thought there'd been an earthquake; others thought there was a train wreck, and the proprietor of the Vanceboro Exchange Hotel thought his boiler had exploded.

Horn made it back to the hotel where he tried, with the help of the owner, to unfreeze his frostbitten hands. He then checked back into the hotel and went to bed. The next morning he was woken up by a pounding on his door and opened it to find two Canadian constables and the local American deputy sheriff, George Ross. The Canadians wanted to extradite Horn immediately, but the Americans instead moved him to a small town south of Vanceboro and charged him with malicious mischief (for breaking the windows) and illegal transportation of explosives. Horn was interrogated by agents of the Bureau of Investigation for several days and freely gave up his story but refused to name anyone else involved, particularly not von Papen. He was finally tried at the US District Court in Boston and convicted of the illegal transportation charges and spent 18 months in the federal penitentiary in Atlanta. When his sentence was up, he was extradited to Canada where he was convicted of sabotage and sentenced to 10 years in prison (Barton 1919, pp. 291–305; Witcover 1989, pp. 33–34; Landau 1937, pp. 20–21).

13.3 The Dark Invader Arrives

By early 1915 German military intelligence was becoming concerned with the increasing number of failures in operations in the United States. So in March 1915 a new intelligence agent was sent to New York to shake things up and get the sabotage efforts back on track. Franz Rintelen (he used a "von" in front of his name but was not an aristocrat) was born in 1878 to a family with many banking interests and connections. For a time he was a German naval officer, serving on the Admiralty Staff. In the early 1900s he spent a couple of years in banking in London. He then crossed the Atlantic and worked in the United States for first Deutsche Bank and then Disconto-Gesellschaft, Germany's second largest bank, from 1906 to 1909. He spoke fluent English, was a member of the New York Yacht Club, and ran in all the right social circles. In 1909 he transferred to Central America for a year, much to the dismay of his friends in New York. Back in Germany in 1910, he married, had a daughter, and rejoined the German Navy at the outbreak of war in 1914. He spent several months on the Admiralty Staff in Berlin working on financial matters before joining naval intelligence. Rintelen was intelligent, good-looking, personable, persuasive, cunning, and very well organized. Needless to say, his arrival in New York on April 3, 1915 with five-hundred thousand dollars and orders to shake up the sabotage efforts against Allied shipping did not sit well with Captains von Papen and Boy-Ed. Even worse was that Rintelen was a loose cannon; he reported directly to Berlin and did not need to go through any of the attachés or von Bernstorff (Fig. 13.4).

Fig. 13.4 Captain Franz Rintelen in 1919 (*Public Domain*. 1919 International Film Service, Inc. From the Library of Congress Prints and Photographs Division)

Rintelen hit the ground running in New York. Working with a now unemployed German export merchant named Max Weiser and the legal help of a shady American lawyer, Bonford Boniface, he opened an export business, E. V. Gibbons, Inc. in Lower Manhattan. He deposited the half a million dollars in a New York bank and then proceeded to hang out at the New York docks during the day and the better clubs at night. At the docks he made contacts with interned German sailors and with Irish nationalists among the stevedores. At night he reacquainted himself with some of his society friends and kept an ear out for gossip. While Rintelen was recruiting potential saboteurs, Weiser was busy trying to buy munitions allegedly to be shipped to Germany. They had no luck getting supplies; the Allies were purchasing the entire American output as fast as it could be manufactured. So Rintelen and Weiser proceeded to do the next best thing, blow up the ships.

Dr. Walter Theodor Scheele had a chemistry degree from the University of Freiberg and had been in the German Army as an artillery officer. In 1893 he was ordered to the United States to investigate the American chemical industry for the German General Staff. He did such a good job at industrial espionage that he stayed for twenty years, marrying an American and opening a pharmacy in Brooklyn, all the while having a second job as a spy. Von Papen knew Scheele and it was he who sent Scheele over to the E. V. Gibbons offices to talk to Rintelen. During the course of his interview with Rintelen, Scheele reached into his trousers pocket and pulled out a small lead pipe and laid it on the desk. The pipe was about 2 in. in diameter and about the length of a large cigar. Scheele told Rintelen that the pipe was what he needed to blow up munitions ships.

The cigar-shaped pipe was divided into two compartments by a copper disk that was soldered into the inside of the pipe. Scheele told Rintelen that one would pour picric acid (also known as 2,4,6-trinitrophenol, an explosive) into one side and, into the other, sulfuric acid. Then each open end would be plugged with wax and sealed. The copper disk acted as a timing device. The two acids on either side would eat their way through the copper, the amount of time it required for this to be accomplished depending on the thickness of the copper disk. Once the copper disk was breached, the acids would mix and quickly generate an intense flame that would shoot out of both sides of the pipe and melt the lead. Of course, anything near the pipe at this juncture would catch fire (Witcover 1989, p. 89). The ideal part was the timing device. Scheele explained that with some experimentation they could cut copper disks of different thicknesses that would set off the incendiary reaction from a few minutes or hours to many days from the time the "cigar bomb" was manufactured. The combination of the cigar bombs and the restless stevedores on the docks loading munitions ships was just what Rintelen was looking for.

Dr. Scheele had already set up a dummy company in Hoboken, the New Jersey Agricultural Chemical Company, where the lab where he would fill and seal the cigar bombs was located. Weiser could buy lead pipe and copper through the E.V. Gibbons Company. All Rintelen needed now was a place to cut the pipe and solder the copper disks without arousing suspicions or conspicuously breaking any American neutrality laws. His attorney, Boniface, had the answer. In New York Harbor in early 1915, there were some 80 German merchant ships interned for the duration. Technically, they were German soil, so by doing the manufacturing on one of the ships they could avoid the American authorities and the neutrality laws. The pipes were cut and the copper disks soldered inside them on an interned ship, the former North German Lloyd liner *Friedrich der Grosse*, where Rintelen hired a retired German merchant ship Captain, Karl von Kleist, to supervise operations. The empty cigar bombs were carried to Scheele's lab in Hoboken to be filled with the two acids. The cigar bombs were then delivered to E.V. Gibbons where they were distributed to sympathetic dockworkers that, for a fee, would plant them on munitions ships bound for Europe. A week or two later, there would be a mysterious fire or explosion on board the ship, and the cargo would be destroyed. The first ship to be targeted in this manner was the *SS Phoebus*, a British merchant vessel carrying munitions to Russia. About 10 days into its crossing, a fire broke out in the ship's hold, and it had to be towed to Liverpool, its cargo a complete loss. The cigar bombs worked and Rintelen ramped up production and recruited as many dockworkers as he could. The fire on the *SS Phoebus* was followed closely by explosions or fires on the *SS Kirk Oswald* on May 2nd, the *SS Bankdale* on May 8th, the *SS Samland*, the *SS Sygna*, and the *SS Ryndam*. The number of ship explosions and fires increased each month. Rintelen was on his way (Witcover 1989, pp. 90–99).

Not content with just causing fires onboard munitions ships, Rintelen's saboteurs also targeted munitions plants. The Anderson Chemical Plant in Wallington, NJ, had an explosion early in May 1915 in which three people were killed. The DuPont Powder Company plant in Carneys Point, NJ, had two explosions during May, and a barge at the Hercules Powder Company plant in Tacoma, Washington, exploded shortly thereafter. The cigar bomb scheme would not end until April 1916 when Karl von Kleist was arrested by New York Bomb Squad detectives, and he told the whole story. Walter Scheele was warned of Kleist's confession by Karl Boy-Ed and fled to Cuba. Eighteen months later Scheele was arrested in Havana and extradited to the United States where he changed sides and gave the US government all the technical details of the cigar bombs.

All the while that the cigar bomb scheme was being implemented, Rintelen was busy making more mischief. Even before his trip to the United States, Rintelen visited former Mexican dictator Victoriano Huerta in Barcelona in February 1915. Huerta had been in Barcelona since 1914, waiting for an opportunity to return to Mexico. Rintelen convinced Huerta that now was the time he should return to Mexico and reclaim the presidency from Carranza. He said that the Germans would supply money and arms for Huerta's cause. The hope on Rintelen's part was that a coup by Huerta would provoke a war between the United States and Mexico that would keep the United States out of the European war and divert American munitions to a Mexican war instead (Tuchman 1958, p. 61). At this point Mexico was still embroiled in a Civil War, with Carranza still fighting Villa's ragtag army and desperately trying to hold onto power. The Americans had just withdrawn from their occupation of Veracruz the previous November, President Wilson had not yet recognized Carranza's government, and relations between Mexico and the United States were at a nadir. It seemed like the right time for a military coup.

Huerta arrived in New York in April 1915 just days after Rintelen himself, and Rintelen wasted no time in meeting with Huerta and in involving von Papen and Boy-Ed in the conspiracy. Unfortunately for the Germans, British Naval Intelligence was also very interested in Huerta's presence in the United States and so was bugging his hotel room and following him all around New York and, of course, noticing who he was meeting with. British Naval Intelligence in the guise of Captain Guy Gaunt, the British naval attaché for naval intelligence, had organized a counterintelligence network around a group of Czech nationalists led by Emanuel Viktor Voska. Voska's agents had infiltrated numerous German and Austro-Hungarian offices around New York and Washington, including a number of agents working at several hotels frequented by members of the German Embassy staff. Voska's agents would tell Gaunt what they'd seen and heard, and Gaunt would tell the New York Police Department Bomb Squad and the US Secret Service. The Americans also began following Huerta around New York.

Rintelen offered Huerta over $800,000 in cash, deposited in a bank in Havana, and supplies of arms and ammunition to be delivered by submarine to Mexico when Huerta returned to Mexico to start his coup. Von Papen traveled to the US-Mexican border on a reconnaissance mission, and Boy-Ed arranged purchases of arms and ammunition for Huerta. These preparations were all observed by Voska's spies. Just to add to the comedy, Mexican President Carranza also sent agents to New York to tail Huerta and discover what he was up to. So the Czechs, under British guidance, the American Secret Service, and the Mexicans were all following Huerta wherever he went. This also led them to connect Rintelen to Huerta because of their frequent meetings and meals while planning Huerta's return to Mexico (Tuchman 1958, pp. 67–76).

Things came to a head on June 24th when Huerta boarded a train for San Francisco so he could attend the Panama-Pacific International Exposition there. Huerta never made it to San Francisco because he changed trains in Kansas City and headed south toward Mexico where he was to meet his Mexican supporters. However, the Americans were informed about his trip and change of trains, and Huerta was removed from the train and arrested in Newman, New Mexico, just 25 miles from the Mexican border. He was charged with sedition and put in prison at Fort Bliss, Texas, just outside of El Paso. Within a few weeks he was released on bail but kept under house arrest at Fort Bliss. He died on January 14, 1916 before he could be brought to trial. While at the time, Huerta's followers claimed the Americans poisoned him, the current view is that he died of cirrhosis of the liver due to his long-term habitual drinking (Tuchman 1958, pp. 84–85).

Rintelen continued to put in place new schemes designed to keep munitions from the Allies. His last effort was the creation of Labor's National Peace Council as an alternative labor organization to the American Federation of Labor led by Samuel Gompers who was stridently pro-British. Because of his views Gompers forbade dock or factory strikes against companies involved with the creation or shipping of munitions. Rintelen and an American crony, the con man David Lamar, known as the "Wolf of Wall Street," organized the National Peace Council to foment strikes in factories in order to stop the shipment of arms and ammunition to the Allies. Rintelen funneled nearly $500,000 to the Peace Council through Lamar, most of which was siphoned off by him. Very few strikes took place, and by December 1915 the National Peace Council had imploded, many of its officers were indicted on fraud charges, and by early 1916 it had ceased to exist (Witcover 1989, pp. 116–117).

By early July 1915 the New York Police Department, the Secret Service, and the Justice Department's Bureau of Investigation were all looking into Rintelen's affairs and activities in New York. In addition, von Papen and Boy-Ed were writing letters to the German Foreign Ministry and the Admiralty complaining about Rintelen and his reckless, high-handed escapades. On July 6th Rintelen received a telegram from the German Admiralty recalling him to Germany. He boarded the Holland-America liner *Noordam* on August 3rd again masquerading as a Swiss national Emile V. Gaché. Ten days later the liner was stopped in the English Channel for a routine inspection, and Rintelen was arrested. By claiming he was a member of the German Navy, he avoided an espionage charge. He spent the next 21 months in a prisoner of war camp in England. While there, he was indicted in New York on December 28, 1915, for conspiracy and fraud involving the National Peace Council and indicted again in March 1916 on federal charges related to the cigar bomb scheme. Once America entered the war in April 1917, he was extradited to the United States and charged sequentially with conspiracy to foment labor agitation, passport fraud, and conspiracy to plant bombs on British ships. He was convicted in a federal court in New York and spent the next 3 years in the federal penitentiary in Atlanta. Rintelen was released on November 19, 1920. After briefly returning to Germany where he was treated as a nobody, rather than as the hero he expected, he moved to England. He volunteered to help the English intelligence services during World War II but was politely declined. He wrote his memoirs, a two-volume set titled *The Dark Invader*, and died in London in 1949 at the age of 72 (Tuchman 1958, p. 77).

13.4 The Minister without Portfolio

While von Papen, Boy-Ed, and later Rintelen were busy concocting ways to keep munitions away from the Allies, Dr. Heinrich Albert, the German commercial attaché, was also quite busy (Fig. 13.5). Heinrich Albert was born in 1874 at Magdeburg in Saxony, the son of a businessman. He graduated from law school and entered the German Interior Ministry in 1904. He was appointed as the commercial attaché to the United States in 1914 and kept that post until the US declaration of war against Germany in 1917. Tall and good-looking, he was not a particularly gregarious or social person, and he did not enjoy his time in the United States. From his rooms in the Hamburg-America shipping line offices at 45 Broadway, he funneled money to a number of pro-German newspapers around the country, kept von Papen, Boy-Ed, and Rintelen funded, and looked for ways to legitimately keep the raw materials for ammunition away from American factories. Albert's most successful adventure was the Great Phenol Plot of 1915 (Mann and Plummer 1991, p. 38).

Dr. Hugo Schweitzer was a German chemist, born in Prussia in 1860, and earned his Ph.D. in chemistry at the University of Freiberg. He came to the United States in the early 1890s, worked for several chemical companies, and became a naturalized American citizen. He eventually became director of the pharmaceutical division of the Bayer Chemical Company, the American subsidiary of Farbenfabrikenvormals Friedrich Bayer & Company of Elberfeld, Germany. He retired from Bayer and continued to consult in the chemical industry and was also on retainer to the German Army. Like Walter Scheele, Hugo Schweitzer was an industrial spy.

Explosives are complex chemical compounds created from simpler materials. Many high explosives used in artillery shells and bullets generally contain a compound called picric acid. Picric acid is made from phenol (hydroxybenzene), also known as carbolic acid. Phenol can also be used to make salicylic acid, the main ingredient in aspirin and TNP (trinitrophenol) another high explosive; in 1914 phenol

Fig. 13.5 Dr. Heinrich Albert (Bundesarchiv bild 102-13486. *Public Domain*. This file is licensed under the Creative Commons Attribution-Share Alike 3.0 Germany license)

was also used by the inventor Thomas Edison to make his high-quality vinyl phonograph records. This is where the story gets complicated. At the beginning of World War I, most of the phenol used in the United States was imported from Great Britain. Once the war began the British government designated the entire output of the phenol industry to be used to make explosives, cutting off the exports to the United States. Because the munitions industry was so lucrative for American companies, the entire, much smaller, phenol production of the United States was also being used for explosives manufacturing. This left the aspirin industry, dominated by the Bayer Company, and Thomas Edison without a regular supply of phenol. Edison, being the inventor and entrepreneur that he was, created a synthetic version of phenol and proceeded to build his own factory to manufacture it at Silver Lake, New Jersey, in Essex County, about 20 miles west of New York City (Mann and Plummer 1991, pp. 39–40).

Enter Dr. Schweitzer, who had two goals in mind: first to find a steady supply of phenol for the Bayer Company that could be converted into salicylic acid to make aspirin and second to buy up as much phenol as possible to keep it out of the munitions industry, the output of which was nearly all going to the Allies to kill Germans. With $100,000 in seed money from Dr. Albert and the German government, Schweitzer created a company called the Chemical Exchange Association. He then entered into an agreement with Edison to buy 3 tons of phenol per day from mid-June 1915 through the end of the year and 2 tons per day through the first quarter of 1916. He contracted with Heyden Chemical Works, Garfield, New Jersey (owned by Chemiske Fabrik von Heyden, Radebeul, Germany) to convert as much of that phenol into salicylic acid as the Bayer Company could use. He then sold the salicylic acid to Bayer and any remaining phenol to other pharmaceutical companies and others for nonmilitary uses. This was all legal, but it had the effect of further reducing the amount of phenol and picric acid on the American market and sent prices skyrocketing. This increased Schweitzer's profits but hurt the American munitions industry and hampered its ability to fill contracts for the Allies. In short, it was everything that von Bernstorff and Albert wanted to do (Mann and Plummer 1991, pp. 40–42). Unfortunately, the good news didn't last long enough.

By the late spring of 1915, the New York Police Department, the Secret Service, and the Justice Department's Bureau of Investigation were all getting suspicious about the activities of Messrs. von Papen, Boy-Ed, and Albert in New York City. The trio was now regularly being followed, and on occasion the new technology of phone tapping was used on their office telephone lines.

On Saturday July 24, 1915, a month after the start of the Great Phenol Plot, Dr. Albert had an afternoon meeting with a pro-German newspaper owner in his office in Lower Manhattan. Outside were two Secret Service agents watching the comings and goings. Around 3:30 pm Albert and his associate left the office at 45 Broadway and headed for the uptown-bound Sixth Avenue elevated train, followed closely by the Secret Service agents. At 33rd Street the newspaper owner got off, followed by one of the agents. Albert settled down for the trip uptown to his own stop. It was a warm, humid summer day in New York and Albert dozed off to the rhythm of the train. The train stopped at the 50th Street station, Albert's stop, and was just about to leave when he awoke and realized he was about to miss his stop. He leapt through the soon to be closing doors but left his briefcase behind. Realizing his mistake he turned just in time to see Secret Service agent Frank Burke grab his portfolio and dash off the train. Albert gave a chase and Burke ran down to the street and hopped an uptown-bound trolley car. Burke then told the conductor that the man chasing him just caused a commotion on the elevated train, and he was just crazy. The conductor conveniently did not stop at the next corner, losing Albert in the process. Burke immediately called his superior, William Flynn the head of the Secret Service, and when they looked inside the briefcase, they found a treasure trove of memos, letters, check stubs, and other documents all related to Albert's activities. They included documents on Albert's propaganda activities, including payoffs to German-American and Irish-American organizations, funds given to von Papen and Boy-Ed, receipts for funds for the Chemical Exchange Association, and documents showing that the Germans had incorporated and were building a munitions plant in Bridgeport, Connecticut.

That evening Flynn traveled to Maine where the Secretary of the Treasury William McAdoo was vacationing. McAdoo looked over the documents and immediately headed for Washington to show them to President Wilson. The documents did not contain enough evidence to charge any of the men, but they showed a pattern of activities that could be traced back to employees of the German Embassy. President Wilson, still hanging on to his dream of American neutrality, was not convinced that the government could or should do anything at that point. McAdoo reasoned that the publication of Albert's documents would embarrass the Germans and put some of their schemes out of business. So the Treasury Department leaked some of the documents to the anti-German *New York World* newspaper. The *World* used them as their lead story across the entire front page in their August 15th edition (Millman 2006, pp. 31–32). The documents were deeply embarrassing to the German Embassy, and von Bernstorff denied all of them out of hand. However, the publication of the documents led to the government intensifying their surveillance and investigation of von Papen, Boy-Ed, and Albert. It also caused the Bridgeport munitions factory scheme to be shut down, and Albert stopped funding the Chemical Exchange Association. From that point on, Dr. Albert was known in the media as the "minister without portfolio." With the end of Albert's funding and the bad publicity generated by the *World's* stories, the Great Phenol Plot collapsed by the end of the year. Schweitzer tried to revive it, but Edison canceled their contract and Schweitzer could not continue to raise enough money to continue the scheme. Schweitzer himself died of pneumonia in late 1917 (Witcover 1989, pp. 120–121).

13.5 Failure and Recall

After the debacle of Albert's papers, there was one last fiasco to befall the Germans during 1915. James J. Archibald was a pro-German American journalist. He had reported on Germany for several years before the beginning of the war, and in 1915 he was back in the United States and on von Bernstorff's payroll, writing pro-German stories for several American newspapers (Millman 2006, p. 54). Archibald was also friendly with the staff at the German Embassy and was a personal friend of Konstantin Graf von Dumba, the Austro-Hungarian Ambassador to the United States. On September 1, 1915, Archibald was on a Dutch steamer, on his way to Germany. On a tip from the Americans, the British boarded the ship when it stopped at Falmouth, England, and searched Archibald's luggage. In his bags they found a number of diplomatic documents and letters from von Dumba to his government and also a number of letters from Franz von Papen to his wife. Archibald was arrested for acting as a courier for an enemy nation. The von Dumba documents, among other things, spelled out the use of Austro-Hungarian funds for dock and munitions plant strikes in the United States. Several of the von Dumba letters were not only critical of US foreign policy, they personally insulted President Wilson. The von Papen letters to his wife were also very critical of the United States. In one of the letters von Papen responds to his wife about news from the battlefront "How splendid on the eastern front! I always say to these idiotic Yankees that they should shut their mouths and better still be full of admiration for all that heroism" (Millman 2006, p. 55). The British gladly turned over the documents to the *Chicago Tribune*, the *Washington Post*, and the *New York World*, who promptly published them on their front pages. From this point on, the attaché's days in the United States were numbered. The United States demanded Ambassador von Dumba's recall on September 6th and the Justice and State departments stepped up their investigations into von Papen and Boy-Ed.

Oblivious to the nationwide public ridicule, von Papen and Boy-Ed in mid-September took a vacation through the American West. Reporters dogged them at every stop. Returning to New York in October, von Papen and Boy-Ed at last figured out that they could no longer be effective in planning and recruiting for acts of sabotage. Von Bernstorff tried to help by defending his attaches to the Secretary of State and claiming no knowledge of any of the contents of Albert's or von Dumba's papers. But by this time it was no use. Finally, on December 8th, Secretary of State Robert Lansing wrote a note to von Bernstorff demanding Von Papen and Boy-Ed's recall. The German government officially recalled the two attaches on December 10th. Von Bernstorff then tried to distance himself from the recall disaster by trying to place all the blame on the now departed Rintelen and wired to the Foreign Office, "Convinced Rintelen was principal reason for recall of attachés. His immediate disavowal necessary" (Tuchman 1958, p. 84).

Von Papen departed the United States on December 21, 1915 and Boy-Ed followed on January 1, 1916. To add insult to injury, when the ship carrying von Papen was stopped by the British, the British authorities claimed that while von Papen had diplomatic immunity and free passage through British lines to Germany, his luggage did not. They then confiscated his luggage in which von Papen had foolishly stored many of his official papers and the bankbooks of the various accounts he had used to pay his agents in the United States. Needless to say, the British shared all this information with the Americans who indicted von Papen in April 1916 on charges of conspiracy in the Welland Canal plot. Von Papen went on to serve on the

Western Front and as a liaison officer with the Ottoman Army in Palestine. He entered politics after the war and became Chancellor of the Weimar Republic in 1932 and Vice-Chancellor for a time in 1933–1934 under Adolf Hitler. He later served as Ambassador to Austria and Turkey before retiring from government. He was tried at Nuremburg in 1945 but was acquitted. Von Papen died in 1969 at age 89.

Karl Boy-Ed returned to Germany and was appointed the head of *Nachrichten-Abteilung N*, German Naval Intelligence. He retired after the war, married an American, and settled in Hamburg, Germany. He died on his 58th birthday in 1930 from a horse-riding accident.

The departures of Rintelen and the two attachés ended the first phase of German espionage in the United States, but it did not stop it by any means. Others were more than ready to pick up the pro-German torch and continue the work.

References

Barton, George. 1919. *Celebrated Spies and Famous Mysteries of the Great War*. Boston, MA: The Page Company. http://books.google.com/books?id=D8QiAAAAMAAJ&printsec=frontcover&source=gbs_ge_summary_r&cad=0#v=onepage&q&f=false.

Keegan, John. 1999. *The First World War*. New York: Knopf.

Landau, Captain Henry. 1937. *The Enemy Within: The Inside Story of German Sabotage in America*. New York: G. P. Putnam's Sons.

Mann, Charles C., and Mark L. Plummer. 1991. *The Aspirin Wars: Money, Medicine, and 100 Years of Rampant Competition*. Boston, MA: Harvard Business School Press.

Millman, Chad. 2006. *The Detonators: The Secret Plot to Destroy America and an Epic Hunt for Justice*. New York: Little, Brown and Company.

Pohlmann, Markus. 2005. "German Intelligence at War, 1914–1918." *Journal of Intelligence History* 5(2): 25–54. doi:dx.doi.org/10.1080/16161262.2005.10555116.

Tuchman, Barbara W. 1958. *The Zimmermann Telegram*. New York: Macmillan Company.

Witcover, Jules. 1989. *Sabotage at Black Tom: Imperial Germany's Secret War in America – 1914–1917*. New York: Algonquin Books.

Chapter 14
Spies Among Us: Baltimore, Germs, Black Tom, and Kingsland (1916–1917)

Abstract While the initial focus of the German espionage attempts was in New York, von Papen, Boy-Ed, Rintelen, and their associates did not limit themselves just to the nation's largest port. Other ports, like New Orleans and San Francisco, were also targeted and agents recruited in and sent to those places for sabotage activities. However, the other main hub of German spy activity was much closer to the home of the German Embassy—Baltimore.

14.1 Baltimore Heats Up

Paul Hilken was an important businessman in Baltimore. He was a first-generation German-American citizen; his father Henry was the honorary German consul in Baltimore and the Baltimore agent for the North German Lloyd shipping line. Paul also worked for the North German Lloyd line out of Baltimore with responsibilities in New York as well. He lived in Bremen, Germany, for 5 months in late 1913 and early 1914, being trained in the duties and skills of the managing director of American operations of North German Lloyd. The advent of the war in August 1914 put a hold on his promotion, but not his ambitions. By 1915, Paul Hilken had effectively taken over his father's work as the Baltimore agent while his father ran a tobacco export company, Schumacher & Company, which the two of them owned (Millman 2006, p. 34). Hilken also made many trips to the North German Lloyd headquarters in Bremen, Germany, with side trips to Berlin. With a large house, an American wife, and three children in the fashionable Roland Park neighborhood of Baltimore, Paul Hilken was the epitome of the American success story. Not particularly good looking, Hilken was of medium height and had a slight build; but he looked intelligent and he was. He had degrees in mechanical engineering from Lehigh University and MIT and had also completed a very difficult course in naval architecture at MIT. Despite being married, he had a wandering eye and tended to exaggerate his accomplishments. He was a difficult boss and generally not a very likable person. German was spoken in his home when he was growing up, and he was fluent in the language and, despite having been born in the United States, was much more of a German patriot than an American one (Messimer 2015, pp. 6–8) (Fig. 14.1).

J.F. Dooley, *Codes, Ciphers and Spies*, DOI 10.1007/978-3-319-29415-5_14

Fig. 14.1 Paul Koenig, Captain of the merchant submarine Deutschland, and Paul Hilken in Baltimore, July 1916 (*Public Domain*. From the Library of Congress Prints and Photographs Division. LC-B2- 3912-10. http://www.loc.gov/pictures/item/ggb2005022220/)

On Sunday, April 18, 1915, Paul Hilken received a phone call at his Roland Park home from Franz Rintelen. Rintelen introduced himself and assured Hilken that he was from Germany and that he wanted Hilken to help him with some work he was doing in the United States. The two men met later that day in Philadelphia where Rintelen showed Hilken a letter of introduction from the German Admiralty. Rintelen talked about his objectives in America and about how he was organizing strikes and ship bombings in New York. He told Hilken that he wanted to expand his operations into other ports. Baltimore was one of the largest ports on the East Coast, and there were already dozens of German ships—and their crews—interned there. The more Rintelen talked about his plans, the more interested and excited Hilken became. Rintelen then asked Hilken to help him set up a sabotage network in Baltimore and to be the paymaster for that network (Landau 1937, p. 46; Messimer 2015, p. 6; Millman 2006, p. 34).[1] Hilken jumped at the chance. He and Rintelen then spent the rest of their meeting talking about details, the transportation of the cigar bombs from New York to Baltimore, and what kinds of targets Rintelen wanted Hilken's organization to focus on.

Paul Hilken was not the type of man to be a field agent or to run a group of field agents; he was a businessman and a manager, not a spy. So at his next meeting with

[1] Note that all three sources have different dates and places for this initial meeting between Rintelen and Hilken. The most likely seems to be Millman's date of mid-April in Philadelphia.

Rintelen at Hilken's home in Roland Park the last weekend in May 1915, he brought along someone who could run the organization on the ground, keep the recruits in line, and do the dirty work. Captain Friedrich Hinsch had commanded the North German Lloyd merchant steamship *SS Neckar* at the outbreak of the war in August 1914. Like all German merchant ship officers, Hinsch was a member of the German naval reserve and so he was anxious to get back to Germany and get into the war. But with the English fleet patrolling the Atlantic, it was not so easy. Hinsch and the *Neckar* spent the first couple of months of the war dodging the English Navy in the Atlantic and meeting up with German naval vessels in need of supplies. However, by the end of September 1914, his ship was in desperate need of engine repairs, was short on fuel and supplies, and was being chased by the British. Hinsch was forced to put in at Baltimore and have his ship interned for the duration of the war. By the time May 1915 came around, Hinsch was bored and looking for something to do that would make a difference in the war. Friedrich Hinsch was tall and heavy set with blonde hair and blue eyes; in his mid-40s, he was quick to anger and quick to use his fists. His men, who still depended on him for food, shelter, and money, feared him. He was the ideal person to recruit and run the group of sailors and stevedores who would plant the cigar bombs on munitions ships in Baltimore. Rintelen was pleased with Friedrich Hinsch. He gave Hilken over $10,000 to start his organization, and they arranged for the delivery of cigar bombs. Hilken was the paymaster and gave funds to Hinsch. Hinsch created the organization, and he paid an African-American stevedore named Edward Felton to recruit more dockworkers to do the dirty work. Hinsch paid Felton who then distributed the money to the saboteurs at the docks who would actually plant the bombs. The organization was tight and efficient. None of the stevedores knew about Hinsch or Hilken; Felton only knew about Hinsch and Hilken but not about Rintelen. Neither Hilken nor Hinsch knew about the saboteurs in New York or who made the cigar bombs, nor did they know any of the stevedores (Millman 2006, pp. 47–48). When Rintelen left the United States in August 1915, he also left a stash of funds with Hilken to continue the Baltimore operations. Unfortunately for the German Admiralty, there is very little evidence that the Baltimore crew blew up many ships during 1915 and 1916. Nearly all the fires and ship accidents over this period attributable to the cigar bombs are traceable back to Scheele and his organization in New York (Messimer 2015, p. 11). That didn't stop the Baltimore group from trying, however, and they were more successful at their next exploit.

In addition to espionage and sabotage in neutral countries, Abteilung IIIB looked to other methods to slow the flow of supplies to the Allies. Section P (for Politik) of Abteilung IIIB, headed by Captain Rudolf Nadolny, was in charge of sabotage in all its forms. In the spring of 1915, Section P became the first military organization to advocate and plan a campaign of biological warfare against both combatants and neutral nations.

14.2 Germs

Anton Dilger was a German-American medical doctor. Dilger's father Hubert was a horse artilleryman in Baden before emigrating to the United States in 1861. He fought for the Union during the American Civil War as a captain of artillery, earning a Medal of Honor at Chancellorsville. Born in 1884, the tenth child of Hubert and his wife Elise, Anton Casimir Dilger grew up on the family farm, Greenfield, in the Shenandoah Valley of rural Virginia riding horses and shooting from an early age (Fig. 14.2). At 10 years old, Anton was shipped off to Germany to live with his older sister, Eda, and her husband in Mannheim and to finish his education. He lived in Mannheim with his sister and her husband for 8 years, finishing secondary school there. Anton Dilger then attended medical school at the University of Heidelberg in Germany, graduating in 1908. In 1909, he attended a prestigious graduate course for doctors at the Johns Hopkins University Medical School in Baltimore and then returned to Germany to finish his training in surgery at Heidelberg. Tall, good looking, cosmopolitan, a music lover, and a ladies' man, with the obligatory dueling scar on his cheek and culturally steeped in everything German, Anton was looking forward to an extremely successful career as a surgeon. By 1912, he was ready to start his career and he just needed to decide whether to stay in Germany or go back to America. Then, as we see over and over again in this narrative, war intervened (Koenig 2009).

In 1912 in the Balkans, the Balkan League, made up of the small independent states of Serbia, Greece, Montenegro, and Bulgaria, had their eyes on expansion and the recovery of the European territories of the now crumbling Ottoman Empire. The proximate excuse for the war was the Ottoman response to a series of revolts in favor of independence in the Ottoman province of Albania. The Balkan League supported the Albanians, and all of them declared war on the Ottoman Empire in October 1912. Anton Dilger was recruited by the Bulgarians to volunteer as a battlefield surgeon and served during the First and Second Balkan Wars in 1912 and 1913. Back in Germany after World War I broke out, Dilger volunteered for service and

Fig. 14.2 Dr. Anton Dilger circa 1916 (*Public Domain*. Origin unknown. Retrieved from https://commons.wikimedia.org/wiki/File:Anton_Dilger.jpg)

was taken on as a civilian surgeon at a Military Reserve Hospital in Rastatt, Baden, in November 1914.

As an American citizen with German relatives and sympathies, Dilger was a prime target for recruitment by the German Secret Service. He was recruited in mid-1915 by Rudolf Nadolny for Section P of Abteilung IIIB to work on biological warfare; the Germans wanted to disrupt the transport of horses and mules to the Allies. With the stagnant front in France, the cavalry on both sides had a very limited role to play in the war. But horses and mules were essential for logistical support, particularly for the artillery forces, transporting supplies and ammunition from railheads to the front lines, for ambulance service, and for messengers between the front lines and the headquarters behind them. The Allies and the Central Powers used approximately 6,000,000 horses during the war. Because of the use of machine guns and artillery and the unsanitary conditions of the trenches, the death rate among the animals was extraordinarily high. Both sides in the war were constantly desperate for more horses and mules, and America was one of the largest suppliers for the Allies. For the Germans, restricting the number of animals shipped to Europe from America became a high-priority goal. Nadolny and his associates in Section P had come up with the idea of infecting horses and mules in transit so that while the animals were cooped up in transport ships crossing the Atlantic, any disease would spread quickly. Two of the most virulent diseases that affect horses and mules are anthrax and glanders, both of which could be cultured fairly easily by someone with knowledge of microbiology and good laboratory techniques. Anton Dilger had both these skills. By this time, Dilger was a complete German patriot and was more than willing to do what he could to help Germany win the war. He signed on to the anthrax scheme and sailed for America on September 29, 1915, with four vials of anthrax and glanders bacteria and a stash of money in his luggage, determined to set up a bacterial laboratory and to work with the German agents already in America to infect as many horses and mules as possible (Messimer 2015, p. 16).

After arriving in New York on October 7, 1915, and taking a train south, Dilger stayed with his sister Jo and her husband Adolf in Washington for a few days. Then after a short visit home to the family farm in Front Royal, Virginia,[2] he rented a house in the new Chevy Chase neighborhood of Washington, DC, just 6 miles from the White House. He moved into the house with his older sister Em and his brother Carl. Em kept house for the threesome, and Carl, who had extensive experience as a brewer, helped Dilger set up the lab in the basement of their house on 33rd Street in Chevy Chase. Dilger then got in touch with Paul Hilken to arrange for more funds and for the bacteria to be distributed; from this point on, Anton Dilger's house was known as "Tony's Lab" (Millman 2006, p. 70).

Dilger used the four vials he had brought from Germany to start the cultures in his lab. Two of them contained the anthrax bacteria, *Bacillus anthracis*, and were labeled "B" for the Latin word for cattle, *bos*. The other two were labeled "E" for

[2] By pure coincidence, Colonel Parker Hitt, the American First Army's Chief Signal Officer and author of the standard text for training Army cryptanalysts, *Manual for the Solution of Military Ciphers*, also lived in Front Royal, VA.

equus, Latin for horse, and contained the glanders bacteria, *Burkholderia mallei*. Under the right conditions, Dilger could grow new cultures of the two types of bacteria every week or so. Shortly after the lab was set up, Hilken sent Friedrich Hinsch over to Tony's Lab to introduce himself to Dilger and to set up the schedule for deliveries of the bacteria (Koenig 2009, pp. 86–90).

Hinsch used stevedores in Baltimore and New York to infect horses and mules destined for the Western Front with the diseases. Stevedore foreman Eddie Felton hired a number of his workers in Baltimore and Newport News, Virginia, to infect the animals using syringes for the anthrax and swabs in the horse's noses for the glanders. Hinsch would pick up two or three dozen vials of germs from Tony's lab and meet Felton at Paul Hilken's offices at the Hansa House in downtown Baltimore, and Felton would distribute them to his men. Occasionally, Hinsch would travel to New York to distribute more vials there. The United States Army had a horse paddock in Van Cortlandt Park in the Bronx, and Hinsch's recruits would infect horses and mules there before the animals were moved down to transport ships in the harbor. The most dangerous part of the operation was that both the anthrax and glanders diseases could easily move from animal to human, so the saboteurs were instructed to wear rubber gloves and be extremely careful in handling the vials of bacteria. Needless to say, not all of them were perfect, and a few human cases of glanders did crop up on the East Coast over the next year or so. The Americans and the Allies, however, did not make the connection between human cases of glanders and sick horses on animal transport ships. Overall, the plan worked well in terms of the saboteurs being able to infect individual horses and mules, but the hoped-for epidemics never really took hold. One reason is that in 1905 a reliable early test for glanders, called the *mallein test*, had been developed and by 1915 was in regular use by the Allied animal divisions. This test could detect glanders very early, and the sick horse would be disposed of before it could infect others.

14.3 Herrmann and Hinsch Divide the Work

By early January 1916, Tony's Lab was up and running, and Carl Dilger had mostly taken over growing and distributing new cultures. Anton Dilger was anxious to get back to Germany and take up his next assignment from Section P. He sailed back to Germany on board the Norwegian passenger liner *Kristianafjord* on January 29, 1916, leaving Carl in charge of Tony's Lab (Koenig 2009, p. 111).

On the ocean voyage across the Atlantic to Norway, Anton Dilger made the acquaintance of another German-American, Frederick Herrmann. Herrmann was tall, blonde, and 22 at the time. Being careful not to reveal any of his associations in the spy network in America or with Abteilung IIIB in Berlin, Dilger became friends with Herrmann, and they spent much time together on the voyage to Norway. Separating once the ship docked, Dilger and Herrmann both took a train to Copenhagen only to run into each other again at the office of German military intelligence in that city. It turned out both Dilger and Herrmann were German spies.

Fred Herrmann had been born in Brooklyn to German immigrant parents. He spoke German fluently, but with an American accent. Although not well educated, he was smart and eager for adventure. In February 1915, on board a Dutch ship heading to Germany for a visit with his grandmother, he was recruited by naval intelligence agent Paul von Dalen. Von Dalen wooed the young Herrmann with exciting stories of spies and derring-do, narrow escapes, wine, women, and song. For a 21-year-old with few prospects, this was too good to pass up. A year later, Herrmann had spied for the Germans in both England and Scotland, mostly keeping an eye out for naval maneuvers and reporting on ships coming in and out of harbors. On his second trip to Scotland, where he had enrolled in an Edinburgh University forestry program, the British Secret Service became aware of his activities and he was put on a ship to New York and expelled from Britain. After this close call, Herrmann had made contact with Captain Karl Boy-Ed in New York and was used as a courier, carrying secret messages from Boy-Ed to naval intelligence in Berlin (Koenig 2009, p. 121). On the trip to Berlin in February 1916 when he met Anton Dilger, Fred Herrmann was switching his employer from the Navy to the Army and was headed for a meeting with Nadolny and Captain Hans Marguerre of Abteilung IIIB, Section P.

In fact, three agents from the United States were converging on Berlin for this meeting at Abteilung IIIB, Paul Hilken, Anton Dilger, and Fred Herrmann. Dilger was being tapped for more biological warfare work and would continue his volunteer labor in surgical units. Hilken was increasing his role as paymaster of the Baltimore cell and was given additional responsibilities. He would be the agent in charge of facilitating the arrival and loading of Germany's first commercial merchant submarine, the *Deutschland,* which was scheduled to arrive in Chesapeake Bay in July 1916. Hilken would create a company, the Eastern Forwarding Company, which would sell the *Deutschland's* cargo and arrange for the purchase and loading of nonmilitary cargo bound for Germany. Captain Hinsch would arrange for the docking of the vessel and for stevedores to offload and load the cargo. This was all possible because America was still a neutral country and business was business.

Herrmann would have the biggest new job. He was to become Hinsch's right-hand man, would join in the work at Tony's Lab, and in addition would begin a new campaign of destruction. By the beginning of 1916, the Germans had decided that just blowing up one munitions ship at a time was not doing enough to hinder the supply of arms and ammunition to the Allies. They had to go bigger and destroy targets that would have a much more long-term effect. So they decided the time had come to scale down the munitions ship destruction scheme and ramp up a plan to blow up munitions factories instead. Fred Herrmann and Friedrich Hinsch would be the linchpins of that new plan. To that end, Herrmann was supplied with a new replacement for Walter Scheele's cigar bombs. These "pencil bombs" were easier to hide, but more dangerous to detonate. Made of a thin piece of glass tubing with two compartments separated by a thin capillary section, each compartment was loaded with different explosive chemicals, for example, sulfuric acid and a mixture of chlorate of potash and sugar. When the tip of the glass tube was broken and the tube set tip down, the effect would be the same as a cigar bomb, a very intense incendiary reaction that would start a fire (Koenig 2009, pp. 126–127; Landau 1937, p. 76).

After a few more days in Berlin, Fred Herrmann, with two boxes of 30 incendiary pencil bombs each, set sail for the United States again. Back in Baltimore, Paul Hilken introduced Herrmann to Friedrich Hinsch, and the two began to work out the plans to blow up munitions factories rather than just individual ships. This would be a much more difficult assignment as the plants were well guarded and everyone entering and leaving was usually searched. Herrmann and Hinsch drew up a list of factories and divided the list between the two of them. One of the factories on Herrmann's list was the Canadian Car and Foundry Company, Ltd., munitions plant in Kingsland, New Jersey, about 20 miles from New York City. Hinsch drew the Lehigh Valley Railroad's Black Tom Munitions Depot in Jersey City, right in New York harbor. Herrmann was also introduced to Carl Dilger and proceeded to learn the processes of creating the anthrax and glanders bacteria cultures.

14.4 Black Tom Explodes

In February 1916, Friedrich Hinsch, using the pseudonym of Francis Graetnor, approached a young man in New York's Penn Station to ask the time. Michael Kristoff was a 23-year-old Austrian immigrant, down on his luck, and using the last of his money to take a train to visit his sister in Columbus, Ohio. He was tall and slim, with reddish hair and blue eyes. Kristoff was forever getting and losing jobs and seemed slightly mentally challenged. Hinsch and Kristoff struck up a conversation and discovered that they had much in common, or so Kristoff thought. Finding Kristoff unemployed, Hinsch hired him to carry his bags on a business trip that he was taking to the Midwest; Hinsch also promised to help Kristoff find a job at a factory when they returned to New York. Over the course of the next several weeks, the two men traveled to Philadelphia, Bridgeport, St. Louis, Detroit, Chicago, Cleveland, Akron, and Columbus. Staying in each city a few days, Hinsch would depart the hotel and be gone all day, leaving Kristoff to watch the luggage, but giving him strict instructions never to look inside. And in each city they visited, there was a chemical, munitions, or projectile plant explosion. Finally, near the end of their trip, Kristoff could contain himself no longer, and while Hinsch was out for the day, he opened one of Hinsch's briefcases. Inside he found cash, factory blueprints, maps, and photos labeled with the names of the factories that had been recently blown up or burned down. When the two men returned to New York, Kristoff rented a room from an aunt in Bayonne, New Jersey. Hinsch was true to his word and found Kristoff a job at the Eagle Iron Works plant forging metal. What Kristoff did not know at the time was that his new job was just up the road from the largest munitions depot in the country, out in New York Harbor on Black Tom Island (Millman 2006, pp. 63–65).

Through the spring of 1916, Hinsch continued to plan the attack on Black Tom. With the recall of Rintelen and the expulsion of Franz von Papen and Karl Boy-Ed, the sabotage organization in New York was mostly shut down, with the NYPD, the

Fig. 14.3 Kurt Jahnke *Public Domain*. From the Library of Congress Prints and Photographs Division. Call Number LC-B2-5000-9 [P&P]

Secret Service, and the Justice Department's Bureau of Investigation continuing to investigate other Germans and German-Americans in the city. Hinsch had to go farther afield to find agents to do his job, and in the end, he had to go all the way to San Francisco to find the right men.

Kurt Jahnke was born in what is now western Poland in 1882. In 1899 at the age of 17, he emigrated to the United States. He joined the Marine Corps and fought in the Philippine-American war in 1900, later becoming a naturalized American citizen. When World War I started, Jahnke was recruited for intelligence work and sabotage by the German consul in San Francisco, Franz Bopp. Jahnke had a talent for intelligence work and explosives and was involved in several ship and factory explosions (Fig. 14.3).

Lothar Witzke was born in Posen in 1895 and joined the German Navy in 1912. He was a junior officer on board the *SMS Dresden*, a German light cruiser that saw action in the South Atlantic and off the Pacific coast of South America early in the war. The *Dresden* was damaged by the British Navy at the Battle of Más a Tierra off the coast of Chile and later scuttled on March 14, 1915. Witzke was interned with the rest of the crew at Valparaiso, Chile. He escaped later in 1915 and made his way via Mexico to San Francisco where he met Jahnke and joined the ranks of the German saboteurs.

In June 1916, Jahnke and Witzke made their way east from San Francisco to New York. They spent at least some time at Martha Held's establishment at 123 West 15th Street in New York, and they met with Friedrich Hinsch and Michael Kristoff. Martha Held was a former opera singer whose townhouse in Manhattan was a favorite meeting place for interned German ship officers and the intelligence agents. While at Martha Held's, several meetings were held to plan out the details of how to blow up the munitions depot at Black Tom. Black Tom Island was, by 1916, a peninsula. Originally an island, the Lehigh Valley Railroad had built a causeway out to the island and laid railroad tracks and built piers where barges would dock. There was no gate at the base of the peninsula, just a guard shack that could easily be bypassed. Black Tom also had seven large brick warehouses on it for storing munitions. By Jersey City ordinance, no barge or railroad car containing explosives was allowed to stay at Black Tom overnight, a law that was routinely flouted. There were guards stationed at the base of the peninsula who regularly made rounds of the depot, but the Germans had bribed at least some of these guards to look the other way. On the night of July 29–30, 1916, there were approximately two million pounds of explosives of various types on Black Tom Island, much of it in the warehouses, but some in railroad boxcars, and about 100,000 lb on the *Johnson Barge No. 17* which was tied up to a dock near the middle of the island. One hundred years later, there is still controversy over who blew up the facilities at Black Tom Island and how they did it. The author Jules Witcover has what is probably the best and most plausible description

> Witzke and Jahnke came in to the Black Tom terminal over water around midnight in a small boat laden with explosives, time fuses and incendiary devices. Kristoff meanwhile infiltrated the depot from the land side. They then set small fires in one or more of the boxcars containing TNT and gunpowder, and placed explosives with time fuses there. They also planted time bombs and incendiary devices on a barge – the Johnson 17 – that was loaded with more explosives and tied up to a pier at another point off the yard. Then Witzke and Jahnke retreated onto the darkened river to await the outcome of their work and Kristoff fled by land. (Witcover 1989, p. 161)

About 12:30 AM, one of the guards discovered a small fire in a boxcar and put it out. Shortly thereafter, several other fires were discovered, and the Jersey City fire department was called. The fires quickly got out of control and both the guards and the firefighters fled in fear of the possibility of explosions. At 2:08 AM, the first titanic explosion occurred as the boxcars and then the warehouses started to go. The explosion blew out windows all across Manhattan and Brooklyn on the New York side and through Jersey City, Bayonne, and Hoboken in New Jersey. The explosion was heard and felt as far away as Philadelphia. Everyone was awake. About 20 minutes after the first blast, the *Johnson 17* barge went up in a second gigantic explosion sending shrapnel across New York Harbor, including dotting the Statue of Liberty, nearly a mile away, and leaving a 300 ft wide crater on one side of the Terminal. For hours, ammunition and charges went off, raining casings and shrapnel down all over the harbor. Miraculously only four people were killed, including the captain of the *Johnson 17* who had made the mistake of sleeping on board that night. Black Tom Island and the depot were a complete wreck; it was estimated that the damage caused by the explosions was in the neighborhood of $20,000,000 in 1916 dollars (Fig. 14.4).

Around 4:00 AM that Sunday morning, Michael Kristoff's landlady, Mrs. Anna Rushnak, heard him pacing back and forth in his room. Going to see what was wrong, she heard him moan over and over "What I do? What I do?" When Mrs. Rushnak related this story to her daughter, the daughter went to the police who immediately started investigating Kristoff. He was finally arrested on August 31, 1916. The Jersey City police interrogated Kristoff over and over with Kristoff repeating the story of his meeting and travels with Graetnor including his discovery of blueprints, maps, and large amounts of cash. He insisted that he wasn't near

Fig. 14.4 Black Tom Pier after the July 30, 1916, explosion (*Public Domain*. From the Library of Congress Prints and Photographs Division. LC-DIG-ggbain-22664 (LC-B2- 3963-9 [P&P]))

Black Tom on the night of July 29–30, but he never gave a plausible story of just where he was. With no real evidence against him, the police were forced to let him go in mid-September. Kristoff then disappeared. The Lehigh Valley Railroad hired private investigators to track him down. Records show he joined the US Army in May 1917, but was discharged for health reasons in September 1917. He finally turned up in a jail in Albany, New York, in 1921, serving time for petty theft, but he added nothing new to his story. When released he disappeared again. Michael Kristoff was finally found in a pauper's grave on Staten Island in 1928. He had apparently died of tuberculosis. Documents found when his body was exhumed claimed it was Michael Kristoff in the grave, but a check of dental records against his US Army records did not match. No one has ever fully explained what happened to Michael Kristoff or what was his exact involvement with the Black Tom explosion (Messimer 2015, p. 80; Landau 1937, pp. 138–139). Jahnke and Witzke apparently got cleanly away and were soon back on the West Coast, still up to no good.

There still remain a couple of problems with this explanation of who blew up Black Tom Island:

1. Jahnke and Witzke were usually based on the West Coast, so it was unusual for them to be in New Jersey. There is also no indication that they traveled to the East Coast for any other meetings or sabotage work, so why just this one job?
2. Kristoff was later diagnosed as slightly mentally challenged; his part in the plot in this version of the story was to access the Terminal from the land side, meet

up with Jahnke and Witzke, and independently start fires and set explosives. Given his mental state, it seems unlikely he would be trusted with a job as important as this. Overall, while Witcover's narrative is believed to be the most plausible, there are still questions about who actually set off the Black Tom explosions.

Friedrich Hinsch was almost certainly the person ultimately behind the planning of the Black Tom explosion, and it was the biggest success the German saboteurs had had up to this point. But they were not done. Fred Herrmann, working under Hinsch, was also busy during the latter half of 1916. While Carl Dilger was gone to Germany visiting his brother in the early summer of 1916, Fred Herrmann took over the creation of the germs at Tony's Lab and also made incendiary pencil bombs in the basement of the house in Chevy Chase. With the new initiative to bomb ammunition plants, the Baltimore crew shut down the Chevy Chase lab in August 1916. Fred Herrmann moved the lab to St. Louis that same September in order to put the germs closer to where the horses and mules were first brought together for shipment. Never as successful or as knowledgeable as either Dilger brother, Herrmann accidentally allowed the germ cultures to die of cold and finally closed shop on the day before Thanksgiving 1916 (Koenig 2009, pp. 174–175). This was fine with Herrmann because he had to be back in New Jersey for his next big job. His target was the Canadian Car and Foundry Company's shell assembly plant at Kingsland, New Jersey, only about 10 miles northwest of Black Tom Island.

14.5 Kingsland Burns

The Canadian Car and Foundry Company was based in Montreal, Canada, and in early 1915, the company signed a contract with Russia for $83,000,000 to supply artillery shells for the Russian Army. Because they didn't have enough capacity at their Canadian plants for this contract and all their other war work, the company built an assembly plant in Kingsland (now Lyndhurst), New Jersey. The plant opened in the spring of 1916 and by early 1917 had 38 buildings on the site, all surrounded by a 6 ft high chain-link fence topped with barbed wire. All the employees entered and exited the plant through a single gate and searches were common. Explosives, shell casings, shell warheads, and fuses were shipped to the plant and assembled there and ultimately loaded on transports in New York harbor and shipped to Russia. By 1917, the plant could produce upward of 3,000,000 shells per month (Landau 1937, p. 93).

Because of the security at the plant, Hinsch and Herrmann decided they needed operatives inside the plant itself. Hinsch had met a man named Carl Thummel, a German national who had emigrated to the United States in 1902. Using the name Charles Thorne, he had joined the US Coast Guard in 1913; shortly after that, he met Hinsch in Baltimore and they became friends. Thorne resigned from the Coast Guard in May 1916, and Hinsch began using him as a courier, sending him back and forth to England several times during the summer of 1916. In September 1916, Paul Hilken arranged for Thorne to get a job as an assistant employment manager at the Canadian Car and Foundry plant in Kingsland. Thorne was responsible for hiring men who would be assembling shells and hired a number of men sent to him by Hinsch. One of these men was Theodore Wozniak, an Austrian national. Wozniak was hired in December 1916. He met regularly with Fred Herrmann who was paying him for information on the Kingsland plant. Herrmann also gave Wozniak several pencil bombs (Mixed Claims Commission 1940, p. 163). Wozniak's job in Building 30 of the plant was to clean out newly arrived shell casings. The cleaning was a multi-step process, involving wiping out the shell casing, cleaning a coating of grease applied to the casing before shipping using rags soaked in gasoline and denatured alcohol, and then drying off the shell casings. There were 48 benches lined up side by side in Building 30 and the gasoline soaked rags piled up during a workers shift.

A little after 3:00 PM on the afternoon of January 11, 1917, Theodore Wozniak was at his bench working on cleaning shells. He apparently spilled some gasoline and suddenly a small fire broke out. The fire spread quickly across Wozniak's bench and leapt to adjacent benches. The men in the building ran as the fire engulfed the entire building within minutes. The fire spread to other buildings, setting off the explosives stored there in a series of titanic explosions heard in New York City and as far north as Westchester County and as far east as Long Island. The fire and explosions went on for more than 4 hours. According to Witcover, "The Kingsland plant itself was completely destroyed, with estimated damages of 17 million dollars. A later inventory indicated that 275,000 loaded shells and more than a million unloaded shells, nearly half a million time fuses, 300,000 cartridge cases, and 100,000 detonators, plus huge amounts of TNT were destroyed in the fire" (Witcover 1989, p. 193). The only saving grace was that no one was killed. All 1,400 workers managed to get through the fence and escape across a frozen marsh to safety (Fig. 14.5).

Suspicion that the fire was intentionally set surfaced immediately. Several of Wozniak's coworkers in Building 30 testified that the fire had started at his workbench. Wozniak himself admitted as much, but claimed that a spark from a rotating machine designed to hold the shell casings while they were being cleaned was the culprit. No one, however, could say whether Wozniak had deliberately started the fire, so he was never charged. Wozniak disappeared shortly after the fire and was not found again until more than a decade later when the American and German Mixed Claims Commission was looking for evidence of complicity in the fire. And it would be a decade after that, in 1939, that the German-American Mixed Claims Commission would finally decide that Hilken, Hinsch, Herrmann, and Wozniak were indeed responsible for the devastation at Kingsland (Mixed Claims Commission 1940, pp. 308–310). The key evidence in their decision was not any documentation written before the event but a message that Fred Herrmann sent to Paul Hilken from Mexico City in April 1917. This message was in code and written in two parts. First Herrmann wrote the text of his message on several consecutive pages of the January 1917 issue of *Blue Book* magazine using lemon juice as an invisible ink. The lemon juice disappears when dry and can be revealed using heat; Hilken used a hot iron to

Fig. 14.5 Aftermath of the Kingsland explosion, January 11, 1917 (*Public Domain*. Original from International Film Service, Inc. Retrieved from https://commons.wikimedia.org/wiki/File:Ruins_of_the_Kingsland_Munitions_Explosion.png)

reveal the message (Macrakis 2014, p. 25). In the message, Herrmann used a numerical book cipher to hide the names of various people mentioned in the message. These numbers embedded in the cipher message were always four digit numbers and were constructed as follows. The first digit of the number was dropped. The remaining three digits were then reversed, and these numbers indicated a page in the magazine. On these pages, over certain letters were tiny holes made with a pin that spelled out the name of the person or place in the message. If one holds the page up to a bright light, the pinpricks can be read and the rest of the message deciphered. The entire deciphered and translated message is:

> Have seen 1755 [Eckardt] he is suspicious of me Can't convince him I come from 1915 [Marguerre] and 1794 [Nadolny] Have told him all reference 2584 [Hinsch] and I 2384 [Deutschland], 7595 [Jersey City Terminal], 3106 [Kingsland], 4526 [Savannah], and 8545 [Tonys Lab] he doubts me on account of my bum 7346 [German] Confirm to him thru your channels all OK and my mission here I have no funds 1755 [Eckardt] claims he is short of money Send by bearer US 25000. Have you heard from Willie. Have wired 2336 [Hildegarde] but no answer Be careful of her and connections. Where are 2584 [Hinsch] and 9107 [Carl Ahrendt} Tell 2584 [Hinsch] to come here I expect to go north but he can locate me thru 1755 [Eckardt] I don't trust 9107 [Carl Ahrendt], 3994 [Kristoff], 1585 [Wolfgang] and that 4776 [Hoboken] bunch, If cornered they might get us in Dutch with authorities See that 2584 [Hinsch] brings with him all who might implicate us. tell him 7386 [Siegel] is with me. Where is 6394 [Carl Dilger] he worries me Remember past experience Has 2584 [Hinsch] seen 1315 [Wozniak] Tell him to fix that up. If you have any difficulties see 8165 [Phil Wirth Nat Arts Club] Tell 2584 [Hinsch] his plan O.K. Am in close touch with major and influential Mexicans Can obtain old 3175 [cruiser] for 50000 West Coast What will you do now with America in the War Are you coming here or going

> to South America Advise you drop everything and leave the States regards to 2784 [Hoppenburg] Sei nicht dum mach doch wieder bumm bumm bumm. Most important send funds Bearer will relate experiences and details Greetings (Landau 1937, p. 245)

The Kingsland explosion was the last big effort of the German spy network in Baltimore. Less than a month after the Kingsland explosion, on February 1, 1917, Germany would resume unrestricted submarine warfare and the United States would break diplomatic relations with Germany. Later, in March, the Zimmerman telegram would be released and the United States would declare war on Germany on April 6, 1917. The saboteurs would all disappear soon after that.

After his meeting in Berlin in early 1916, Anton Dilger stayed in Germany until June 1917 working at military hospitals in Rastatt and Karlsruhe as a surgeon. He allegedly worked on more biological warfare schemes and was linked to anthrax and glanders outbreaks in Romania after it declared war on Austria-Hungary in August 1916, followed immediately by Germany's declaration of war against Romania (Koenig 2009, p. 111). Dilger returned to the United States on July 4, 1917, but America was only a waypost to his next assignment from Abteilung IIIB in Mexico. Unknown to all of them, this would be the last time that Anton Dilger would see his family and his family farm in Virginia. From Front Royal, Virginia, Dilger drove across country, stopping in St. Louis and Kansas City and then headed south and crossed into Mexico in early August 1917. By the middle of the month, Dilger was in Mexico City along with most of the other German spies who had been active in espionage and sabotage in the United States during 1914–1917.

Once the United States declared war on Germany, sabotage by enemy agents meant the death penalty instead of just a prison term for the likes of Anton Dilger, Friedrich Hinsch, Fred Herrmann, Kurt Jahnke, and Lothar Witzke. So as soon as the United States declared war, most of the German spies in America headed for Mexico and safety. When Anton Dilger arrived in Mexico City, he went right to work trying to negotiate deals where the German military would support the Mexican government with arms, ammunition, and money. The Imperial German government's continuing hope was to convince Mexico to invade the southern United States and draw American forces to the border area and away from Europe. The biggest problem the German spies had in Mexico was not the Mexican government or its continuing Civil War; it was themselves. The various agents from New York (Paul Koenig, who had organized dockworkers and worked for von Papen and Rintelen), Baltimore (Dilger, Hinsch, and Herrmann), and San Francisco (Jahnke and Witzke) all thought they should be in charge. Also, half of them represented the German Army through Abteilung IIIB and half represented the German Admiralty's Naval Intelligence Service, *Nachrichtenabteilung N*, so there were interservice rivalries involved as well. The German Ambassador to Mexico, Heinrich von Eckardt, was singularly unprepared to manage these groups or negotiate between them. Hence, all the groups were ineffective and nothing got done, each one peppering von Eckardt and Berlin with letters and telegrams denigrating their competition. Dilger traveled from Mexico to Madrid, Spain, in mid-December 1917 to be closer to Berlin and where he thought he would have better communications and influence in getting money for the Mexican plots. His proximity to Berlin didn't help; but it did give his enemies in Mexico the opportunity to plot against Dilger while he was away. His cover identity was also blown by careless remarks by several people in the Mexican Foreign Ministry, and by February 1918, French, British, and American agents were on his trail. Dilger rapidly became useless to the German military intelligence establishment, and by the summer of 1918, he was cut off and isolated in Spain. While in Madrid, Dilger became ill, most likely from the constant travel and the stress of his undercover work. In February, he was hospitalized for a blood clot in his lung, and then in the fall of 1918, he contracted the Spanish flu, which was at that time in the process of killing 50 million people worldwide. Ironically, the man who was behind the first germ warfare assault in modern history died of the Spanish flu on October 17, 1918 (Koenig 2009, p. 264).

Or did he? Henry Landau in his 1937 book *The Enemy Within* says this about Anton Dilger's death in Madrid, "It was whispered that he knew too much. It was a deadly poison that removed him – at least so it was later intimated by a former German agent" (Koenig 2009, p. 264; Landau 1937, p. 194). That former German agent was likely either Fred Herrmann or Anton's brother Carl, who both testified under oath before the German-American Mixed Claims Commission in the 1930s that German agents in Madrid poisoned Anton Dilger (Koenig 2009, p. 266). No other evidence has arisen to confirm how Dilger met his death. We will likely never know.

References

Koenig, Robert. 2009. *The Fourth Horseman: One Man's Mission to Wage the Great War in America*. New York: Public Affairs/Perseus Group.

Landau, Captain Henry. 1937. *The Enemy Within: The Inside Story of German Sabotage in America*. New York: G. P. Putnam's Sons.

Macrakis, Kristie. 2014. *Prisoners, Lovers, & Spies: The Story of Invisible Ink from Herodotus to Al-Qaeda*. New Haven, CT: Yale University Press.

Messimer, Dwight R. 2015. *The Baltimore Sabotage Cell*. Annapolis, MD: Naval Institute Press.

Millman, Chad. 2006. *The Detonators: The Secret Plot to Destroy America and an Epic Hunt for Justice*. New York: Little, Brown and Company.

Mixed claims commission (United States and Germany). 1940. *Opinions and Decisions in the Sabotage Claims Handed down June 15, 1939, and October 30, 1939 and Appendix*. Washington, DC: U. S. Government Printing Office. http://hdl.handle.net/2027/mdp.39015073384821.

Witcover, Jules. 1989. *Sabotage at Black Tom: Imperial Germany's Secret War in America – 1914–1917*. New York: Algonquin Books.

Chapter 15
John Manly and the Waberski Cipher Solution

Abstract The high point of John Manly's service in MI-8 was his decipherment of the "Pablo Waberski" cipher message and his subsequent testimony at Lothar Witzke's court martial in August 1918. This chapter contains Manly's own report of the decipherment. It is a fascinating tale and a very good explanation of the decipherment of a double transposition cipher message.

When last we left Lothar Witzke, he had fled from the Black Tom explosion in New Jersey back to San Francisco in August 1916. It is probable that Witzke and Kurt Jahnke continued their sabotage in the San Francisco area and they are likely to have been involved in the Mare Island Naval Station explosion in March 1917. When America declared war on Germany on April 6, 1917, Witzke, Jahnke, and most of the other German agents in the United States quickly left for Mexico because crimes committed against a neutral America that would bring about a relatively short prison sentence would bring the death penalty when committed against a belligerent America. However, just because the German agents had decamped to Mexico City did not mean that they weren't planning on returning and continuing their sabotage activities later (Fig. 15.1).

By December 1917, Kurt Jahnke was one of the agents running the German Secret Service operation out of Mexico City. Jahnke and his operatives were planning to send agents back into the United States to foment dissent within labor unions and the Army and to blow up more munitions factories if possible. Over the course of the summer and fall of 1917, Kurt Jahnke sent Lothar Witzke back into the United States at least twice on reconnaissance missions. On January 16, 1918, Jahnke sent Lothar Witzke along with two other agents to cross the United States border at Nogales, Arizona. Unfortunately for Jahnke and especially for Witzke, both of the other agents were also Allied spies. William Graves, a black Canadian dockworker who had lived in the United States, was working for British intelligence, and his job was to disrupt German intelligence operations in Mexico. Paul Altendorf, a Pole with a medical degree from the University of Krakow and who had been in the Mexican Army, had been recruited by the Treasury Department's Bureau of Investigation to do the same thing. These two joined Witzke on his travels north from Mexico City toward the United States border. The three made their way north at a leisurely pace and finally got to Nogales at the end of January. Along the way, Altendorf claimed he'd had a change of heart and left the group. In reality, he headed to Nogales on his own and met up with his contact from Treasury, Special Agent

J.F. Dooley, *Codes, Ciphers and Spies*, DOI 10.1007/978-3-319-29415-5_15

Fig. 15.1 Lothar Witzke (alias Pablo Waberski) in 1918 (Copyright Brown Brothers, Inc. (http://brownbrothersusa.com) Used with permission)

Byron S. Butcher to report and to set a trap for Witzke. Witzke hung around Nogales for a couple of days, crossing the border, but always going back to his hotel on the Mexican side. Finally, the Americans thought they knew enough, and when Witzke crossed the border on February 1, 1918, Butcher and his men were waiting for him. On searching Witzke, a cryptogram was found folded up and sewed in his jacket (Gilbert 2012, pp. 93–95). This was dispatched to MI-8 and then languished on Herbert Yardley's desk for many weeks as just one of a large number of messages that MI-8 needed to decrypt. Finally, John Manly got hold of it, and he and Edith Rickert set to work deciphering the Waberski cipher.

Manly's work on decrypting the Waberski cipher is masterful. His essay is a classic explanation of how a gifted cryptanalyst approaches an unknown message and solves it. Manly begins by determining the language of the cipher message, a crucial step in gaining information about the message. He then does a frequency analysis to give himself hints on the type of cipher system used and to provide data on which letters are used and which are not. Once he has a guess on the natural language and the cipher system type, he can take his knowledge of that system and the language characteristics and begin to make educated guesses about how different parts of the message relate to each other. In this case, Manly guesses that the cipher is a columnar transposition cipher, and using his knowledge of German, he begins to organize the message into columns that would make sense for a German language message. He moves back and forth between the original message and the table he is constructing, making changes in the table to accommodate language characteristics and his knowledge of formulaic German diplomatic messages. In the end, he comes up with a brilliant solution. Note, though, that John Manly's decryption is not a general solution of transposition ciphers. Even after his solution, he doesn't know all the details of how the cipher message was constructed. Confronted with another message of this type, he would follow roughly the same procedure to tease out a decryption.

Here is John Manly's description of the decipherment of the Waberski cipher and a recreation of his testimony at Lothar Witzke's court martial at Fort Sam Houston in San Antonio, TX, in August 1918 (Manly 1927).

****

Waberski: John Matthews Manly

The following is the exact form in which this famous message appeared when it was found upon that master spy who went under the name of Pablo Waberski when he was seized by the agents of the United States government:

```
                    15-01-18
seofnatupk asihelhbbn uersdausnn
lrsegglesn nkleznsimn ehneshmppb
asueasriht hteurmvnsm eaincouasi
insnrnvegi esnbtnnrcn dtdrzbemuk
kolseizdnn auebfkbpsa tasecisdgt
ihuktnaeie tiebaeuera thnoieaeen
hsdaoaiakn ethnnneecd ckdkonesdu
eszadehpea bbilsesooe etnouzkdml
neuilurnrn zwhneegvcr eodhicsiac
niuanrdnso drgsurriec egrcsuassp
eatgrsheho etruseelca umlpaatlee
clcxrnprga awsutemair nasnutedea
errreoheim eahktmuhdt cokdtgceio
eefighlhre litfiueunl eelserunma
znai
```

The original of this document was promptly forwarded to Washington by the Military Intelligence Department of the Southwest at Fort Sam Houston, but it was far from receiving prompt attention when it reached Washington. As several times happened in our experience, the officer who received it was so impressed with its importance and with the secrecy, which should be preserved with regard to everything concerning it, that he sent it in to Captain Yardley, who was then Chief of the Section of Code and Ciphers, with no indication of its nature, its source, or its importance. Unfortunately at the time, the Code and Cipher Staff was excessively busy with deciphering a stream of documents that turned out to have little or no importance. Consequently, this nameless and undescribed document was laid aside for attention at some more convenient season, and somehow in the press of routine business, it was sidetracked for a long time. Had MI-8 been properly informed of its source and importance, it would undoubtedly, as later events proved, have been deciphered almost immediately but at the time the Staff of MI-8 was totally inadequate to the volume of cipher messages which had to be deciphered every day, and it was not until this important message had been in our hands for several weeks that we were asked whether we had deciphered it.

When this question was asked, no one remembered anything about the message, as it had attracted no special attention. Then for the first time, we were informed of its importance and immediately began work upon it, as we should have done weeks before had we been told where it came from.

The cipher is so important in itself and so representative of a type of ciphers extensively used for official German communications during the war that the reader will undoubtedly be interested in following in detail the steps of the processes by which it was deciphered.

There are two main classes of ciphers. The essential feature of one consists of substituting for the ordinary letters of the alphabet either arbitrary signs representing these letters or the usual letters with different values. This is the type most familiar to the public because it is the one most commonly used in fiction. In the other type of cipher, the letters retain their normal value but are shifted out of their normal positions. Ciphers of this type are called transposition ciphers.

Transposition ciphers are of all degrees of difficulty, from the simplest to the most complex. If a message were simply written backward, it would be a form of transposition cipher, but of course it would be read very easily, as the following example will show:

```
egassem elpmaxe na si siht
```

Anyone can see almost at a glance that this is a reversed or inverted form of the following:

```
this is an example message
```

Perhaps the most common form of transposition cipher is that known as the columnar transposition. In this case, the correspondents would agree that the letters forming the words of the cipher should be written in a certain number of columns and then transposed by reading the letters down the columns, taking the columns in some fixed order that had been agreed upon. The number of columns and the fixed order would together constitute the key. For example, two correspondents might agree to use a key of nine columns and to take the columns in the following order 7, 6, 9, 5, 2, 3, 1, 4, 8. Suppose the message to be, "Immediately upon the receipt of this you will proceed to Vanceboro, Maine and dynamite the bridge spanning the St. Croix River." This would be first written out in the following form:

```
7 6 9 5 2 3 1 4 8
i m m e d i a t e
l y u p o n t h e
r e c e i p t o f
t h i s y o u w i
l l p r o c e e d
t o v a n c e b o
r o m a i n e a n
d d y n a m i t e
t h e b r i d g e
s p a n n i n g t
h e s t c r o i x
r i v e r
```

This is the first step of the process of encipherment. The second step consists in taking the letters of the vertical columns in the order agreed upon. This would result in giving the message the following form:

```
attueeeidno
doiyoniarncr
inpoccnmiir
thowebatggi
epesraanbnte
myehloodhpei
ilrtltrdtshr
eefidoneetx
mucipvmyeasv
```

If the message were to be sent by telegraph, it would then be divided into groups of five letters, because the telegraph companies have agreed to count a group of five letters in a cipher message as a single word. The message as sent and received by telegraph would then read as follows:

```
attue eeidn odoiy oniar ncrin poccn miirt howeb atggi epesr
aanbn temye hlood hpeii lrtlt rdtsh reefi donee txmuc ipvmy
easv
```

The correspondent who received this message would prepare a sheet of paper with nine columns headed by the numbers of his key. He would then count the number of letters in the message and divide this number by nine to ascertain the length of the columns. In this case, he would find that there were 104 letters and since nine into 104 goes eleven times and five over, all the columns but five would contain eleven letters and that five, which naturally would be the first five, would contain twelve. Then beginning at the column headed with Fig. 15.1, he would write the first eleven letters of the message in regular order down that column. The second, with twelve, would be written down the column headed 2; then columns 3 and 4 would have eleven letters each. Columns 5, 6, and 7 would have twelve, column 8 would have eleven, and finally column 9 would be the last one with twelve letters. It would now be in the exact form in which it had been first written by his correspondent and could easily be read by taking the letters horizontally.[1]

Since it is difficult to remember accurately a series of several numbers, it is customary to use as the key some word that can easily be remembered and translated into a series of figures by assigning to the letters their relative order in the alphabet. For example, the word Wilson would give the key 6, 1, 2, 5, 4, 3; the word Hindenburg would give the series 5, 6, 7, 2, 3, 8, 1, 10, 9, 4. The keyword is very easy to translate into the corresponding figures.

The first step in the process of attack upon a cipher is to make a count of the frequency with which each letter or sign appears. This is for the purpose of ascertaining what type of encipherment was used. If the letters of the alphabet appear in their normal frequency, it is pretty certain that the cipher belongs to the transposi-

[1] In the original document, Manly erroneously puts the length of his sample cryptogram at 109 characters, rather than 104. All the rest of his figures for this example were based on 109 characters and were thus incorrect. The example has been changed to use the correct value of 104, and the subsequent values were also corrected.

Table 10.1 Frequency table of the Waberski cipher message

Letter	Count
A	34
B	10
C	15
D	17
E	63
F	4
G	11
H	20
I	27
J	0
K	12
L	16
M	11
N	42
O	15
P	8
Q	0
R	26
S	34
T	21
U	25
V	3
W	2
X	1
Y	0
Z	7

tion type. Of course the frequency of the letters varies considerably with the language in which the message was originally written. But the normal frequency of any language is easily ascertained, and every expert knows frequency tables for the principal European languages.

An actual count of the letters of the Waberski message gives Table 10.1[2]

Any cipher expert would at a glance recognize this as a normal frequency table for one of the west European languages. As a matter of fact, he would probably recognize it immediately as German, for it is more like the German frequency table than either the English or the Spanish, and it was of course highly probable that the message was written in one of these three languages (Fig. 15.2).

[Ed. Fig. 15.2, a frequency chart of Table 10.1 will give the reader a better view that this message is written in German.]

However, assurance on this point was in this case very easy. As the message contains not a single *q*, it was clearly not written in Spanish, in which *q* is a very

[2] In Manly's original document from the Friedman Collection at the George Marshall Research Library, there are no tables or figures in the text. The editor has added the tables, copies of the cryptogram, its solution and translation, and the frequency chart in this version. For a similar explanation of the Waberski solution, see Yardley (1931, pp. 140–171).

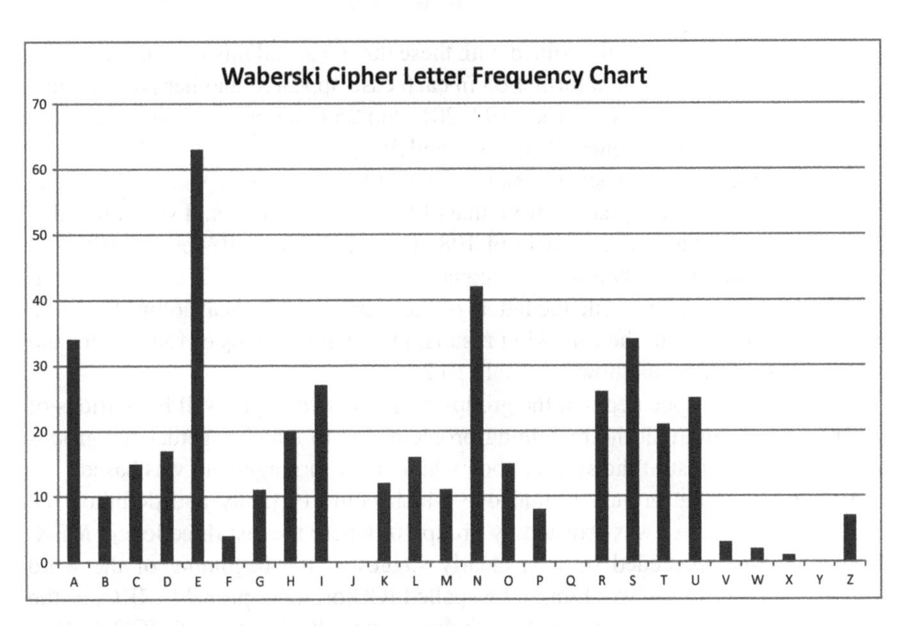

Fig. 15.2 Frequency chart of the Waberski cipher

frequent letter. That it was not English was indicated with equal clearness by the fact that it contained only two *w*'s, whereas a message of this length would normally contain six *w*'s. The comparatively small number of *t*'s and the comparatively large number of *e*'s and *n*'s confirmed this conclusion.

Having decided tentatively but with considerable confidence that the message was in German, the next step was to try to bring together the letters in the order that they occupied in the original message. This of course could not be done all at once. The first step was to make a beginning by bringing together letters that certainly were joined in the original message. It is a peculiarity of the German language that the letter *c* never occurs in native words except before an *h* or a *k*. A glance at the frequency table shows that the message contains *c*'s, *h*'s, and *k*'s, but at present we do not know which *c* goes with which *h* or *k*. This can be ascertained only experimentally. In the present case, each letter in the message was numbered, and it was discovered that the positions occupied by the letter *c* were numbers 85, 109, 145, 199, 201, 259, 266, 270, 290, 294, 319, 331, 333, 381, 387. The *h*'s occupied the following positions: 14, 17, 52, 56, 69, 71, 152, 172, 181, 193, 217, 253, 264, 307, 309, 367, 373, 378, 396, 398. It was next discovered that the intervals between certain occurrences of the letter *c* matched exactly with the intervals between occurrences of the letter *h*. Thus, for the *c*'s the interval between 85 and 109 is 24, and the interval between 109 and 145 is 36.[3] In the list of *h*'s, these same intervals appear between numbers 193, 217, and 253. It would seem probable, then, that

[3] In Manly's original document, he erroneously uses the pairs 109 to 145 and 145 to 199, which don't match with differences of 24 and 36. These have been corrected.

these three *c*'s were originally joined with these three *h*'s, and this was confirmed by the fact that fifty-four letters further on in each case appeared another pair of equal intervals, that is, in the *c*'s between 199, 201, and 259 and in the *h*'s between 307, 309, and 367, the pair of intervals being 2 and 58.

It was quite clear that six *c*'s had been correctly matched, with six *h*'s, and subtracting the number of each *c* from that of the corresponding *h*, it was found that there was an interval between them of 108. Thus, 217 minus 109 equals 108, 253 minus 145 equals 108, 309 minus 201 equals 108, and 367 minus 259 equals 108. It was therefore decided to write the letters of the message in vertical columns of 108, and this was done with the following results, giving 100 groups of four letters and eight groups of three, as shown in Table 10.2.

It immediately appeared that the groups thus obtained might well be portions of German words and that the remaining problem was to match together the groups and ascertain the rest of the system upon which the disarrangement was based.

Inspection of the groups to find one which could certainly and definitely be matched with another was arrested by group 10, where the last three letters MEX, since they were preceded by a *k*, clearly suggested the beginning of the word Mexico. The rest of the word might be spelled ICO or, more probably, IKO, as the message was clearly in German. Search for a group beginning with ICO or IKO proceeded as far as group 13, which was noted as a possible continuation of group 10. Further examination of the message showed that this was the only group beginning with these letters and, therefore, almost beyond a doubt the group that must be joined with group 10.

After a good many groups had been matched in this way, it became clear that the system was very complicated indeed and that the best way to discover it was to bring to light as many uniformities as possible in the intervals between the groups. Since the factors of 108 most likely to be used for a columnar arrangement are 9 and 12, it was regarded as probable that these columns of fours would be divided either into 12 blocks of 9s or 9 blocks of 12s. The arrangement of 12 9s was first tried and proved to be correct.

Arranged in this way, the blocks present the following appearance:

```
scha enpa odet ftal ndbe arbe tzic ubli pesc
kmex ausr skon ikop hoef eleg ista hena blow
bzus ndzu unkt ende ramm sula deni aber ufun
skia nbis npun lsru ramt stre eand gszr gewa
iche einr sser nder ngge ktvo lich ehre zuei
nkom stde inha maih neck eist heim ntau eich
send hbit mauc peso punk berd ardt sang utsc
ehoe andi soro rige iese hauf teri herg tnih
ehei usch rder mage verl naci sist mauf ekai
ansu iese ntpu chen onal unte ange serl iess
iche nder schu nkon rdem nkta vorz enun gesa
dsei ede  sul  nec  bsa  tzu  nam  ndt  rep
```

The group KMEX, with which we started, stands in the first line of the second block, and the group IKOP, which we decided must follow it, is in the fourth line of the same block. If, now, we try to put together the first and fourth lines of each

Table 10.2 First step in arranging the columns of the Waberski cipher

1	scha	37	iche	73	ehei
2	enpa	38	einr	74	usch
3	odet	39	sser	75	rder
4	ftal	40	nder	76	mage
5	ndbe	41	ngge	77	verl
6	arbe	42	ktvo	78	naci
7	tzic	43	lich	79	sist
8	ubli	44	ehre	80	mauf
9	pesc	45	zuei	81	ekai
10	*kmex*	46	nkom	82	ansu
11	ausr	47	stde	83	iese
12	skon	48	inha	84	ntpu
13	*ikop*	49	maih	85	chen
14	hoef	50	neck	86	onal
15	eleg	51	eist	87	unte
16	ista	52	*heim*	88	ange
17	hena	53	ntau	89	serl
18	blow	54	eich	90	iess
19	bzus	55	send	91	iche
20	ndzu	56	hbit	92	nder
21	unkt	57	mauc	93	schu
22	ende	58	*peso*	94	nkon
23	ramm	59	*punk*	95	rdem
24	sula	60	berd	96	nkta
25	deni	61	ardt	97	vorz
26	aber	62	sang	98	enun
27	ufun	63	utsc	99	gesa
28	skia	64	ehoe	100	dsei
29	nbis	65	andi	101	ede
30	npun	66	soro	102	sul
31	lsru	67	rige	103	nec
32	ramt	68	iese	104	bsa
33	stre	69	hauf	105	tzu
34	eand	70	teri	106	nam
35	gsze	71	herg	107	ndt
36	gewa	72	tnih	108	rep

block, we find that they give us the following groups: 1. SCHAFT/AL; 2. K/MEXIKO/P; 3. B/ZUSENDE; 4. SKI/ALS/RU; 5, ICHEN/DER; 6. N/KOMMA/IH; 7. SEND/PESO; 8. EHOERIGE; 9. EHEIM/AGE; 10. ANSUCHEN; 11. ICHEN/KON; and 12. D/SEINE/C. All of these groups are obviously fitted to form parts of a German sentence. Let us look for the group that should follow each.

The SUL in the third line of block 12 would obviously make a good continuation of the KON of group 11. Let us see if the third line of each block is a continuation of the fourth line of the preceding block. On experiment, it immediately appears that this is true. The continuation of block 12 is naturally found in block 1.

Our last series when thus extended becomes D/SEINE/CODE/T, which clearly gives us the German words meaning "his code" and the beginning of a word that may well be "telegram." Searching in block 2 for the rest of the word "telegram," we find it in line 6: ELEG, and a still further continuation appears in line 5 of block 3, RAMM. Experiment shows that in every case the sequence of lines thus established gives an intelligible and correct sequence of letters and words. We have obviously obtained the beginning of the system, and the same process continued will give us the whole system. In order to show this to the eye, it is only necessary to shift the lines so that those that come in sequence are placed in sequence. This will result in placing line 2 at the top and line 7 at the bottom of our pattern. The whole sequence of the vertical arrangement of the lines will be 298143657. The whole message, as now rearranged, presents the following appearance:

```
enpa ausr ndzu nbis einr stde hbit andi usch iese nder ede
pesc blow ufun gewa zuci eich utsc tnih ekai iess gesa rep
ubli hena aber gsze chre ntau sang herg mauf serl enun ndt
scha kmex bzus skia iche nkom send ehoe ehei ansa iche dsei
ftal ikop ende lsru nder maih peso rige mage chen nkon nec
odet skon unkt npun sser inha mauc soro rder ntpu schu sul
arbe eleg sula stre ktvo eist berd hauf naci unte nkta tzu
ndbe hoer ramm ramt ngge neck punk iese verl onal rdem bsa
tzic ista deni eand lich heim ardt teri sist auge vorz nam
```

Inspection of the groups as now arranged shows that if we get 12 exactly similar series by beginning with the top group in each block and moving diagonally one space to the left and downward except between the fourth and fifth groups, where the movement is downward in the same column.

These twelve similarly constructed series are obviously parts of the original message. It still remains to determine the order in which these parts are to be placed. This in turn admits of no possible doubt, as there is only one way in which they can be arranged to make good sense, good spelling, and good grammar at each joint between the parts. The final order in which the groups at the heads of the columns, that is, the first groups of the series, are to be taken is 8, 11, 2, 5, 1, 6, 7, 3, 4, 9, 10, 12. This is shown as follows:

```
andi ekai serl iche nkon sul  arbe hoer deni
nder rep  ubli kmex ikop unkt stre ngge heim
ausr ufun gsze iche nder inha berd iese sist
einr eich sang ehoe rige rder unte rdem nam
enpa blow aber skia lsru sser eist punk teri
stde utsc herg ehei mage ntpu nkta bsa  tzic
hbit tnih mauf ansu chen schu tzu  ndbe ista
```

```
ndzu gewa ehre nkom maih mauc hauf verl ange
nbis zuei ntau send peso soro naci onal vorz
usch iess enun dsei nec  odet eleg ramm eand
iese gesa ndt  scha ftal skon sula ramt lich
ede  pesc hena bzus ende npun ktvo neck ardt
```

Taken in this order, the groups form the following message in German:

An die Kaiserlichen Konsular-Behoerden in
Der Republic Mexiko Punkt
Strenggheim Ausrufungszeichen
Der Inhaber dieses ist ein Reichsangehoeriger
der unter dem namen Pablo Waberski
als Russe reist punkt er ist deutscher geheim
agent punkt Absatz ich bitte ihm auf ansuchen
schutz und Beistand zu gewaehren komma ihm
auch auf, Verlangen bis zu ein tausend pesos
oro nacional vorzuschiessen und seine Code
telegramme an diese Gesandtschaft als
konsularamtliche Depeschen abzusenden punkt
Von Eckardt

which was undoubtedly the message as originally composed. A literal translation of it is as follows: (the punctuation marks, which are spelled out in the German, are represented by the marks in the translation; thus, punkt means "period," ausrufungszeichen means "exclamation point"):

To The Imperial Consular Authorities in
the Republic of Mexico.
Strictly Secret!
The bearer of this is a subject of the Empire who travels as a Russian under the name of Pablo Waberski. He is a German secret agent.
Please furnish him on request protection and assistance, also advance him on demand up to one thousand pesos of Mexican gold and send his code telegrams to this embassy as official consular dispatches.
Von Eckardt

The actual work of securing this decipherment was very exhilarating. It was at noon on a Saturday that we began to search for the system underlying the arrangement of the groups of four letters.[4] The success in forming these four-letter groups by bringing together letters separated in the original message by an interval of 108 assured us that we were on the right track and that the complete solution of the message was merely a matter of careful, scientific investigation. There was a special stimulus to complete the solution that very day for Colonel Van Deman had informed Captain Yardley that the Chief of Staff accompanied by members of the congressional committee would make a visit of inspection on Sunday morning. Two members of the code and cipher group, therefore, devoted all their energies to discovering

[4] The "we" in this case was most likely John Manly and Dr. Edith Rickert. Herbert Yardley, in his book *The American Black Chamber*, Chapter VII, implies that it was he who came up with the solution to the Waberski cipher. This is not the case. As Manly's document shows, it was he and likely Dr. Rickert who worked through the weekend to solve the cipher.

the system and completing the solution of the cipher. In spite of various interruptions, the work went on successfully, but it was not completed at six o'clock, the usual close of the working day. The two experts could see that success was just at hand and, therefore, decided to take dinner near the office and continue their work in the evening. To cut a long story short, the complete solution was obtained, a translation made, and numerous type-written copies of the cipher message as received, the solution of it, and the translation were prepared before the experts left the office late in the evening.

The contents of the message showed clearly that it was of the highest importance. The bearer was confessedly an official agent of the German Secret Service, and the cipher message was a credential testifying to his status and conferring upon him almost unlimited powers. A real spy, obviously one greatly trusted and relied upon, had beyond a doubt been captured, and this document, which was found upon his person, furnished the most indisputable and damning evidence against him.

The triumphant feelings of Colonel Van Deman as he awaited the visit of inspection can easily be imagined. Not only was he able to point to a well-organized, smoothly working division, he could cite a fresh achievement of the Code and Cipher Section, which, in the opinion of the Chief of Staff and the other members of the visiting committee, would alone have justified the whole organization.

The decipherment and translation of the cipher were quickly communicated to the Military Intelligence Office at Fort Sam Houston, where Pablo Waberski was in prison awaiting his trial. The importance of the message as evidence was immediately recognized, and preparations were made for the trial. But at once there arose a serious legal question, the decision of which interposed a long delay before the trial actually occurred. This question was whether under the laws of the United States an enemy agent, operating in the country three thousand miles from the actual seat of war, should be tried by a military tribunal under martial law, or whether the case must be tried by a judge and jury in the district criminal court. The question was naturally submitted to the highest legal authorities, and it was determined that according to the eighty-second article of war as interpreted by the whole series of its legal antecedents as well as by the general principles of international law, the proper tribunal for the place was a court martial.[5]

The trial was held at Fort Sam Houston beginning August 14, 1918, and the two members of the Code and Cipher Section who had been most active in securing the solution[6] were ordered to proceed from Washington to Fort Sam Houston to testify at the trial if called upon concerning the correctness of the solution and translation of the cipher and to explain to the court the method by which the solution had been obtained.

The journey from Washington to San Antonio at the season was not only long but also hot. The whole country was suffering from one of the most extreme waves of heat known in many years. The trains were crowded with passengers in all stages of

[5] Lothar Witzke, a.k.a. Pablo Waberski, was also an officer in the German Navy.

[6] This was John Manly and Edith Rickert.

disarray and with tempers exasperated by the clouds of dust and the scorching heat. But the arrival in San Antonio brought a welcome relief. The days were hot, but at sunset a strong, cool breeze swept in from the Gulf of Mexico, bringing a relief that was not only welcome but also positively delicious after the torture of the hot and dusty days.

The visitors from Washington, therefore, awaited with some tranquility their turn to appear as witnesses before the court martial. Only one of them was called, but for him the occasion was not without its excitement. He felt that he was summoned to a task of unusual importance and solemnity. A fellow man was being tried for his life before one of the highest special tribunals in the world. Only the president of the United States himself would review the decision reached. The prisoner was charged not only with the mere fact of being a spy but with having been a leader in several of the most disastrous and destructive plots from which this country had suffered since the beginning of the war. The witness had never before appeared in a court of any kind, but he felt not only the seriousness of his obligations but an absolute certainty concerning the validity of the damning testimony which he was called upon to give.

The room in which the court martial sat was a large bare room on the second floor of one of the hastily constructed wooden buildings of the great Army post that had been developed in the suburbs of San Antonio and named Fort Sam Houston. As the witness entered to give his testimony, he was surprised at the group that confronted him. There was no large body of tense and eager spectators such as crowd in upon every sensational trial in the civil courts, but only a small group of six or eight quiet and serious men. The group consisted of the presiding officer of the court martial and his associates, the legal officer conducting the prosecution for the government, the counsel for the prisoner, and the prisoner, pale and nervous, but quiet as a stone.

The witness from MI-8 was duly sworn and asked the usual questions concerning himself, his career, and his competency to testify about codes and ciphers. The attorney for the government then showed the witness two papers, and the examination proceeded.

****

Manly then answered the prosecution and defense questions, elaborating on the process used to decrypt the Waberski transposition cipher message. He verified that the copy the prosecutor had was accurate and that there was no other way to decrypt the cipher message other than the one he described, ending with "… there is no possibility of its being deciphered to show anything else. There might be a conceivable variation in which the particular form for these same results could be secured, just as if you were going from one place to another; you can go north and then go west, or you can go west and then go north, and arrive at the same point." The end result is the same.

Lothar Witzke was convicted of espionage and sentenced to death largely on the strength of Manly's evidence. He was the only German spy given the death sentence during the war. In 1920, President Wilson commuted Witzke's sentence to life in prison, and in 1923, clemency was granted and Witzke was released and allowed to return to Germany. As the last German spy to return home, Witzke was greeted as a hero and awarded the Iron Cross First and Second Class.

References

Gilbert, James L. *World War I and the Origins of U.S. Military Intelligence*. Lanham, MD: Scarecrow Press, Inc., 2012

Manly, John M. "Waberski." Item 811. Friedman Collection, George Marshall Foundation Research Library, Lexington, VA, 1927.

Yardley, Herbert O. *The American Black Chamber*. Indianapolis: Bobbs-Merrill, 1931.

Chapter 16
Madame Victorica Arrives in New York

John Matthews Manly

Abstract This article, number IV in his sequence, is the first of four Manly articles on the German spy Madame Marie de Victorica. It includes background material on Madame Victorica and a several page discussion on secret inks and scientific methods of cryptology and steganography. The rest of the article is a chronological description of Madame Victorica's first few days in New York City in early 1917 in which she is making contacts with the German community in New York and with other German agents already in the city. The article ends abruptly after page 9; possibly a page is missing. Herbert Yardley has an entire chapter in *The American Black Chamber* on Madame Victorica that contains some of the same details as these four articles.

On New Year's Day, 1917, in a certain street in Berlin, a richly dressed lady accompanied by a German aristocrat entered a house that was apparently the private residence of a well-to-do family. There was nothing in the appearance of either to attract any attention other than that naturally accorded to a well-dressed man and a lady of unusually attractive appearance and bearings. But the man was a high official of the German Foreign Office, and the lady was the ablest and most important of the three adventurous women whom Germany sent to the United States as secret agents during the progress of the war.[1]

The lady had all the qualifications that E. Phillips Oppenheim[2] or any of the masters of romantic fiction could have desired. In person she was of the subtle blond type often developed as a result of international marriages. There were strains of German, French, and Spanish blood in her ancestry. Her manners had been formed by association from childhood with the best society of official circles in Continental Europe and South America. Her accomplishments were many. She wrote and spoke

[1] The other two women were possibly Mme. Despina Davidovitch Storch and Mrs. Elizabeth Charlotte Nix, both arrested in New York on March 12, 1918, under suspicion of espionage. Mme. Storch died on March 30, 1918, allegedly of pneumonia. The list of female spies might also include the Baroness Ida Leonie von Seidlitz, who arrived in New York on June 1, 1915, and ran a German spy ring that was also involved in fomenting Irish plots against England. The Baroness was also arrested in March 1918 and eventually deported.

[2] E. Phillips Oppenheim (1866–1946) was an English novelist. He wrote 100 novels and many short stories. His most famous novel was *The Great Impersonation* (1920).

J.F. Dooley, *Codes, Ciphers and Spies*, DOI 10.1007/978-3-319-29415-5_16

fluently German, French, Spanish, and English. She was an accomplished newspaper woman and writer for the silent drama.

Her Christian name was Marie, and the British believed that she was the famous Marie, "la femme blonde" of Antwerp,[3] head of the Central Espionage Bureau at Antwerp, whose duty was to recruit and instruct persons for Secret Service. Whether this be true or not, there is abundant evidence, not only that she was fitted for Secret Service work but that she was not without experience in the kind of intrigues which it involves.

Certain facts in regard to her family and career are well ascertained. She was born in the city of Posen in 1878, and her maiden name was Marie Else von Kretschmann. The family was said to be related to the Kaiser. Her father was Baron Hans von Kretschmann, a general in the Franco-Prussian War and author of well-known works on military science.[4] According to Madame Victorica, it was to her father that Maréchal Bazaine surrendered his sword.[5] Her mother was the Countess Jennie von Gustedt,[6] daughter of a Prussian diplomat accredited to South American republics, and partly of French descent—the name De Vussière later assumed by Madame Victorica as one of her aliases, having belonged to her family of her mother's side. She had an elder brother, Carlos, a Jesuit priest and chaplain in the Austrian Army,[7] and a sister who died in 1917, the wife of Count Esterhazy. This sister was a prominent writer of socialistic and women suffrage books under the name of Lily Braun.[8]

---

[3] In 1934 Herbert Yardley published a short story in Liberty magazine titled *H-27, The Blonde Woman from Antwerp*. Like many publications of the day, Yardley and Manly get the association between Madame Victorica and the real *Blonde Woman from Antwerp* wrong. The female German spy in Antwerp was Dr. Elsbeth Schragmüller. The British called her Fräulein Doktor because she had a Ph.D. in political economy. She was not a spy but ran the French section of the spy center and school out of German-occupied Antwerp, Belgium. Schragmüller tells her own story in her memoir published in 1930. (See also Kahn 2004, p. 113.)

[4] Hans von Kretschmann was a major during the Franco-Prussian war and was possibly a Baron by marriage.

[5] This is unlikely. Madame Victorica's father, Hans von Kretschmann (1832–1899), was not made a general until 1890. During the Franco-Prussian War, he was a Major, so would probably not have received Marshal Bazaine's sword on his surrender of the Army of the Rhine to Prussian forces at Metz on October 27, 1870.

[6] Madame Victorica's grandmother was the Baroness Jeromée Catharina Jenny von Gustedt (1811–1890), an illegitimate daughter of Jerome Bonaparte, Napoleon's youngest brother and the King of Westphalia from 1807 to 1813. Her mother was Jenny Auguste Frieda Karoline von Gustedt (1843–1903), nee von Pappenheim.

[7] There are no known records of Madame Victorica having a brother.

[8] Madame Victorica's older sister Lily was never married to Count Esterhazy. Lily Braun (1865–1916) was a German feminist writer, socialist, and supporter of women's suffrage. She was born Amalie von Kretschmann. Braun wrote many political tracts and novels espousing women's rights. She married Georg von Gizycki, her mentor and 14 years her senior, in the summer of 1893. He died in 1895. In 1896, Lily married Heinrich Braun (1854–1927), who was a Social Democratic politician and a publicist. After 1914, Lily became increasingly nationalistic and alienated many of her socialist colleagues. She died on August 8, 1916, after suffering a stroke (Lischke 2000, pp. 25–30).

Educated mainly in Germany, she passed much of her early life in South America, her father having been naturalized as a citizen of the Argentine Republic. She, therefore, learned both German and Spanish as a child and spoke like a native.[9]

She had been married three times. First to a Chilean mining engineer, who is said to have died of tuberculosis; then, in 1904, to a physician, Pablo Aliro Montero de Siomont, who also died shortly after their marriage; and finally, in 1913, to Manuel Gustave Victorica, an Argentinian.[10] According to Madame Victorica, her third husband deserted her a few months after marriage and was a "crook." She professed to have had no communications with him for several years and to be seeking evidence to obtain a divorce from him; but this is not very probable, partly because she was a Roman Catholic and partly because there is some evidence that she was connected with his activities as an agent for the Austrian government in reporting the sailing of food ships from Brazil.

After the death of her first husband, she began writing political articles for German newspapers, specializing on subjects connected with Chile. In 1907, she was sent to Chile to write on industrial and agricultural conditions there. She took part in the campaign against the separation of church and state—she was in all things an ardent conservative—and was engaged in intrigues against the German Minister to the Country.

Her activities in Chile seem to have attracted favorable official attention at home, for on her return to Germany she was invited by Prince von Bülow to enter the Secret Service, but declined.[11]

She continued her work as foreign correspondent and was associated for some time with Dr. E. J. Dillon, the noted English journalist and political writer in St.

---

[9] Not true. The Kretschmann's spent their time in Germany and never lived in South America as a family. Marie Kretschmann's father was never a naturalized Argentine citizen. Manly is likely either getting this part of the story from Yardley (who repeats this in *The American Black Chamber*) or from contemporaneous newspaper accounts which got many of the details of Madame Victorica's case incorrect.

[10] Marie Else von Kretschmann's first husband was the German painter and graphic artist Otto Eckmann (1865–1902). They were married in 1898 and he died of tuberculosis in 1902, probably having infected her as well. Her second husband was Pablo Aliro Montero de Siomont, a Chilean-French physician, whom she married in 1904 while on a visit to South America. Nine months into their marriage, he had a complete mental breakdown and claimed to have killed all his patients. (Other reports are that they divorced in 1909.) Marie left him and lived in Chile for 5 years, working as a journalist. Her third husband was José Manuel Victorica (1880–1950) who was also known as Manuel Gustave Victorica. They were married in 1913 in Heidelberg. He was an Argentine citizen, which enabled Marie to acquire an Argentine passport. Theirs was a marriage of convenience most likely set up by the German military intelligence office, *Abteilung IIIB*. He was also likely a German spy. He was arrested in Pontarlier, France, on January 10, 1917. He was tried and convicted of espionage at Besancon and sentenced to life imprisonment on April 28, 1918; after the war his, sentence was commuted (Kahn 2004, pp. 33–34; Macrakis 2014, pp. 166–167).

[11] Bernhard Heinrich Karl Martin von Bülow was German Chancellor from 1900 until July 1909. It was not von Bülow who recruited Marie de Victorica but more probably Colonel Walter Nicolai (1873–1947), the head of *Abteilung IIIB*, the department within the German Grand General Staff with responsibility for Intelligence, Press, Propaganda and counterintelligence from 1912 through 1919.

Petersburg.[12] We shall later see that she paid Dr. Dillon the compliment of using his name as an alias for one of the most important German secret agents in America. From her journalistic work and from scenarios for the moving pictures, she earned a handsome income. At the outbreak of the war, she was in Bulgaria reporting on the efforts to reconcile Bulgaria to the Holy See. She barely got through Belgrade (Serbia) before the bombardment, losing all her personal effects. During the Russian invasion and until after the Battle of Tannenberg, she was on her estate in East Prussia.

Later she volunteered for war service and was at first employed by the Bureau of Public Information, working in the Catholic Department under Erzberger on translations and on scenarios for war propaganda pictures.[13] She was then drafted into the Secret Service and sent on missions to Switzerland and Holland.

Such had been the career of Madame Marie Victorica up to the time that she entered the house in Berlin accompanied by Secretary Baron von dem Bussche-Haddenhausen of the German Foreign Office.[14] She was about to start on a mission far exceeding in importance, difficulty, and danger, any that she had previously undertaken. She had been commissioned to visit the United States and was on this fateful day receiving her final instructions and her equipment for the mission.

The house she had entered—to all outward appearances a quiet and unromantic residence—was in reality the invisible ink laboratory of the German Secret Service.

The arts of transmitting messages secretly are manifold, and many of them are of great antiquity. Aeneas Tacticus, a Greek writer on military affairs of the fourth century B.C., enumerates 20 different methods of secret communication. Some of the methods in early use have been made famous by the anecdotes of Plutarch and other ancient writers. The most famous of these devices, perhaps, is that known as the skytale, which consisted in writing a message on a narrow strip of parchment wound spirally around a wooden rod. When unwound, the letters of the words of the message were so scattered that they did not form intelligible words and could not be read. The method of reading was to rewind the parchment around a rod that was of the same diameter as the first. This was successfully used by Athenian generals.

---

[12] Emile Joseph Dillon (1854–1933) was an English-Irish author, journalist, and linguist. Dillon had three doctorates in philology and oriental languages. He was the Russian correspondent for the *Daily Telegraph* newspaper from 1887 to 1914. He reported on such events as the 1894–1895 Turkish massacres of Armenians, the Dreyfus trial of 1899, the Boxer Rebellion in 1900, the Russian revolution of 1905, the Portsmouth Peace Conference of 1905 to end the Russo-Japanese War, the Mexican revolution in 1920, and the Versailles Peace Conference in 1919.

[13] Matthias Erzberger (1875–1921) was a German politician and was Reich Minister of Finance from 1919 to 1921. During the war he was in the Department of Information and was in charge of foreign propaganda targeted at Catholic groups. Erzberger was prominent in the Catholic Center Party and was a member of the Reichstag. He was one of the authorized representatives of the German government at the signing of the armistice and was assassinated for his trouble.

[14] This was Hilmar von dem Bussche-Haddenhausen (1867–1939), a German diplomat who served in the United States, Argentina, and Romania. From 1916 to 1918, he was Under Secretary of Foreign Affairs in Berlin.

But the incident of greatest interest to us is one reported of the great statesman, Themistocles. While he was in exile from his native country, one of his friends, wishing to communicate to him the news that the time was ripe for a revolt and a return to power, shaved the head of one of his slaves, tattooed the message in indelible ink on the bare skin, and, having waited long enough for the growing hair to cover the writing, sent the slave to Themistocles with the sole oral message that he was to shave the slave's head. The writing in this instance was not done with invisible ink, but it is rumored that in more recent times, invisible inks have been used for writing on the bare skin and messages afterward brought out by the use of a developer.[15]

Notwithstanding the long history of secret inks, the arts of producing and developing them were at the beginning of the war in as backward a stage of development as were the arts of code and cipher construction and attack or any of the other arts of military value. It had long been known that writing produced with the juice of lemons, onion, and various other fruits and vegetables, or with milk or a solution of alum, could be developed by various simple processes, such as the application of heat, but this was practically all that was known on the subject even by the Germans at the beginning of the Great War.

As has so often happened in the history of the world, in this case also necessity was the mother of invention. Numerous German agents who had been making use of milk, alum, or the juices mentioned to produce invisible writing were seized by the British and the writing easily developed. Two cases in particular—that of Eva de Bournonville[16] and that of Courtenay de Rysbach[17]—brought vividly home to the Germans the defects of these means of secret writing.

With the thoroughness characteristic of all their methods, and with their knowledge of the enormous military and political value of the means of secret communication, they set to work to devise improvements in these as in so many of the other arts. For them this was not difficult. Whatever may be thought of the relative claims of the English, French, and Germans in the field of pure science, there can be no doubt that in 1914 the Germans led the world in the application to practical uses of

[15] Manly is making a distinction between *steganography*, where the very existence of a message is hidden, and *cryptology*, where the message is visible but unreadable. Secret inks are an example of steganography. The message itself is written in plaintext but is hidden from view, just like the tattoo on the slave's head.

[16] Eva de Bournonville was a Swedish citizen who was spying for Germany in the United Kingdom. She arrived in September 1915 and was arrested and charged with espionage in November 1915. Convicted and sentenced to death, her sentence was commuted after a plea for clemency from the Swedish government. She was released and deported to Sweden in 1922. De Bournonville wrote a number of letters using secret ink including some relating to antiaircraft defenses and a possible air raid on Croydon (West 2013, pp. 38 – 39).

[17] Courtenay de Rysbach, a British citizen of Austrian descent, was a German spy in the United Kingdom. He was caught in July 1915 when British censors intercepted four sheets of music headed for Norway that were impregnated with a message written in secret ink. The message gave detailed information about the British war effort, including possible military targets. De Rysbach was sentenced to life imprisonment for espionage in October 1915. He was released from Dartmoor Prison in 1924 (Macrakis 2014, pp. 139–142).

the more subtle refinements of modern scientific thinking. Their work in the fields of applied chemistry and physics was especially advanced; and with the close official connection between the government and the men of science, it was natural that when the need for improved types of secret writing was recognized, the German leaders should impose upon their chemists the task of devising forms of invisible writing that would become visible, not under some general physical reagent, like heat, but only under one particular chemical reagent.[18]

The chemists responded with several inventions of great value. And for a time, although the British and French knew that certain papers that they had seized contained invisible writing, they were entirely unable to develop and read it. Then began a battle of wits between the chemists of Germany and those of the Allied countries, very similar to that which we have discussed in connection with the activities in code and cipher.

On the one side, the effort was to devise a substance safely and easily transportable which would produce writing that would become visible only when there was applied to it in a proper manner some special reagent not difficult to procure but not likely to be thought of by the enemy or applied accidentally. The task on the other side was to devise methods or reagents that would disclose the presence and develop the visibility of secret writing of any character whatever. Like the contest between the makers of armor plate and the makers of big guns, or like that between the constructors of code and cipher systems and the code and cipher analysts, this contest was one in which victory perched momentarily now on the banner of one side and now on that of the other.

The conditions of the contest were exacting and difficult. To be easily and safely transportable, the secret ink or the substance for making it must be such that it would not lose its valuable qualities by the mere passage of a brief period of time and that it could be carried on the person of the agent in some form which would escape suspicion and elude even the most searching examination, if suspicion had once been aroused. The former of these conditions was less difficult to meet than the latter, but the latter was solved in two different ways.

---

[18]Many German scientists worked for the government during World War I. A number of them signed the famous "Manifesto of the Ninety-Three (German men of letters)" on October 23, 1914. The manifesto, a denial of German atrocities in Belgium during August and September 1914, included signatures of not fewer than 12 German Nobel laureates in science: Adolf von Baeyer, 1905 recipient of the Nobel Prize in Chemistry; Emil Adolf von Behring, received the 1901 Nobel Prize in Physiology or Medicine; Paul Ehrlich, awarded the 1908 Nobel Prize in Physiology or Medicine; Hermann Emil Fischer, 1902 recipient of the Nobel Prize in Chemistry; Fritz Haber, received the 1918 Nobel Prize in Chemistry (Haber was later called the father of chemical warfare for his work in creating chlorine and mustard gas); Philipp Lenard, winner of the 1905 Nobel Prize for Physics; Walther Hermann Nernst, won the 1920 Nobel Prize in chemistry; Wilhelm Ostwald, received the 1909 Nobel Prize in Chemistry; Max Planck, awarded the Nobel Prize in Physics in 1918; Wilhelm Roentgen, awarded 1901 Nobel Prize in Physics; Wilhelm Wien, received the 1911 Nobel Prize in Physics; and Richard Willstätter, won the 1915 Nobel Prize for Chemistry. (See http://en.wikipedia.org/wiki/Manifesto_of_the_Ninety-Three accessed on August 14, 2014.) Most of the signatories later repudiated the Manifesto.

The first of these ways was to disguise the substances for secret writing by incorporating them in such things as perfumes, tooth washes, tooth pastes, soaps, gargles, and other medicinal or cosmetic preparations. Sometimes, of course, the only disguise attempted was the use of a false label. The other method was to make use of substances that could be soaked up and retained by ordinary articles of wearing apparel in sufficient quantities to be recovered as needed for use in writing.

The articles most commonly used were silk scarfs, handkerchiefs, ties, collars, stockings, underclothing, and night clothing, boot laces, washing gloves, sponges, and the like. Some of these were best adapted for carrying the inks that made an entirely colorless fluid. Others, such as boot laces and cloth covered buttons, were used especially for carrying a certain secret ink which would have discolored the white articles.[19]

The usual method of using these substances was to soak a bit of the cloth in distilled water and then wring it out into a glass tube used for writing. In the case of the famous ink known as F ink, the solution necessary for producing good results had a strength of only one hundredth of one percent, but when the proper reagent was applied, the writing became clear and easily legible.

To prepare for use the extremely subtle ink known as P ink—with which one of the agents arrested in Seattle was provided—it was necessary only to soak one inch of a bootlace in cold water. Other types of ink required soaking in boiling water.

The number of inks used was very great. They were distinguished by various letters of the alphabet, for example, E, F, G, H, N, P, R-1, and R-2, and are referred to by these designations in the secret message themselves.

No doubt the reason for having so many inks was twofold. In the first place, new inventions were made from time to time and more refined methods introduced. In the second place, the use of different inks in corresponding with different agents was essential to security. Otherwise, the arrest of a single agent or the treachery of one would put the enemy in possession of all the secrets of this means of communication.

In some instances, considerable care had to be exercised in regard to the inks. Agents who used the type of ink carried in black bootlaces were warned that the laces must not be exposed to direct sunshine. They, therefore, could not be used but could only be carried. On one occasion, an agent who was bringing a letter to America to be posted in New York to Madame Victorica found that the secret writing had begun to develop because he had carried the letter next to his skin, and for the sake of safety, he was obliged to destroy it.

This hasty survey of some of the outstanding points in the use of invisible inks will give the reader some idea of the difficulty of the task that confronted the chemists of the Allies. One of the greatest difficulties was that if a paper suspected of containing invisible writing were tested with the wrong reagent for developing the writing, the portion tested was often either disfigured or so effected that it would afterward fail to respond even to the right reagent. It was necessary, therefore, not

[19]When arrested in 1918, Madame Victorica had in her luggage two silk scarfs impregnated with invisible ink.

only to discover what reagents might be used for developing the various kinds of invisible writing but to invent such a technique of examination as made it possible to carry out the test to success without spoiling the writing to be tested. How this was done, though a wonderful story, is a story that perhaps will never be known.

Although Madame Victorica met five of the German chemists at the secret ink laboratory, she said later that she was not taken into the laboratory itself. As a matter of fact, she needed no instruction in the methods of using the ink with which she was provided or the developer which was necessary to render legible the invisible writings which she was later to receive. She had already served on secret missions to Holland and Switzerland.

"I was conducted," she said later when telling the story of this visit, "into a large room in the laboratory entirely filled with clothing of all kinds – scarfs, handkerchiefs, ties, stockings, night gowns, and all sorts of things. I was told to select any article I liked, as they were all soaked in secret inks. I took two white silk scarfs."

In this, as in so many other instances, Madame Victorica told the truth but apparently not all the truth, for when she was arrested at the Hotel Nassau, overlooking the entrance to New York Harbor,[20] and her effects seized and examined, it was found that invisible inks could be prepared not only from the two scarfs—which, by the way, were the same pattern as those seized by the British on Hagen, one of the chief German agents arrested soon after the beginning of the war—but also from several handkerchiefs. All of these were impregnated with a substance which, when dissolved in water, produced the famous F ink.

Before she left Berlin on January 5, 1917, Madame Victorica received instructions as to the purpose of her mission. If we could believe all her statements, she was only to indulge in propaganda.

She said, "I was ordered to encourage pacifist activities and Irish unrest and to attempt to stir up the Catholics to resent the interference by Italy with the freedom of the Pope."

But again it is clear that she was not telling the whole truth.

It appears clearly from secret writing, which were developed in the laboratory of MI-8, that she was to arrange an ingenious method for importing into America a new and powerful explosive called theta, for placing bombs on ships, to locate submarine bases, and after her arrival she was ordered to ascertain what would be done about the Panama Canal. It is also clear from the evidence that she was more or less concerned in the plot against the Welland Canal.[21]

---

[20] The original text says "at Long Beach Hotel," but this can't be true because the Long Beach Hotel burned to the ground on July 29, 1907 and was never rebuilt. (See http://captfxco.homestead.com/LongBeachHotel.html accessed on August 14, 2014) The hotel that Manly means is probably the Hotel Nassau in Long Beach, NY. Other articles refer to Madame Victorica staying at the Hotel Nassau.

[21] This would be the "Von Papen Plot." Franz von Papen, the German military attaché assigned to the German embassy in the United States, was indicted by a federal grand jury in April 1916 for plotting to blow up the Welland Canal that links Lakes Erie and Ontario in Canada. At the time, von Papen was already back in Germany, having been declared *persona non grata* by the US government in December 1915 and expelled. It is unlikely that Madame Victorica had anything to do

She left Berlin as planned, traveling via Christiania and Bergen. In Christiania, she went to the office of the Argentine Consul to obtain a visa for her Argentine passport. She, of course, did not inform him of the purpose of her mission, but seems to have reported that she was visiting the United States to gather evidence to secure a divorce from her third husband. The consul called her attention to the numerous notations of the German military police on the back of the passport and warned her that she would never get through the British cordon at Kirkwall with such a passport. He, therefore, gave her a new one.

One of the instructions given to Madame Victorica before she left Berlin was that immediately upon her arrival in New York, she should go to see a certain German chemist of worldwide reputation. She was told that he would supply her with the reagent necessary to develop the letters she was to receive in secret writing and to supply a certain ingredient necessary for the manufacture of a powerful explosive named theta, which it was planned to use.[22]

She arrived in New York on January 21 and registered at the Hotel Knickerbocker. Abundant funds had been provided for her by cable so that on her arrival there was $5,000 to her credit at one of the principal German banks and shortly afterward an additional $3,500 was cabled for her.[23] Quite naturally, one of her first acts after her arrival was to do a little shopping. She secured the services of a Mrs. Hunt, a professional shopper recommended by the Hotel Knickerbocker, and promptly spent something over $4,000 for clothes.

She attempted immediately to see the famous chemist, but he was out and she left her card. Accompanied by Mrs. Hunt, she then attempted to see Albert Paul Fricke.[24] Fricke was the manager of Richter and Company, manufacturers of toys and medicine with headquarters in Berlin.[25] Fricke was at Lakewood, but she left a message,

---

with the plot to blow up the Welland Canal, as she did not arrive in the United States until January 1917.

[22] This may have been Dr. Hugo Schweitzer, a renowned chemist and for a time head of the pharmaceutical division Bayer Company in America. Schweitzer was known to have been involved with sympathetic German-Americans, including Franz von Papen and Heinrich Albert. He was behind the *Great Phenol Plot*, attempting to corner the American market on various chemicals, particularly phenol, that were to be used to make ammunition for the Allies, and helping to facilitate the leak of the formula for mustard gas to the Germans. He was indicted, but died suddenly in November 1917. (See http://en.wikipedia.org/wiki/Great_Phenol_Plot accessed on August 14, 2014.) The "theta" explosive Manly refers to may be pentaerythritol tetranitrate, also known as "tetra" or PETN. Tetra was invented in Germany in 1894 and was widely used by the Germans during World War I.

[23] Upon her arrival, she deposited a check for $35,000 in a Manhattan bank.

[24] Albert Paul Fricke was a naturalized American citizen employed as the American director of a German toy company, Richter and Company, who acted as the distributor for payments among the German spy ring. He was arrested in June 1918 and charged with treason. Further charges were made in December 1918. He was tried—and acquitted by a jury—in April 1919.

[25] Richter and Company was a German manufacturer of toy building blocks (*Anker Stones*), chocolate, and patent medicines. The US subsidiary was founded in 1887 and was confiscated by the US government in 1918 as a German-owned company. During the war, the European arm of Richter and Company produced artillery shells for the German Army.

and the next day he telephoned her and they had luncheon together. She identified herself to him and attempted to get him to send the cable announcing her arrival. This he declined to do so it was sent through the hotel. The telegram was addressed to H. Fels, Berlin, N.W., Dorotheen Str. 21, and read: "Happily arrived. Fine passage."

Fels (rock) is a synonym for stein (stone) and was the official designation of the chief of the Secret Service to whom she was to report. She says, "'Fine passage' means I got through Kirkwall without any trouble."

She told Fricke that she wanted the address of (Herman) Wessels, whom she called Rodiger,[26] a name under which Fricke also knew him. At first Fricke refused to give her this address, but she convinced him of her connection with the Foreign Office and the Naval Secret Service, and he finally did so.

On January 25, Fricke took Madame Victorica to the Kaiserhof[27] for luncheon and Wessels joined them.

> "At first," she says, "Wessels was suspicious of me but I answered all his questions and gave him details about the naval office in Berlin, and used the names of officers and clerks. All this did not satisfy, so I finally showed him my passport with the visa of Mr. Samburger, Chief of the Berlin Secret Service. The reason this convinced him was that passports for ordinary travelers are issued and visaed by the Pass Bureau, but in order to avoid publicity, the passports of secret agents are visaed by the secret service and generally signed by Samburger himself. I also wrote down for him the new secret numbers to be used in translating the ABC code."[28]

The passport here mentioned was not the new one visaed in Christiania but the old one that she had hidden and brought with her for just such purpose as that to which she now put it.

Fricke left Wessels and Madame Victorica at the Kaiserhof, and they continued their conversation. She then gave him a message that Lieutenant Eisenhafer was coming to help him and told him of the new secret ink, promising to give him some for his own use.

Fricke says that Madame Victorica gave Wessels a new fountain pen at this interview. This, no doubt, contained some of the ink that she had extracted from one of her scarfs.

Victorica then asked Wessels about the firm to which certain shipments of Swedish artificial marble containing the materials for making explosives could be sent, and this helped to show him that she really was what she professed to be. She

[26] This was German Navy Lt. Commander Herman Wessels, a German spy who arrived in the United States in early 1917 to set up an extensive spy ring in the eastern United States during 1917 and 1918. He went by several aliases, Carl Rodiger being the most common. Wessels was arrested in May 1918 and charged with espionage and treason. He was tried in a naval court martial in 1920.

[27] The Kaiserhof Restaurant and hotel was a German restaurant at the corner of Broadway and 39th Street near the Casino Theatre and the old New York Metropolitan Opera House.

[28] This is likely the cipher system that Manly describes in the next chapter—Article V. He is referring to the 5th edition of William Clausen-Thue's ABC Universal Commercial Electric Telegraphic Code, at the time the most popular non-secret commercial code used (Kahn 1967, p. 838).

told him also that she wished to see Wünnenberg,[29] and although he advised her to keep away from Wünnenberg as he thought Wünnenberg was under observation, she insisted on meeting him. Wessels told her that she could do this through Emil Kipper, a retired manufacturer, and that Kipper would also introduce her to an Irish leader with whom she had been instructed to communicate.

[Ed. The article ends abruptly here. Possibly there is a page is missing.]

References

Kahn, David. 2004. *The Reader of Gentlemen's Mail: Herbert O. Yardley and the Birth of American Codebreaking*. New Haven, CT: Yale University Press.

Lischke, Ute. 2000. *Lily Braun: 1865–1916, German Writer, Feminist, Socialist*. Rochester, NY: Camden House.

Macrakis, Kristie. 2014. *Prisoners, Lovers, & Spies: The Story of Invisible Ink from Herodotus to Al-Qaeda*. New Haven, CT: Yale University Press.

West, Nigel. 2013. *Historical Dictionary of World War I Intelligence*. Historical Dictionaries of Intelligence and Counterintelligence. New York: Scarecrow Press.

Yardley, Herbert O. 1931. *The American Black Chamber*. Indianapolis, IN: Bobbs-Merrill.

Yardley, Herbert O. 1934. "H-27, The Blonde Woman from Antwerp." *Liberty Magazine*, April 21.

[29] This is Karl Wünnenberg, also known as Charlie "Dynamite" Wünnenberg, a German agent and explosives expert. He was implicated in the Black Tom explosion. Wünnenberg was arrested, tried, and convicted in the summer of 1917 and sentenced to 2 years in prison.

Chapter 17
Madame Victorica and German Agents in the United States

John Matthews Manly

Abstract This chapter contains the second of Manly's Madame Victorica articles, number V. The majority of the article is about German agents in the United States, especially Charles Nicholas Wünnenberg, known as "Dynamite Charley," and Hermann Wessels (aka Carl Rodiger), and not so much about Madame Victorica. The article spends quite a bit of time talking about the various aliases used by the German agents. It also introduces the attempts of Madame Victorica to enlist the aid of Irish-Americans to work against the British in the United States and in Ireland.

When Madame Victorica arrived in New York on her sinister mission, there were already several German agents at work with very similar purposes and plans. Of those with whom we are concerned, the earliest to arrive was Charles Nicholas Wünnenberg (Yardley 1931, pp. 113–114; Landau 1937, pp. 86–91; Macrakis 2014, pp. 154–155), more picturesquely known as "Dynamite Charley." He was a man about 39 years old and was lame. He had been a marine engineer and had come to this country in 1915 as a German agent. During 1916, he had been especially engaged in efforts to induce American newspapermen to go to England as spies. The most famous of those who yielded to his solicitations was George Vaux Bacon (Macrakis 2014, pp. 153–160), whose arrest and trial in October 1916 were of worldwide notoriety.

The coming of the second of the agents with whom we are concerned was announced in October 1916, in the following cablegram from the office of Richter and Company in Copenhagen.

> Rodiger of Olten branch will arrive on S.S. Christianiafjord. Notify Binder of Behrens and Company in reference to will.

The manager of neither the New York branch of Richter and Company, who received this cablegram, nor Binder of Behrens and Company understood the meaning of the mention of the "will," but later Fricke[1] explained that it was undoubtedly

[1]Albert Paul Fricke was the New York manager of Richter and Company, mentioned in Article IV. Richter and Company was a German manufacturer of toy building blocks (Anker Stones), chocolate, and patent medicines. The US subsidiary was founded in 1887 and was confiscated by the US government in 1918 as a German-owned company. During the war, the European arm of Richter and Company produced artillery shells for the German Army. Fricke was arrested along with Madame Victorica and 6 others in 1918. See http://www.ankerstein.ch/downloads/CVA/Book-PC.pdf (accessed on August 14, 2014).

J.F. Dooley, *Codes, Ciphers and Spies*, DOI 10.1007/978-3-319-29415-5_17

inserted for the purpose of inducing the British cable censor to pass the message as an ordinary business message.

The real name of the agent thus announced was not Rodiger but Hermann Wessels (O'Leary & Kelly 1920, pp. 240, 525, 529, 540). A Swiss passport for Carl Rodiger, manager of Richter and Company's branch house at Olten, Switzerland, had been procured, and the photograph of the real Rodiger removed and one of Wessels substituted. This was a very frequent practice during the war and was not difficult. The British attempted, but without success, to induce neutral countries to devise and adopt a method for printing an indelible photograph directly upon the passport, in order to prevent just such substitutions as this.

This alias "Rodiger" was, however, the name by which Wessels was chiefly known after his arrival in the United States, and most of the German agents and sympathizers knew him by no other name. He had, however, a number of aliases. One of them was "Haro Schroejers," a name under which a Dutch passport was procured for him in November or December 1916. Another was "C. Stamm," of 19 North Broadway, Yonkers. That Rodiger, Schroejers, and Stamm were one and the same is shown by the fact that on March 22, 1917, about 6 weeks after Madame Victorica left Berlin, each of them received a letter signed "Richards"[2] and absolutely identical in wording. On the same date, three letters, different from these in the wording of the clear text, but clearly containing in secret ink the same messages as this contained, were sent to Madame Victorica. The three letters to Wessels (addressed to Rodiger, Schroejers, and Stamm) read as follows:

> Dear friend,
>
> It is a very long time since I heard directly from you, what is the matter? In the meantime I was very pleased to have some news from our friend Clarke. By same mail I will send you the wanted documents in duplicate for his business, as it seems that the originals are lost. How is your catarrh going on? I can recommend very much my special medicine Kalium Iodatum also to you. Please ask Clarke about further details I have given to our friend before his departure. Do use it immediately after receiving this letter and the result will be very satisfactory.
>
> I hope that you received my last remittance by wireless and did you get the bond from Clarke?
>
> In a week or two my cousin Oscar will call upon Clarke and no doubt you will meet him, too. He will give you full report about my business, please give him your assistance in everything. He will go to the South as soon as he can.
>
> With the Kindest regards I am
> Yours truly,
> Richards.

"Clarke," as we learn from other documents, was a cover name for Madame Victorica. The documents wanted for Clarke's business are known to have been the letters of introduction that she was to use in establishing connections that might aid in carrying out her plans. The originals were to have been brought over on the sub-

[2] "Richards" is also known as Frank Richards and was a shadowy character who weaved in and out of this espionage story. Apparently, whoever Richards was, he was never caught by either American or British counterespionage authorities. Richards may have been the cousin of one of the Irish-American conspirators, John T. Ryan (Literary Digest 1919; Weyl 1950, p. 313).

marine *Deutschland*,[3] but they never arrived and are said to have been lost. Copies were promised in the present letter, but for some reason they were not enclosed, and Madame Victorica was obliged to form acquaintances without their aid. The recommendation of "my special medicine, Kalium Iodatum,"[4] which is repeated in similar terms in so many of the letters to their agents is a means of informing them that this is the proper chemical reagent for developing the invisible writing on the blank pages of the letter containing the recommendation. "Clarke" had been thoroughly instructed in the use of this "medicine" before she left Berlin and is here ordered to pass the instructions on to Wessels. The next sentence of the letter implies that this ink had to be developed soon in order to come out clearly. "Did you get the bond from Clarke?" apparently means, "Did you get a part of the $35,000 recently sent to Madame Victorica?" Finally, "cousin Oscar" is the agent variously known as "L. A. 7," "Ernesto Escobar," "Harry Wood," and "Oscar Eiyelhardt."[5] He was a sort of "supervisor" or "district manager" of agents in North and South America and later furnished Wessels with a new code and new ink and offered to send him $20,000 or $25,000 immediately upon his return to South America.[6] "Richards," the signer of the letters to Wessels, was the official designation of one of his superiors.

The other important aliases of Wessels were "Dillon" and "Hudson." Madame Victorica said that "Dillon" was the name used for Wessels in messages from Germany and "Hudson" that used in letters from the United States; but both names were used in the letters from Germany—"Hudson" in the earlier ones and "Dillon"

---

[3] The German submarine *Deutschland* was one of two unarmed merchant submarines used as blockade runners during World War I. The *Deutschland* made two trips to the United States during 1916; however, a third trip in January 1917 (possibly the trip mentioned here) was aborted because of a worsening of German-US relations. The *Deutschland* was taken over by the German Imperial Navy on February 19, 1917, and converted into the U-155. As a U-boat, the *Deutschland* made three successful cruises, sinking 42 Allied ships. She was surrendered under terms of the Armistice on November 24, 1918 (see http://en.wikipedia.org/wiki/German_submarine_Deutschland Accessed on August 14, 2014).

[4] *Kalium Iodatum* is potassium iodide (or iodide of potash), which is sometimes used as a homeopathic medicine. In this case, it comes in tablet form that is dissolved in water and used to develop the secret ink messages.

[5] No one by any of these names was arrested in the period 1917–1918 for espionage. But see the footnote below.

[6] Oscar/Harry Wood may have been one of the several German agents—Karl Frick, or Kurt Jahnke, or Franz Rintelen—who were all supposedly involved in the Black Tom explosion in 1916. Rintelen, a German naval officer who went under several pseudonyms, including Emil Gasche, E. V. Gibbons, and Frederick Hansen, was involved with the attempt to foment a coup d'etat in Mexico by former Mexican President José Victoriano Huerta Márquez (1850–1916) in 1915. Huerta was arrested near the Mexican border in late June 1915 and died in of cirrhosis of the liver while still in custody in January 1916. Rintelen was recalled to Germany in July 1915 but was arrested by the English when his ship was stopped and searched. He ultimately spent the rest of the war in English and later American prisons, being freed finally on November 19, 1920. He settled in Germany, but later moved to England and died in London in 1949. Given this chronology, it is unlikely that Rintelen was involved in Black Tom. (See O'Donnell 2004, pp. 44–45; Tuchman 1958, pp. 63–81.)

in the later. It will be remembered that the use of the name Dillon is probably due to Madame Victorica's associations with Dr. E. J. Dillon in St. Petersburg.

Wessels was born in Papenburg in 1877 and had for many years been a seaman and officer of the merchant marine and reserve officer in the German Navy. In the summer of 1916, he was summoned to Berlin and sounded out concerning a trip to America. The mission was frankly one of sabotage and particularly the placing of incendiary bombs on American and Allied ships. He was instructed to see "what could be done against England" and to make use, if possible, of Irish sailors, hiring them to enlist in the British Navy for the purpose of placing bombs containing solution fachlig and tetra,[7] a comparatively recent explosive. Wessels later declared that he was instructed not to do anything against the United States. This is doubtless true, as the German government was then endeavoring to keep this country neutral.

The plan that Rodiger, or Wessels, had for introducing the explosives into this country was to have the ingredients for making them enclosed in toy blocks of artificial stone, which were then to be shipped in marked boxes to the address of Richter and Company in New York, importers of children's toys and of similar goods.[8]

Soon after his arrival, Wessels called on Fricke, the manager of Richter and Company, introducing himself as Rodiger and showing the Swiss passport. He told Fricke that letters would come addressed to the firm but intended for him. They would be recognizable by the fact that they contained the word "aunt" or "tante." He told Fricke his plan and said that he was ready to pay $10,000 to $50,000 for each ship destroyed. Fricke replied that if Richter and Company chose to send such blocks to the New York branch, he (Fricke) could not prevent it but that it would be difficult to get them away without suspicion. He refused to aid Wessels in procuring tetra for use in making bombs. Apparently, Fricke attempted to discourage all these schemes as much as possible.

Wessels also went to see the famous chemist mentioned in the article preceding this, but later said that he did not find him very sympathetic; in fact, "he practically threw me out of the office, and I never saw him again except on the street."[9]

There can be little doubt that his contacts with Wünnenberg and Wessels were among the reasons why this chemist declined to receive Madame Victorica and cooperate with her.

From six to ten letters in secret writing were received by Fricke for Wessels, four of which were developed either in Fricke's office or in the bathroom of his house, and Wessels attempted through Fricke to get in touch with Madame de Chaudière,

[7] This is likely pentaerythritol tetranitrate (PETN), an explosive that was first created in 1894. PETN was patented by the German government in 1912 and was used by the German Army in World War I.

[8] These were Richter and Company's *Anchor Stone Building Sets*.

[9] This was likely Dr. Hugo Schweitzer, a renowned chemist and head of the pharmaceutical division of the Bayer Company in America. Schweitzer was heavily involved along with Heinrich Albert in the Great Phenol Plot to corner the American market on the chemical phenol that was used in the creation of explosives and thus deny the Allies American-made munitions. He was arrested and indicted, but died in 1917 before he could be brought to trial.

who he said was receiving money and messages from Germany and who later was the addressee of several letters intended for Madame Victorica.

About December 1916, according to Wessels,

> a man came to see me who certainly mistook me for someone else. He called me doctor. He was a dark man like an Italian or a Jew, with long hair, but he spoke German. He told me that he was working in a munitions factory, which he knew was not right because he was a good German. He said he wanted to help Germany and he was willing to blow something up. I did not trust him and sent him away.[10]

Wessels attempted through several intermediaries to establish contact with the leaders of the movement for the freedom of Ireland, but for some time was unsuccessful. Finally, Fricke introduced him to Emil Kipper,[11] a retired manufacturer who was able and willing to help him. Kipper asked Wessels about his business, but Wessels refused to tell him anything except that he wanted to reach some of the Irish leaders. Kipper then asked him if he had been told on the other side about Wünnenberg. When he admitted this, Kipper insisted upon calling Wünnenberg on the phone and inviting him to join them at the club. But Wessels and Wünnenberg were not friendly. Wessels had kept away from Wünnenberg, and Wünnenberg suspected that Wessels had been sent over to investigate him and his relations with his associate, Sander[12] ____, which had, in fact, been a part of the mission with which Wessels was charged. Madame Victorica says that on the occasion of her only meeting with Wünnenberg, he tried to impress her with his own importance and "boasted and said that Rodiger (Wessels) was a green horn. Rodiger came and stayed just a moment; he got away as soon as he saw Wünnenberg."

Early in January 1917, Wessels succeeded in his desire to meet the Irish leaders. By an arrangement made by Kipper, Jeremiah O'Leary and John Devoy[13] called for Wessels and took him to a private house, where they had a conference. Wessels said:

> I was sent by the German Government to find out if you can do anything against England through your people.

He then explained his plan and told him they wanted men to put bombs on ships. Devoy took no part in the conversation and O'Leary became angry and refused to have anything to do with the plan. "You want to get us into trouble," said he, "while you Germans sit behind."

[10] This could have been Theodore Wozniak, who was later implicated in the fire and explosion of the Kingsland Munitions Factory on January 11, 1917.

[11] Kipper was indicted along with Marie Victorica, Herman Wessels, Albert Paul Fricke, Jeremiah O'Leary, John Ryan, and J. Willard Robinson on treason charges on June 8, 1918 (New York Times 1918).

[12] This is Albert O. Sander.

[13] John Devoy (1842–1928) was an Irish patriot and rebel leader who lived in the United States for a number of years after his exile by the British from Ireland in 1871. He was a leader of the *Clan na Gael* organization, a founder of the *Friends of Irish Freedom*, a reporter for the New York Herald, a friend and associate of Sir Roger Casement, and the publisher and editor of the Gaelic American newspaper. There is no evidence of Devoy's collaboration with Wessels. (See Golway 1998.)

A little before this, Wessels had received a visit from Alexander Victor Kircheisen,[14] alias Charles Nelson, alias L. A. 7, who arrived in the United States on the *S.S. Columbia*. He met Wessels through Fricke and told him that he had been sent by the German Navy to find out why Wessels had not written. The chief to whom Wessels was to report, the head of the naval secret service, was a very mysterious person known as "Meister Wilhelm."[15] Apparently, he did not communicate directly with his agents, but only through intermediaries. Wessels replied that his reason for not writing was that his stock of secret ink had given out and he did not know how to get any more. He told Kircheisen of the difficulty of getting anyone to work with him, of his lack of success in establishing favorable connections, and he wrote a report in "a white fluid that looked like water" given him by Kircheisen and delivered it to Kircheisen to be taken back to Germany. How many times Kircheisen and Wessels met is unknown. Kircheisen sailed for Germany on January 10, 1917, carrying the report of Wessels and promising to return with help and explosives.

Madame Victorica made her first report on January 25, 4 days after she landed in New York. It was written in open code and secret ink and sent through the medium of her uncle, Count von Alvensleben, of Zurich, Switzerland.[16] She said later that in this letter she reported on her trip and on her meeting with Wünnenberg, but her memory was at fault for she had not yet met Wünnenberg. The date of her letter is fixed by that of one of the letters in secret ink acknowledging it.

It will be remembered that although Wessels had advised Madame Victorica to keep away from Wünnenberg, she insisted upon meeting him and arranged to do so through the medium of Kipper. The meeting, in fact, made her acquainted not only with Wünnenberg but also with Frey, through whom she became acquainted with the Irish leaders whom she was desirous of meeting.

The meeting with Wünnenberg and Frey occurred at Kipper's house. Kipper told Frey that Madame Victorica had come highly recommended by Cardinal Hartman[17] and Frey agreed to introduce her to leading Catholics in this country when her

---

[14] Alexander Victor Kircheisen (alias Charles Nelson) was born in Germany and became a naturalized American citizen in 1914. He worked as a seaman and quartermaster on merchant vessels traveling between the United States and the Netherlands. Kircheisen was on board the steamer *Annie Larson* when it was captured for running guns to Hindu revolutionaries in June 1915. These are the same Hindus that William Friedman testified against in San Francisco. Friedman had broken the book cipher that they were using. In June 1917, Kircheisen was arrested in Copenhagen, Denmark, as a German spy and deported to Sweden. He made his way back to the United States in 1918. He was also known as agent "K-17." (See Kahn 1967, pp. 371–373; Tunney and Hollister 1919, pp. 69–107.)

[15] This is probably either Fritz von Prieger or Karl Boy-Ed who were at different times during the war the head of *Nachrichtenabteilung N*—the German Office of Naval Intelligence.

[16] This may be Werner Ludwig Alvo Count of Alvensleben-Neugattersleben (1840–1929) or one of his sons. However, it is not known whether Count Alvensleben lived in Zurich during World War I.

[17] This is likely Felix von Hartmann, who was the Cardinal Archbishop of Cologne, Germany, from 1912 to 1919. Von Hartmann was an ardent German nationalist and a supporter of the war. How he might have been associated with Marie de Victorica is unknown. The identity of "Frey" is also unknown.

letters should arrive. She says she did not tell Frey of her connection with the German government, but imagined that he guessed her mission "in a way."

> We talked about a food embargo—of which he was strongly in favor—and about the possibility of a war with the United States, and Frey declared that in that event he was for America first.

Wünnenberg was at first very suspicious and cautious. He had been told that a dark lady was coming to help him, bringing messages and information.[18] He thought Madame Victorica was posing as this lady, and although it is said that gentlemen prefer blondes, Madame Victorica's style of beauty was in this instance not in her favor.

Two days later, the dark lady expected by Wünnenberg called at the Hotel Knickerbocker to see Madame Victorica. She was in considerable difficulty, for she had brought over messages in secret ink for Wünnenberg and Sander, and neither they nor she could get the proper reagents to develop them. Thinking that Madame Victorica might help her, she called and explained to her the cause of Wünnenberg's suspicions. Unfortunately, Madame Victorica did not know of any other method of developing the secret writing than the one that had already been tried. Wünnenberg later succeeded in developing the letters himself.

On February 2 and 3, Madame Victorica sent two reports to Germany, in one of which she told what she had learned about Wünnenberg. Receipt of these letters is acknowledged in letters to her dated March 22. It appears that the secret writing in her letters of this date could not be developed as well as in that in the letter of January 25.

On February 3 occurred an event that not only greatly increased the dramatic intensity of the situation but also was a portent of an even more serious event that was soon to occur. It was on this day that diplomatic relations were severed between Germany and the United States and Secretary Lansing handed Ambassador von Bernstorff his passports.

The next day, February 4, the spirits of Madame Victorica were, perhaps, somewhat cheered by the reception of the following wireless message from Germany to a firm in New York:

> From Germany
> To: Schulz Ruckgaber, New York, February 4, 1917
> Give Victorica following message from her lawyers: lower terms impossible will give further instruction earliest and leave nothing untried. Very poor market will quote however soonest our terms want meanwhile bond have already obtained license.
> Disconto.[19]

---

[18] This "dark lady" may have been Madame Despina Davidovitch Storch (1894–1918), a possible German spy of Turkish ancestry who had been in America for some time, likely since 1916. She was arrested on March 12, 1918, but died suddenly of pneumonia while in custody on March 30, 1918, before she could be deported.

[19] *Schulz & Rukgraber* was a New York bank and *Disconto-Gesellschaft* Bank of Mannheim was a large German commercial bank.

At the first glance, this message does not seem especially cheering. It seems to concern itself with some lawsuit and with some contemplated sale, but Madame Victorica was not interested in the words themselves. They meant nothing to her. She had no lawsuit, she was not trying to sell anything, and she did not care anything about either a bond or a license. What she was interested in was the secret message that this telegram conveyed. This was very important to her, for it told her, as she afterward explained, that a remittance of $35,000 was on the way but that she must be very careful with the money and not leave it in any bank, for a declaration of war was expected at any minute and there was danger that it might be confiscated.

The method of enclosing a secret message in an innocent looking business telegram or cablegram which was a favorite one with the Germans and with various modifications was in constant use between Germany and America between the outbreak of the war in 1914 and the entry of America into the war in April 1917. The method was frequently used by von Papen,[20] von Igel,[21] and other important German agents.

Madame Victorica had been instructed in the general methods before she left Germany and had been given detailed information which she memorized and which she afterward repeated in messages written to her in secret ink.

The method is very simple. The only things in the messages that have any meaning are the initial consonants of the words of the message. Each one of these consonants represents a number or figure, according to a set of values agreed upon beforehand. Words that begin with vowels have no significance for the secret message and are of value only because they make it easier to compose a telegram apparently concerned with some entirely innocent affair. In the particular message before us, the significant letters begin with the word "lower" and, taken in order of occurrence, are L, T, W, G, F, L, N, V, P, M, W, Q, H, S, T, W, M, B, H, and L. It had been agreed beforehand that d or t should represent Figure 1; n, z, or y, Figure 2; m or w, Figure 3; q or r, Figure 4; s or sh, Figure 5; b or p, Figure 6; f, ph, or v, Figure 7; h, c, ch, or j, Figure 8; g, k, or x, Figure 9; and l or c, Figure 0.

Translating the initial consonants into figures with the aid of this table of values, which she had memorized before leaving Germany, she obtained a series of numbers.

---

[20] Franz von Papen was the military attaché to the German ambassador in the United States from 1913 to 1916. He was expelled for alleged espionage activities in December 1915 and returned to Germany. In April 1916, he was indicted for conspiring to blow up the Welland Canal linking Lakes Erie and Ontario in Canada. Later in his career, he was the Chancellor of Germany in 1932 just before Adolph Hitler took power. Von Papen and von Igel used a 10,000 group one-part code to communicate with Berlin and superenciphered their messages in much the same way as described here for Madame Victorica.

[21] Wolf von Igel was von Papen's assistant and successor as military attaché to Count von Bernstorff, the German Ambassador to the United States. He took over von Papen and Boy-Ed's spy ring after they were expelled from the United States. However, von Igel himself was arrested on April 19, 1916, along with a trove of documents that proved the complicity of the German government in the plot to blow up the Welland Canal.

These she divided into groups of five, because she was going to use them with a commercial codebook of five-letter groups: 01397, 02763, 34851, and 33680.

Having obtained these groups, she knew that she must find out what they meant by looking them up in the ABC Codebook.[22] This was one of several commercial codes which business firms were allowed to use for telegraph and cable messages during the war. The primary purpose of commercial code is not secrecy but the reduction of the cost of cable messages. The code consists of arbitrary groups of letters or figures to which the codebook assigns special meanings. A group of such figures—in the ABC Code, there would be five figures in a group—may mean a whole sentence, and in this way, a very long message can be sent by the use of a very small number of code groups. When a person wishes to send a message in code, he first finds in the codebook the sentences that express what he wished to say and then writes down and sends the groups of figures that he finds in the book as representing these sentences. When his correspondent receives the message, he has only to look up each group of figures in his codebook and find next to it the sentence, phrase, or word that his correspondent intended.

This was the process that Madame Victorica had to go through after translating the initial consonants of her message into figures, but there was one additional thing that had been agreed upon between her and the German Secret Service. This was that the message should be read backward, that is, that the last group of figures should be taken first. She, therefore, took her last group, 33680, and upon looking it up in the ABC Code found that it meant "Remittance sent today." The next to the last group, 34851, when looked up on the codebook, gave the meaning "Safe as possible." The next group, 02763, she found to mean "You must arrange immediately or it is useless." And finally the group, 01397, meant "On account of political affairs."[23]

The whole message then read "Remittance sent today safe as possible. You must arrange immediately or it is useless on account of political affairs."[24]

Knowing what she did about the relations of the two countries, and having doubtless already discussed the possibilities of the future, she understood that the vague statements in the second sentence meant that war was likely to be declared at any moment and that in this event money deposited in any bank to the credit of a German citizen was likely to be seized and held.

Of course, at another time an entirely different arrangement would be made about the values of the letters, so that even a person who knew in a general way what

---

[22] *The ABC Universal Commercial Electric Telegraphic Code*, Fifth Edition (1901), by William Clauson-Thue was the most popular of the commercial codes of the day. It was a two-part code with approximately 103,000 entries, in a codebook of about 1400 pages. There was a smaller paperback pocket edition and an update in 1915 that allowed for five-letter artificial code words (the Fifth Edition used only pronounceable code words). It is likely that Mme. Victorica used the original 1901 Fifth Edition.

[23] The whole decryption process Victorica used is also told in Yardley (1931, pp. 184–186) and Mendelsohn (1937, p. 104).

[24] Verified as the correct decryption from the ABC Code book, Fifth Edition, by the editor.

the system was would be unable to find out what were the figures composing the code groups and consequently be unable to read the message.

Two days after Madame Victorica received this message, her first meeting with one of the Irish leaders occurred. Wessels called for her in the evening about 9 o'clock and took her to a restaurant at Thirty-First Street and Broadway called the Hofbrau.[25] There they met Kipper and Jeremiah O'Leary. O'Leary was introduced as a journalist.

> We talked of Irish affairs and of my plan for working with the clericals. Mr. O'Leary did not like this idea and tried to interest me in the proposed food embargo. We talked a long time and I learned quite a lot. Kipper and Wessels just listened.

About a week later, O'Leary invited her to lunch at the Hofbrau and introduced to her as Captain West a man who later turned out to be John T. Ryan of Buffalo, Chairman of the Executive Committee of the Friends of Irish Freedom in this country.[26] During the next 2 months, she had many meetings with these two men, sometimes in New York restaurants and sometimes in drives to suburban restaurants. She was attempting to enlist the whole body of Irish sympathizers in favor of Germany while, they, on their part, were apparently playing the game for Ireland only.

On February 20, she had a bad scare. She learned that Wünnenberg and Sander had been arrested and indicted for conspiracy to violate the neutrality laws. Although she and Miss Lesen, "the dark lady," in a luncheon conversation at the Waldorf, had agreed that Wünnenberg was "somewhat of a crook," she felt obliged to contribute $3,000 to the bail put up for Wünnenberg and Sander. She said she did so because she was afraid that one or both of them would turn state's evidence and implicate her.

She felt quite sure that they had mentioned her name and, consequently, that she was being watched. She, therefore, left the Waldorf and took an apartment at the Spencer Arms, paying rent for 4 months in advance. But here she suspected that she was being followed and observed, and within 2 weeks, she left this hotel for the Hotel Netherlands, where she registered under the name of Marie de Vussiere.

Shortly after she went to the Netherlands, on the recommendation of O'Leary, she engaged as maid and companion Miss Margaret Sullivan, who remained constantly with her until her arrest, and gradually became more and more of a sympathizer and trusted confidant, finally acting as the medium for the transfer of considerable sums of money.[27]

---

[25] The Hofbrauhaus Restaurant 31st and Broadway, NYC, August Janssen, Proprietor (actual address was 1214 Broadway). Over the years, the restaurant has moved and morphed into Janssen Graybar Hofbrau, Graybar Building, 420 Lexington Avenue (at 44th Street).

[26] John T. Ryan was an attorney from Buffalo, NY, who was heavily involved in the *Friends of Irish Freedom,* John Devoy's *Clan-na-Gael* organization, and the Irish revolutionary movement in the United States. He was indicted in June 1918 along with Madame Victorica, Jeremiah O'Leary, Herman Wessels, and others. Ryan fled the country and remained in exile for more than 4 years.

[27] Margaret Sullivan was recommended to Mme. Victorica by Jeremiah O'Leary and became her maid and later confidante early in 1917. Sullivan was arrested as a material witness at the same time as Mme. Victorica and was held in the Tombs jail in New York City for a year. She refused to testify against her employer (O'Leary and Kelly 1920, pp. 294–297, 461, 481).

O'Leary visited her often, discussing the Irish situation in detail and the activities of Gaffney[28] in Germany. O'Leary was clearly disturbed that the Irish had never received from Germany a definite reply to certain proposals contained in what was confidentially called "the third document." Apparently, the Irish wanted to know what the German government was still ready to do in the way of supplying arms and troops for a repetition of the plans of Sir Roger Casement.[29] As the Irish were much concerned about getting an answer to these proposals, and as Madame Victorica was extremely anxious to send a confidential messenger to report on the general situation and outlook, it was agreed that Ryan and O'Leary should procure a messenger and she should pay the expense of sending him to Europe. It was finally arranged that this messenger should carry her report, ask for an answer to "the third document," and also should present to the German government through Madame Victorica's chief a request that a U-boat should be sent to Newport, R. I., to take on board three Irishmen who were very anxious to go to Ireland. This messenger had considerable difficulty in getting on a ship, but finally took passage as second cook on the Standard Oil tanker *American*, which sailed from New York about March 17. This trip cost Madame Victorica more than $1,000, and as it turned out, the messenger did not return until June 30.[30]

Meanwhile, events were moving very rapidly toward the final break between Germany and the United States, an event which did not change the nature of the activities of Madame Victorica and Wessels, but greatly increased their danger and, for a time at any rate, seemed likely to cause an entire break with their Irish friends. Madame Victorica, herself, gave a vivid account of this critical moment:

> About 8 o'clock on April 5th Mr. O'Leary called in his car and we drove up Riverside Drive. We were talking about the coming war and the draft and volunteering. I maintained that there was so much pro-German feeling in this country that a draft would be necessary. Mr. O'Leary thought this was not true. Suddenly we heard the newsboys calling 'extras' and Mr. O'Leary arose and said, 'Well, Madame Victorica, now we are enemies.' At first I thought he was joking, but I found he was entirely serious. I cannot describe the situation, it was so incredible.

---

[28] This was Thomas St. John Gaffney, the American General Consul in Dresden for 8 years and then in Munich where he was serving at the outbreak of the war. Mr. Gaffney was a German sympathizer and an Irish nationalist. He was asked to resign his post in October 1915. Released from his duties to the American government, Gaffney then became an ardent and outspoken supporter of the Irish nationalist movement. He was a friend and associate of Sir Roger Casement. It was Gaffney who arranged for Casement to take a submarine from Germany to Ireland in April 1916. (See Gaffney 1930.)

[29] Manly is probably referring to the Easter Uprising in 1916, but Casement had no hand in planning the Easter Rebellion and was, in fact, against it, thinking that the Irish were not yet strong enough to fight the British. This was particularly true because of the sinking of the *Aud*, a German freighter carrying guns and munitions to the Irish just off the Irish coast in the days before the Easter Rebellion.

[30] The messenger was Willard J. Robinson (also reported as J. Willard Robinson), O'Leary's former private secretary and a member of the Irish-American revolutionary group. Robinson was arrested along with the others in April 1918 and charged with treason, but was acquitted at trial in 1919.

References

Gaffney, T. St. John. 1930. *Breaking the Silence: England, Ireland, Wilson, and the War*. New York: Horace Liveright. https://ia600505.us.archive.org/33/items/breakingsilencee00thom/breaking-silencee00thom.pdf.

Golway, Terry. 1998. *Irish Rebel: John Devoy and America's Fight for Ireland's Freedom*. New York: St. Martin's Press.

Kahn, David. 1967. *The Codebreakers; The Story of Secret Writing*. New York: Macmillan.

Landau, Captain Henry. 1937. *The Enemy Within: The Inside Story of German Sabotage in America*. New York: G. P. Putnam's Sons.

Literary Digest, The. 1919. "'Frank Richards' American Head of German Spies." *The Literary Digest*, March 1.

Macrakis, Kristie. 2014. *Prisoners, Lovers, & Spies: The Story of Invisible Ink from Herodotus to Al-Qaeda*. New Haven, CT: Yale University Press.

Mendelsohn, Charles. 1937. *Studies in German Diplomatic Codes Used During the World War*. Register No. 191. War Department, Washington, DC: Office of the Chief Signal Officer, Government Printing Office.

New York Times. 1918. "Seven Indicted as German Spies Face Execution." *New York Times*, June 8, sec. News. http://query.nytimes.com/mem/archive-free/pdf?res=9A01E1D8163EE433A257 5BC0A9609C946996D6CF.

O'Donnell, Patrick K. 2004. *Operatives, Spies, and Saboteurs: The Unknown Story of the Men and Women of World War II's OSS*. New York: Free Press.

O'Leary, Jeremiah A., and Michael A. Kelly. 1920. *My Political Trial and Experiences*. New York: Ulan Press.

Tuchman, Barbara W. 1958. *The Zimmermann Telegram*. New York: Macmillan Company.

Tunney, Thomas J., and Paul Merrick Hollister. 1919. *Throttled! The Detection of the German and Anarchist Bomb Plotters*. Boston, MA: Small, Maynard and Company. https://play.google.com/books/reader?id=bNcLAAAAYAAJ&printsec=frontcover&output=reader&authuser=0&hl=en&pg=GBS.PR8.

Weyl, Nathaniel. 1950. *Treason: The Story of Disloyalty and Betrayal in American History*. Washington, DC: Public Affairs Press.

Yardley, Herbert O. 1931. *The American Black Chamber*. Indianapolis, IN: Bobbs-Merrill.

Chapter 18
More German Spies

Abstract The two German spy networks in New York and Baltimore were not the only ones in the United States between 1914 and 1918. *Abteilung IIIB*, military intelligence, and *Nachrichtenabteilung N*, naval intelligence, both sent independent agents to the United States during this period and also recruited Americans to spy in Europe.

18.1 The Journalist

George Vaux Bacon was a Midwest-born American who ended up spying for the Germans. Born on April 30, 1888, in St. Paul, Minnesota, Bacon was of medium height, with brown hair and eyes, and myopic. He was a journalist and wrote primarily for magazines, notably *PhotoPlay*; he was quite interested in the film industry and also wrote for film tabloid magazines. He was recruited during the summer of 1916, at age 28, by two German agents, the German journalist Albert O. Sander and a naval reserve officer Karl Wünnenberg (the infamous "Dynamite Charley"). Sander had lived in the United States for several years, working for German language newspapers. Wünnenberg had been in the United States since about 1900 and claimed to be a naturalized citizen. At the outbreak of the war in 1914, both men were instructed to set up an organization to engage in espionage and sabotage in the United States. They were also instructed to recruit Americans—particularly journalists—to spy on England. They were not connected with von Bernstorff's organization out of the German Embassy, but reported directly to the *Kriegsnachrichtenstelle* (war intelligence agency) in Antwerp (Boghardt 2004, p. 136). Fraulein Doktor Elsbeth Schragmüller, the famous "beautiful blonde from Antwerp," headed the French section of *Kriegsnachrichtenstelle Antwerpen* and all training (Hieber 2005, p. 92). Wünnenberg underwent training in Antwerp in 1915, including the use of secret inks and explosives.

Wünnenberg and Sander set up a dummy company, the Central Powers War Film Exchange, to provide a cover for their work. The Film Exchange is what lured Bacon in. Wünnenberg met Bacon randomly at a bar in New York, and the two started talking about photography and films, and Wünnenberg mentioned the Film Exchange. Eventually, Wünnenberg led the conversation around to the excitement of espionage and how he needed someone to do some investigations in England and

J.F. Dooley, *Codes, Ciphers and Spies*, DOI 10.1007/978-3-319-29415-5_18

Ireland. When Wünnenberg offered Bacon $125 a week plus expenses, he was hooked. Sander found Bacon a job with the Central Press Association, a syndication company that supplied features, stories, and photographs to newspapers around the country. Bacon headed to England, arriving on September 16, 1916. His job was to observe military posts and naval bases and maneuvers and report back to Sander and Wünnenberg through a contact in the Netherlands, one Denis Meisner, a retired German naval lieutenant. Unfortunately for both Bacon and Meisner, the British Secret Service was suspicious of Meisner and was intercepting and reading all his mail as well as watching two safe houses he frequented in Amsterdam. So as soon as Bacon wrote to Meisner and then traveled to Amsterdam late in September, the British began to watch him as well. Because the British had failed to send on one of the intercepted letters from Bacon to Meisner, the two of them suspected they were being watched.

Bacon managed to elude his shadows and returned to England early in November. He was picked up again by the Secret Service when he made a side trip to Ireland later in the month to meet with a member of Sinn Fein. He was detained on December 9 and his belongings were searched. The British discovered the addresses of the safe houses, letters from Meisner, a bottle of Argyrol, and a pair of black wool socks impregnated with a sympathetic secret ink that was a chemical variant of Argyrol, called the "P" ink by the British and Americans. Argyrol, a silver salt sold as a powder, is an antibacterial and antiseptic and at the time was a common treatment for gonorrhea (Macrakis 2014, p. 157). With these discoveries, Bacon knew he had no defense left and made a complete confession on February 9, but claimed he never sent any information about British military or naval operations to the Germans and was just stringing them along to get a good spy story. Bacon was tried and convicted of espionage on February 26, 1917, and sentenced to death (Boghardt 2004, p. 138).

Dr. Stanley Collins related much of the story of George Vaux Bacon to the staff of MI-8 in the summer of 1918. Dr. Collins was a chemist and secret ink expert of the British Postal Censorship office, and it was Dr. Collins' testimony about his evaluation of the secret ink found in Bacon's belongings that earned Bacon the death penalty. Collins gave a series of lectures to the officers of MI-8 in July 1918 when he was in the United States training MI-8 personnel on secret ink techniques:

> When our authorities arrested Bacon, it was a bottle marked 'Argyrol,' found in his medicine-chest, which was responsible for his bad luck. Analysis of the contents revealed a small silver content, but Bacon protested. He said that he carried the Argyrol as a medicinal remedy and antiseptic. But when the 'P' ink was discovered in his socks, he confessed.
>
> As a matter of fact Bacon was entirely sincere in protesting the Argyrol. Having been given no information as to the chemical constitution of 'P' ink, he did not know its similarity to collargol or argyrol and to him the bottle so labeled was, in truth, nothing more than an antiseptic.
>
> I made an examination of all Bacon's possessions and found that the concentration was so low in the solution of the socks that it defied chemical analysis. I made a final test by spectroscopic analysis. The test revealed the presence of silver.

> George Vaux Bacon, who was condemned to death in February, 1917, told in his confession that he had never developed secret ink and that he did not know its composition. He stated that while in Sander's office in New York he saw some of the secret writing from Denmark developed. The letters were placed in a photographic dish and the colorless contents of two brown bottles were poured over them. In ten seconds, he said, the writing appeared, clear and very black. When the solutions were mixed heavy white fumes appeared. Bacon did not confess to the presence of 'P' ink in the dinner-jacket buttons. These were not discovered until after the trial. (Yardley 1931, pp. 67–69)

However, George Vaux Bacon got lucky. First, the British delayed his execution because he was an American citizen. They also sent Bacon's confession to the Bureau of Investigation in Washington. Meanwhile, Bacon was in contact with people back in the United States trying to get the government to grant him a pardon. He wrote the Home Secretary imploring him for a pardon as well. In his letter, Bacon says, "The adventure for which I was punished was a foolish and theatrical one, undertaken as I have always maintained, and this is the truth, not through malice, hatred or dislike for Great Britain or her allies, but simply for excitement. I did not even realize until it was too late, that I was holding myself up for execration by my own race." He also points out that he is related to many Bacons in England, including both Roger Bacon and Sir Francis Bacon (Simpson 2014). This didn't work.

With the United States close to declaring war on Germany in March 1917, the Department of Justice's Bureau of Investigation arrested Sander and Wünnenberg. Bacon then got his second stroke of luck; the Justice Department requested Bacon's extradition to the United States so that he could testify against his former employers. Bacon's sentence was commuted, and he was extradited to the United States where he did testify against Sander and Wünnenberg. His testimony was a key element in their conviction. Bacon himself was also tried and convicted and spent a year in the Atlanta federal penitentiary (Boghardt 2004, p. 138).

A footnote to the story of George Vaux Bacon occurred some fourteen years later. Herbert Yardley, formerly the head of MI-8 and the subsequent War-State Department Cipher Bureau, wrote an article on *Secret Inks* for the *Saturday Evening Post* that appeared on April 4, 1931. Several weeks later, Yardley received a letter in response to that article from none other than George Vaux Bacon:

> Dear Yardley:
> Your article on Secret Inks in *The Saturday Evening Post* for April fourth was very interesting, although there are one or two very slight inaccuracies in Collins' report regarding me.
>
> By the way, I still have the coat of that evening ensemble and my wife recently made a pair of knickerbockers for my eldest son from the trousers. If it is at all possible, I'd like to be informed whether or not my letters from England to Holland were ever brought out and read. In writing them, it had been my desire to make them "hot" enough to keep the Heinies thinking I was sincere, while at the same time not giving them any real information. What the British did not know was that I knew a great deal which I never sent. I am an American, but, after all, am of English descent and had no intention of allowing any act of mine to endanger anyone of my own blood, except possibly myself-and my adventurousness certainly almost cost me my life and has haunted me like a spectre for years.
>
> It was nothing but a fantastic stunt designed to produce an exclusive story on espionage, if I had gotten away with it. However, I had to swear to keep what I knew to myself, so the story was never written.

> After leaving Atlanta in January 1918, I tried to enlist in the Army at Chicago under the name of George Brown, but was rejected on account of severe myopic-astigmatism.
> At the time I got mixed up with Sir Basil Thomson[1] et al. in London, I was 28 and an impractical and rather careless young fellow. When the crash to my plans for a big scoop came, I was so overwhelmed by a feeling of disgrace that I was utterly unable to defend myself properly. I feel that only my mother's plea to Theodore Roosevelt and that grand old lion's insistence on clemency, together with the good heartedness of the British, made the continuance of life on this planet possible for me.
>
> I thought you might be interested in hearing from one of the leading characters in your story. It has created great interest here and I find myself a sort of local historical character, for the time being – a somewhat sinister historical character, however.
>
> Cordially,
> "(Signed)" GEORGE VAUX BACON (Yardley 1931, pp. 70–71)

After serving his time in the Atlanta federal penitentiary, George Vaux Bacon married (three times) and divorced (twice); had at least two children, a boy and a girl; and fashioned a career writing marketing sales brochures in Los Angeles. He registered for the draft during both World Wars I and II, but was rejected both times because of his severe myopia. The rest of his life was quiet and uneventful—at least from the point of view of being a spy. Bacon passed away in Los Angeles in November 1972, having outlived most of the other German spies.

18.2 The Turkish Beauty

While Madame Marie Victorica is the best known of the female German spies that operated in the United States, she was not the only one. Other women were even more adept than Victorica at espionage.

Madame Despina Davidovitch Storch was born in Constantinople in 1895. Raised in Turkey, her childhood and upbringing are murky. She is described as a beautiful brunette who spoke several languages including French, English, and German. It is sure that by the time she was 17 in 1912, she was in France where she married for the first time. Her espionage career apparently began around the time she turned 18. By the time she got to New York in 1916, she was said to be alluring, sophisticated, an excellent dancer, and a serial seducer of men who ensnared them in her charms and led them astray; in short, she was a femme fatale or vamp. Who exactly she was spying for is unclear. It's very likely she was spying for German military intelligence because of documentation found when she was finally caught. She may also have been spying for the Ottoman Empire. She was known, among other things, as "the Turkish Beauty" (Barton 1919, pp. 189–190). Few real facts are known about Despina Storch except for her last days as a spy in New York City. There are sensationalized news magazine articles and newspaper stories about her

[1] Sir Basil Home Thomson (1861–1939) was head of the Criminal Investigation Division (CID) of New Scotland Yard during World War I. Because the Secret Service Bureaus did not have any arrest powers, it was CIDs job to investigate and arrest suspected spies during the war.

in both the news sections and on the society pages. The most over the top being an eleven-part series run in the *American Weekly* section of the *Washington Times* (Washington Times 1918). She was said to have great wealth—at least she spent a lot of money—and ran in the best social and diplomatic circles. As with many things about Despina Storch, her marital status was somewhat ambiguous. One report has her marrying a Frenchman, Paul Storch, in 1912, at age 17. They divorced and he was later an officer in the French army during World War I. Another report has her marrying an English army officer, James Hasketh, in 1915, but separating from him soon afterward and traveling to Paris. Regardless of the truth of these accounts, Despina used the names Storch and Hasketh as aliases in numerous places during her travels.

The period from 1914 through 1916 saw Despina Storch travel through most of the war-torn countries of Europe. Receipts in a safe deposit box opened upon her arrest in New York show stays during this period at sumptuous hotels in Berlin, St. Petersburg, Paris, London, Madrid, Washington, and New York. She and a companion, the Baron de Beville, were briefly detained in Madrid in late 1915 at the request of the French government. However, no charges were ever brought in Spain, and she and the Baron were released after a short time. They embarked from Barcelona to Havana almost immediately after their release and after a brief stay in Cuba headed for the United States.

Mme. Storch's American stay began in 1916 at the Shoreham hotel in Washington, DC, where she stayed for several weeks, ingratiating herself with diplomats including the German Ambassador Johann von Bernstorff. Moving on to New York City, Mme. Storch lived for a year at the Waldorf-Astoria hotel, spending somewhere in the neighborhood of $1000 per month. Once again she was the hit of the social scene, befriending all the upper crust of New York, but also paying particular attention to diplomats and military men. After the United States entered the war in April 1917, Mme. Storch moved to the Biltmore Hotel and checked in as Mme. Nexie, one of her European aliases. At this time, she may also have been in touch with Mme. Marie Victorica. Her coterie now included not only Baron Henri de Beville but also two other acquaintances, Mrs. Elizabeth Charlotte Nix, a fortyish German widow who had arrived from Berlin in 1916, and Count Robert de Clairmont, a 38-year-old Frenchman who may or may not have really been a Count. The quartet were practically inseparable and went to all the right parties and made sure they continued to engage with all the military officers in town. These included all the newly arrived American officers waiting to be shipped out to France. Mme. Storch continued to have an ample supply of funds, but it was never determined where the money came from.

By early 1918, the Secret Service and the Bureau of Investigation of the US Justice Department were more than interested in what Mme. Storch and her friends were up to, but despite their nearly constant surveillance, they had no evidence at all of any espionage activities on the part of the group. Finally, the Secret Service decided that they needed an agent on the inside of Mme. Storch's group, so they enlisted a prominent and wealthy New Yorker, W. H. Vanderpoel as their mole. Vanderpoel already knew the Count de Beville, and he used that association to arrange an introduction to Mme. Storch. Within just a couple of weeks, the two were nearly inseparable and Mme. Storch was trying to seduce Vanderpoel to her side of the conflict. Finally, on about March 12, 1918, the Bureau of Investigation had enough evidence to move, fortunately just in time. Mme. Storch and the Baron de Beville had traveled to Key West, Florida, and were about to take a steamer to Havana "for a vacation" when federal agents swooped in and arrested them before they could board the ship. They were brought back to New York and put under house arrest at their respective hotels. The next day, Mrs. Nix and the Count de Clairmont were also arrested (New York Times 1918a). It turned out, however, that at that time the United States did not have a law that would let the government try and intern a female enemy alien for espionage. That law wouldn't be passed for another 3 weeks. So, on March 19, the Justice Department's attorney asked that the four alleged spies be deported as "undesirable aliens," and President Wilson wasted no time in signing the deportation order. The group, except for de Clairmont who was ill, was moved to a prison on Ellis Island and scheduled for interrogation and deportation. During her interrogation, Mme. Storch steadfastly denied all connections to the German government and military intelligence. Her veneer of confidence cracked a bit when a safe deposit box she had at a New York bank was opened and dozens of cablegrams to Germany, letters to various correspondents including Mrs. Nix and Count de Clairmont, and a codebook were found. Still she continued to deny any involvement in espionage. Her interrogation ended abruptly on March 30, 1918, when she died of a sudden onset of pneumonia. She was just twenty three. Rumors were that she had committed suicide, but the Justice Department and the Secret Service both vehemently denied this.

Mme. Storch was buried at Mt. Olivet Cemetery in Queens, NY, on April 1, 1918, her burial paid for by the Baron de Beville. The next day, *The Sun* newspaper of New York had an interesting piece on the funeral:

> An exquisitely carved white coffin containing the body of Madame Despina Davidovitch Storch, the most romantic spy suspect America has yet known, was placed in a vault on the east slope of Mount Olivet Cemetery, Maspeth, Queens, yesterday afternoon. Thus was drawn the curtain on a life which in twenty-three years knew more diplomatic intrigue than even the popular fiction spy heroine is given by Oppenheim and others. The burial was simpler than those of people who never reached the prominence of the Beautiful Turk. Only one limousine rolled up to the vault after the hearse. It contained the grief-stricken Baron, his parents and a secret service man, who accompanied the French nobleman from Ellis Island. The five knelt on the soft earth about the grave, and James F. Fallen, the undertaker, said a short prayer. The Baron, whose infatuation for the Turkish spy suspect entangled him in the web of her intrigues, wept silently and cast a last look upon the vault as he was led back to the car....
>
> A morbidly inquisitive crowd circled the doorway of the funeral church an hour before the scheduled time for the services. They lined the sidewalk six deep in front of the Hotel Touraine, opposite the Fallon place. They climbed on trucks and pushed around the hearse; many lined the windows of the lofty buildings across the street. None was allowed to enter the funeral parlors, which were guarded by a secret service man. A little after two o'clock the white casket, carried by two undertakers, came out of the building. The chatter of the crowd hushed, and all that stirred the quiet was the music of 'The Girl I Left Behind Me' which echoed into the street, ... (Barton 1919, pp. 196–197)

Aside from the contents of her safe deposit box, which are now missing, the government really had no evidence at all that Mme. Despina Storch was a German agent. She did spend a lot of money and had no visible means of support, but she continued to travel for the better part of two years with the Baron de Beville who may have supported her surreptitiously. The only hint we have is from a newspaper report in the *New York Times* on March 20, 1918, which says, in part, "Mme. Storch confesses she is Mme. Hasketh when confronted with bogus official papers which reveal that she gathered munitions data which she sent to Berlin possibly through a neutral embassy" (New York Times 1918b). Even so, no representative of the United States Government has ever confirmed this account.

18.3 Fraülein Doktor

Elisabeth Franziska Catherina Anna Schragmüller was born on August 7, 1887, in the Westphalian town of Schlusselburg, the first child of Carl Anton Schragmüller and Valesca Cramer von Clausbruch. Most of what we know about her is from an 18-page autobiographical sketch published in 1929 in the popular anthology *Was wir vom Weltkrieg nicht wissen* (What we did not know about the World War).

Her father was a disabled army veteran who became a senior civil servant. Elisabeth, who as an adult began calling herself Elsbeth, was a very bright child and always anxious to learn new things. She was in the first generation of German women able to attend university, and she took advantage of that opportunity. After attending a private women's preparatory school run by her grandmother, she received her school-leaving certificate in 1908 and enrolled in the University of Freiburg. At Freiburg, she studied political economy, earning her doctorate in 1913, just before the outbreak of World War I. After her graduation, Schragmüller briefly held two different jobs. One was as a teacher for the *Lette-Verein*, a not-for-profit organization dedicated to business education for women. Her second job was as a social worker for the *Zentralstelle für Volkswohlfahrt* (Central Office for People's Welfare), where she helped primarily working class people in need. (Hieber 2005, p.98)

Once the war broke out in August 1914, Elsbeth wanted to make a major contribution to the war effort. Against her parent's wishes, she arranged passage to Brussels, Belgium, in late August, just after the occupation. This was no easy task. She then talked her way into the office of the Governor-General, Field Marshal Colmar Baron von der Goltz and convinced him to give her a job. After a brief stint in the censorship office, she was assigned to *Kriegsnachrichtenstelle Brussels* (military intelligence agency Brussels) where she began organizing the human intelligence efforts for France. According to Hieber,

> Her new colleagues were educated (reserve-) officers from different professional backgrounds, and she found herself treated as equal and appropriate to her social background. She got to know the director of *Abteilung IIIb*, Major Nicolai, whom she had to convince about her being the right person for such a dangerous assignment. With Captain Kefer (the

> head of *Kriegsnachrichtenstelle Brussels*), she moved to Lille where she received her training. In the meantime the *Kriegsnachrichtenstelle* moved from Brussels to Antwerp. Having concluded her training in early 1915, Schragmüller was given responsibility for the post's section that organized military intelligence against France, a position she held until the end of the war. In 1936 she described her area of responsibility as: the organization of systematic intelligence on the large Western theater, finally reaching as far as America; the recruitment of contacts [i.e. agents], their instruction, the securing of their lines of communication, personal debriefing, verification of their statements, and the production of reports for the general headquarters. (Hieber 2005, p. 100)

Schragmüller proved to be very competent at selecting, training, and handling field agents. She recruited a number of French soldiers as so-called deserter agents, usually by resorting to blackmail. Her agents all performed well, and her reports were very well received by the higher-ups at Abteilung IIIB, particularly Major Nicolai, who said of her "It is significant that in the German intelligence service a cavalry officer from an old noble family and an extraordinary well-educated woman knew best about handling agents, even the most difficult and sly ones" (Hieber 2005, p. 101).

Schragmüller believed that the tales of her personality and allure that circulated through France were largely from the stories of captured German field agents, although she was, in fact, tall, slender, and blonde. She gave herself the nickname "Mademoiselle Docteur," however, because that was how she introduced herself to prospective agents. Her competence even extended to training the most famous female German undercover agent of all, Mata Hari.

> "In the second half of March 1916 (around 20 March)," Schragmüller remembered, "a meeting between the director of IIIb (Nicolai) and Mata Hari took place in Cologne. He ordered Captain Roepell and Doctor Schragmüller to carry out an additional screening with regard to her suitability for intelligence service, especially concerning her abilities [...] to establish personal relationships and to move in society circles." During the briefing, Schragmüller lived in the same hotel as Mata Hari for about two weeks, and she reported on her: "The behavior of Mata Hari at the hotel, at lunch and at the theater is that of a "Grande Dame", although her eccentric personality and her exquisite elegance attracts general attention. It is an effort for her to obey to the restrictions of her freedom of movement which have to be imposed for her new role. Yet, she is unable to stick to prescribed regulations. (Hieber 2005, pp. 104–105)

In his tell-all book about his days in MI-8 during the war and as head of the joint War-State Department Cipher Bureau during the 1920s, *The American Black Chamber*, Herbert O. Yardley manages to conflate the stories of Elsbeth Schragmüller and Madame Marie Victorica. In the chapter that he devotes to Madame Victorica, we have "Her bankers had described her as a "stunning blonde, about thirty-five years of age." The British cabled: Believe Victorica to be beautiful blonde woman of Antwerp for whom we have searched since 1914" (Yardley 1931, p. 102). Unfortunately for Yardley, Marie Victorica wasn't a blonde, she wasn't stunning, she was closer to forty than thirty five, and as far as can be determined she'd never been to Antwerp. Otherwise he gets it right.

Herbert Yardley used the idea of the Blonde from Antwerp in his fiction as well. He also continued to ignore that he'd gotten all the details wrong. In his 1934 novel

The Blonde Countess, Yardley describes a German secret agent who is playing a cat and mouse game with Nathaniel Greenleaf, his protagonist:

> Her hair, thought Joel, casting another quick glance, is genuine blonde. But she does something to it to keep it nice and bright. Scandinavian of course. Blonde hair and blue eyes and a lovely skin, white as parchment. Probably Mr. Greenleaf thinks she is beautiful. She is beautiful. I must not be catty. But her eyes are really too close together and he should not be deceived when she lowers the lids and looks at him underneath him in that provocative way. He ought to know that is a trick. I will tell him, so he can be on his guard. She is out to wheedle him, trying to make him think he is fascinated her. Men are such chumps when a beautiful woman looks at them that way. I could do it myself only I would not. Never. Not at him. (Yardley 1934a)

Yardley is more up front and direct in trying to create the German enchantress in his short story *H-27, The Blonde Woman from Antwerp*. Greenleaf receives a cablegram from the British warning him of the arrival in Washington of a new German agent, "H-27, traveling on Swedish passport as Amelia Alverson, should arrive about February 22, with plan to assist submarine campaign as result of new discovery. Description: blonde, 29, 5 ft 4. Identifications: left second molar extracted; gunshot wound over right hop; brown birthmark, inch in diameter, below left shoulder blade. Password, Indigo. She brings new code and secret ink" (Yardley 1934b, pp. 22–30). Aside from the fact that she never was a field agent, Elsbeth Schragmüller might have been flattered.

After the war, Elsbeth Schragmüller moved back into her parents home to care for them. She also began a teaching career at the University of Freiburg, the first woman to be a full-time faculty member there. In the late 1920s, the family moved to Munich, and Elsbeth switched from academia to a new career as a public speaker and writer. She published her autobiographical sketch in 1929 and went on to give regular talks about her war experiences. Despite pressure from the government during the 1930s, she refused to become a Nazi. She died on February 24, 1940, most likely from tuberculosis. Despite his best efforts, Walter Nicolai was never able to have Schragmüller granted an Iron Cross (Hieber 2005, p. 108).

References

Barton, George. 1919. *Celebrated Spies and Famous Mysteries of the Great War*. Boston, MA: The Page Company. http://books.google.com/books?id=D8QiAAAAMAAJ&printsec=frontcover&source=gbs_ge_summary_r&cad=0#v=onepage&q&f=false.

Boghardt, Thomas. 2004. *Spies of the Kaiser: German Covert Operations in Great Britain during the First World War Era*. St. Antony's Series. Houndmills, Basingstoke, Hampshire, UK: Palgrave Macmillan.

Hieber, Hanne. 2005. "'Mademoiselle Docteur': The Life and Service of Imperial Germany's Only Female Intelligence Officer." *Journal of Intelligence History* 5 (2): 91–108. doi:10.1080/16161262.2005.10555119.

Macrakis, Kristie. 2014. *Prisoners, Lovers, & Spies: The Story of Invisible Ink from Herodotus to Al-Qaeda*. New Haven, CT: Yale University Press.

New York Times. 1918a. "Spy Net Yields 2 Women Here; Men Also Taken." *New York Times*, March 19, sec. News. http://query.nytimes.com/search/sitesearch/.
New York Times. 1918b. "President Orders Spies Deported." *New York Times*, March 20, sec. News.
Simpson, Rebecca. 2014. "Secrets and Spies of the First World War." UK National Archives. *National Archives Blog*. April 11. http://blog.nationalarchives.gov.uk/blog/secrets-spies-first-world-war/.
Washington Times. 1918. "Mme. Storch – Vampire and German Spy." *Washington Times*, July, sec. The American Weekly Magazine. Library of Congress. http://chroniclingamerica.loc.gov/lccn/sn84026749/1918-06-02/ed-1/seq-32/.
Yardley, Herbert O. 1931. *The American Black Chamber*. Indianapolis: Bobbs-Merrill.
Yardley, Herbert O. 1934a. *The Blonde Countess*. New York: A. L. Burt Company.
Yardley, Herbert O. 1934b. "H-27, The Blonde Woman from Antwerp." *Liberty Magazine*, April 21.

Chapter 19
Madame Victorica and Invisible Inks

John Matthews Manly

Abstract The third of the Madame Victorica articles, number VI, includes discussions of invisible inks and their solutions. Also, this article finally gets down to some of Madame Victorica's plans while in the United States. It contains texts of several cover letters and the translated versions of several solved secret ink messages relating to her plans to acquire explosives and blow up Allied ships in harbor.

When Madame Victorica heard the announcement that the United States had entered the war, and the declaration of Jeremiah O'Leary that henceforth they were enemies, she may well have been discouraged. She had apparently not succeeded in establishing the sort of connections she desired with representatives of the church; she appeared about to lose the connection she had made with the *Friends of Irish Freedom*; and for some time her associate, Wessels or Rodiger, had been discouraged and spiritless. She says of him that he was "stale" and was merely "looking out the window."

This situation, however, was only temporary. O'Leary and Ryan were still expecting a reply to the questions and requests which had been sent to Germany through Madame Victorica's messenger; and however clearly they may have realized the danger of further association with her, they could not sever their relations. Messages and orders were reaching Madame Victorica and Wessels from the other side, and they were expecting the early coming of a messenger—L.A.7—who would bring new supplies of materials and of money. The entrance of the United States into the war indeed made it even more imperative than it had been that they should carry out the plans with which they had come to New York or which had been communicated to them in secret writings since their arrival. What these plans were has been stated in the preceding article, and abundant details in regard to them will appear in the secret messages of certain letters that were intercepted by officers of the United States Government. Some of these letters were dated immediately before the entrance of the United States into the war, but the purposes and plans remained unchanged and the desire to carry them out was only intensified.

Before giving copies of these secret writings, it may be well to explain the general methods of secret correspondence between Germany and its agents in the United States. The Germans well understood the difficulties and attempted to overcome them in very ingenious ways.

J.F. Dooley, *Codes, Ciphers and Spies*, DOI 10.1007/978-3-319-29415-5_19

In the first place, throughout the whole course of the war practically all mails from points in Northern Europe to the United States had to pass at some time under the eyes of the British censorship, and letters which for any reason excited the suspicion of the examiner were held for further examination. For various other reasons letters often failed to reach their destination. As the business of these agents was very important, both these difficulties were met by sending several copies of each secret message. To insure secrecy, the message was written in invisible ink on the blank pages of a letter or crosswise of the lines of the visible writing. Sometimes three, four, or half a dozen letters in which the copies of the secret message were incorporated were identical in form. Sometimes they were entirely different. Sometimes they were sent to the same addressee and signed with the same name. Sometimes they were sent to different addresses or even different addressees and signed with different signatures. They might, of course, in any event, be posted from widely separated points. For example, all four copies of secret letter no. 3, itself dated February 20, 1917, were addressed to Madame Victorica at the same New York address; bore the same date, February 25; and were identical in form and signature. The secret letter no. 7 of March 15th, intended for Wessels, was written on the blank pages of four separate letters, three dated March 22 and addressed to Haro Schroejers, Carl Roediger, Esquire, and C. Stamm; and a fourth of the same date addressed to Madame Victorica and entirely different in its visible contents. A single copy of secret letter no. 2, dated February 3, 1917, was developed in a letter dated March 7. An earlier copy was developed in a letter dated February 13, and that letter indicated that still an earlier copy probably dated February 3 had also been sent.

Some of these letters, all of which were, of course, entirely innocent in appearance, were posted in various neutral countries, but inasmuch as such letters stood a chance of being examined by the British postal censorship, and after [it entered] the war by the US postal censorship, it was regarded as safer either to deliver them by hand or to post them in America, as we had no censorship of domestic mail. The agents employed for this work were usually stewards or sailors on boats plying between European ports and New York who received considerable sums for posting the letters at Hoboken or in New York. In some cases they were furnished considerably greater sums if they brought replies.

Besides the difficulties and dangers connected with the transmission of the letters, the German agents in this country had to contend with certain disadvantages connected with the use of the secret inks. Occasionally, owing to the passage of time before a letter was received, or to some misunderstanding about the proper reagent for developing it, secret writings which had been safely received could not be read. Sometimes the German Secret Service learned or inferred that the Allied experts were on the watch for the use of certain inks, and it was therefore necessary that their use be discontinued. As we shall see later, the important series of documents brought back from Europe to Wessels by Madame Victorica's messenger[1] could not be read, because the messenger bringing the developer never arrived, having been arrested before he reached this country.

[1] J. Willard Robinson.

The earliest of the letters signed by H. Fels, the name assumed by Stein, the chief of the Secret Service of the Foreign Office, was written only three days after Madame Victorica sailed from Germany. It is dated January 8 and was obviously intended to serve as a memorandum of instructions given to her before she left:

> My dear friend,
>
> How are you? Did you have a good passage and how did you find our mutual friends there? No doubt, they are all glad to see you. What about the pills 'Kalium Iodatum' from Burrough, Wellcome & Co. London; can you recommend them? –
>
> Did you call upon old Fricke 74 Washington Street and have you seen Binder from Behrens, 95 Broad Street?
>
> How did you find my aunt Chaudiere 320 W. 111 Street, is she well?
>
> I hope you will write as soon as possible to my cousin Ensler, Schiedamsche Singel 76. Have you written to my friend Jacobsen, Gammel Kongevey 102. Hoping you will receive this letter in best health, I am with kindest regards,
>
> Yours truly,
> H. A. Fels.

The clear text of this letter, like that of nearly all the others, contains what is known as open code—that is, it is expressed in language which on the surface has an entirely innocent meaning but which has also a secondary meaning that can really be understood only by one familiar with the subjects referred to in the obscure expressions. This can easily be illustrated by the present letter.

In it the question, "Did you have a good passage?" really means, "Did you have trouble getting through the British blockage at Kirkwall?"

"How did you find our mutual friends there?" means, "How are our other agents in New York succeeding in their plans?"

The next question means, "Do not forget that the secret writing in this letter is to be developed by applying to it a solution made from the pills mentioned." "Can you recommend them?" means, "Report whether you had any difficulty in using them." These pills contained iodate of potassium, the reagent used for developing the invisible ink known as F ink, which had been agreed upon for messages at this time to Madame Victorica and some of the other agents in the United States.

The inquiries concerning persons in New York are made, not because Fels wished to hear about them but for the purpose of giving their addresses to Madame Victorica. These were the persons to whom letters meant for Madame Victorica might be sent from time to time. Such addresses are known as cover addresses; and it is, of course, very necessary that a secret agent should have several so that if one of them should fall under suspicion letters can be received through one or more of the others. These addresses had undoubtedly been given to Madame Victorica before she left Berlin, but her superiors were afraid that she might have forgotten them. As a matter of fact, she had forgotten the address of Mrs. Chaudière and her wish to know it was one of the principal reasons why she attempted to see the famous chemist, as related in the preceding article, and why she was so much disturbed by his refusal to see her. Of course, the chemist may have known nothing of her reasons for wishing to see him, and Mrs. Chaudière may have been entirely ignorant of the real contents of the letters which came in her care or even addressed to herself but obviously intended for another. Cover addressees would naturally suspect that some sort of secret correspondence was being carried on with their aid, but they need not know the exact nature of it.

It is difficult to express the astonishment with which we read in the clear text of this letter the question about the pills of Kalium Iodatum. It was well known to the chemists of the secret ink laboratories of England, France, and the United States that this chemical was in common use as a developer for a certain sort of secret ink. That the Germans could have hoped to disguise the purpose of their reference by talking about pills is a curious indication either of the ostrich-like quality of their own thinking or of the low opinion they had of the knowledge possessed by the Allies on the subjects of secret inks and their developers. The fact is that the only chance such a clumsy sentence had of failing to arouse suspicion would have been that the letter containing it should have entirely escaped examination and even if we had been entirely ignorant of the properties and uses of Kalium Iodatum, the recurrence of a mention of it in so many letters with the same signatures would inevitably suggest either that it was a matter to be investigated or that the writer was demented.

As a matter of fact, Captain Carver and Lieutenant McGrail,[2] the chemists of the MI-8 laboratory, were thoroughly equipped with instruments and knowledge concerning the very latest discoveries in the field of invisible writing. Soon after the Code and Cipher Section of military intelligence was organized, an incident occurred related elsewhere in these articles which led us to urge the establishment of a secret ink laboratory, and General Van Deman (then Major) had secured the services of two brilliant young chemists and had commissioned them and sent them to England and France to familiarize themselves with the new forms of invisible ink and the latest methods of developing them.[3] These officers had been cordially welcomed by our Allies, had been furnished with all the information in their possession, had been given every opportunity for study and research, and upon their return to America had promptly secured the necessary equipment and begun their work.

Ultimately it had been found necessary to establish two laboratories. One was at Washington where the central office of MI-8 was situated; for as has already been said, it was to this center that all mysterious documents of every kind obtained by any of the protective agencies of the United States were sent. In fact a number of documents came to it also from the military intelligence officers in Canada, as the Allies were working in close cooperation. In addition to this laboratory, it was deemed wise to establish another in New York in order to handle without delay documents seized from international mails by the postal censorship or obtained by any of the other agencies from the strangely mixed population of this great city,

[2] Captain (Dr.) Emmett K. Carver. – ran the MI-8 secret ink laboratory in the postal censorship office in New York City. Lieutenant (Dr.) A.J. McGrail ran the lab in Washington. Carver traveled to England in August 1918 with Dr. Stanley Collins (the British chemist and secret ink expert) to learn more about secret ink recovery techniques. Collins had been in the United States since May 1918, instructing the Americans on secret inks and recovery techniques (Macrakis 2014, pp. 145–147; Kahn 2004, pp. 32–33).

[3] Carver and McGrail.

consisting as it did of representatives of almost every nation in the world with the sympathies and connections which it was not always safe to predict.

These two laboratories worked in the closest cooperation and really formed a single unit. Both were fitted up with all the apparatus and reagents necessary for their work and, in addition to the officers in charge, were supplied with highly trained and enthusiastic personnel.

As might have been inferred from the question about the pills, the letter just discussed did, in fact, contain a message written in invisible ink, and when treated in the laboratory with Kalium Iodatum, the writing became clear and legible.

The most important part of the secret writing was an explanation of the method for reading the secret messages contained in telegrams. How this was done with the aid of the ABC Code Book and the table of numbers and letters was explained in the preceding article in connection with the wireless telegram received by Madame Victorica on February 4 about the remittance of $35,000. The rest of the message is taken up with cover addresses, including those already given and with another reminder of Kalium Iodatum pills for developing the secret writing.

The next letter sent to Madame Victorica at the same address and signed by Fels is dated Nxia (Christiania) February 13, 1917. The clear text of it reads as follows:

> My dear friend,
>
> I hope that you received in the meantime my last letters from 3/2/17. inst and also my remittance by wire, although I was informed today by the bank that you asked for further money, but I hope that everything is settled now. I cabled also on the 4th inst to your bank asking to give you the message from your lawyers and no doubt the message must have reached you. How are you going on? How is the market there? Did you meet Hudson and how is he, his last letter is dated 28/12 and since then no message from him. Now, good bye, with the kindest regards I am
>
> Yours truly,
> Fels.

This letter contains interesting references to the letter just read and also to a letter dated February 3 which had not yet been received but which clearly contained the same secret message as a later letter of March 7. It also confirms the remittance of $35,000 on February 3 and the reading of the secret meaning of the message contained in it. Hudson is a cover name for Wessels. It was not permitted to refer to agents either by their real names or by the aliases under which they were operating.

The messages contained in these letters, both those in the clear text and those in the invisible writing, were in themselves innocent enough. They show clearly, to be sure, that the writer and the addressee are engaged in some sort of business calling for great secrecy, but they give no indication of the nature of that business, strongly as they suggest the importance of it, at least to the persons concerned. These letters were, however, only preliminary and served mainly as reminders of arrangements that had already been made. The real purposes and the dangerous plans which brought Madame Victorica and her associates to the United States are, however, clearly disclosed in the letters which follow.

The first of these, a letter from Fels to Madame Victorica, dated March 3, 1917 is not very interesting in its clear text, as it largely repeats the questions and remarks of the preceding ones. It contains really only one new item, namely, that he has "heard from her uncle," which means that he has received the letter written by Madame Victorica to her uncle, Count von Alvensleben of Zurich.[4]

The message in invisible ink is, however, extremely interesting and important. It is headed "Letter No. 2" and dated February 3, 1917, being obviously a duplicate of other copies sent earlier but lost. After explaining that letter no. 1 did not contain the letters of introduction as had been intended, and confirming the telegram instructing her to put her money in a safe place on account of political conditions, it repeats the telegraph code and several of the cover addresses, but it further contains instructions concerning two plans; one of which had apparently been discussed before she left Germany but the other of which was perhaps entirely new to her.

The earlier plan is expressed in words accurately translated as follows:

> Order as soon as possible through a trustworthy clergyman by wire from (firm and address given) by the following telegram – 'Altar containing three holy figures, four columns about two meters in height … to match six meters wide and three meter high. Style Renaissance … painting, also receivers' addresses.

In other secret messages there are references to this plan, which was later thoroughly explained by Madame Victorica herself.

The plan was to import into this country, enclosed in the hollow spaces of these specially prepared columns and the altar, materials for manufacturing explosive bombs to be used in blowing up ships—"little square, black things containing theta," Madame Victorica called them. Some of the ingredients could not be produced in this country as they were recent German inventions.

To understand this plan thoroughly, it is necessary to bear in mind the conditions of international commerce at this time. Not only was direct trade with Germany interrupted, but trade with neutral countries was very closely supervised. Not only must the articles ordered be of an apparently innocent character, but they must be sent in response to a genuine order by a firm not on the black list.

A church altar with holy figures was not very likely to be suspected of containing ingredients for explosives, but in order that they should come through safely, it was necessary that they should be shipped from a well-known firm in a neutral country and the shipping firm should be able to exhibit an order by letter of telegram from some person who would normally and naturally have a use for such objects—in this case a clergyman—whose address might well include the name of his church. The German Secret Service gave explicit instructions in regard to the firm to which the order should be addressed but left it to Madame Victorica to find the trustworthy clergyman. How far he was to be in the secret does not appear. Possibly it was thought that Madame Victorica could find some clergyman who would approve of her real purpose; possibly it was thought that she could deceive him with some story

[4]This may be Werner Ludwig Alvo Count of Alvensleben-Neugattersleben (1840–1929) or one of his sons. However, it is not known whether Count Alvensleben lived in Zurich during World War I.

of wanting the figures for use in a private chapel. In any event this was one of the most important plans that Madame Victorica had when she came to this country and thoroughly explains her efforts to establish connections with members of the clergy.

This plan obviously superseded the plans for procuring explosives for use on ships upon which Wessels had been working. It was learned that there were two of these. In the first place, he had been trying to obtain the chemicals needed for the manufacture of explosives from the famous chemist previously mentioned, but both he and Madame Victorica had failed to enlist his aid. Wessels declared, "He practically kicked me out of his office."

Wessels had then proposed that the ingredients should be imported in children's blocks of artificial marble which were to be addressed to a German-American firm in New York engaged in importing children's toys.[5] But the New York manager of this firm had shown no sympathy with the plan and had done all that he could to point out its difficulties and dangers and to discourage it. Nothing more is heard of either of the plans after Madame Victorica's plan appears in the secret message.

The very terms in which Madame Victorica's plan is mentioned in this and later secret messages prove clearly that it had been thoroughly discussed with her before she left Berlin and that the only things that remained to be settled were details which were dependent upon the arrangements that she could make on this side of the water. The receiver in New York and the date of shipment had to be arranged for by her, and the date would in turn necessarily determine the choice of the shipper in Europe. The principal message takes it for granted that she has had some success in making arrangements and urges immediate action.

Of equal interest and significance is the other plan discussed in this secret message. It is expressed in terms that may be thus translated:

> Advise immediately where U-boat or sail-boat can sink sacks of (an illegible word) material on American coast—perhaps between New York and Cape Hatteras. Position must be free of currents, so that sacks will not be lost. Water depth not more than twenty meters. Can a messenger be landed there? The marking with buoys of positions of sunken sacks has been successfully carried out in Spain. Wire agreement and where material is to be sunk. Confide in as few persons a possible. Indicate in writing at once how plan is to be carried out.

These questions with reference to locating a submarine base on the Atlantic Coast are of special interest. Everyone will recall the great excitement on the subject in the United States at this time and the contradictory opinions as to whether attempts of this sort were really being made or not. For many weeks scarcely a day passed without a call upon MI-8 to solve the mystery of flashing lights or strange sounds supposed to be signaling to German submarines. Hysterical persons all along the coast from Maine to Carolina reported having seen lights flashed by ones, twos, and threes from hilltops near the coast. Radio enthusiasts heard strange and unintelligible signals from their buzzers. Scarcely one of these observers had made an accurate record of what they had seen or heard, but all thought that experts who could read code and cipher ought to be able to tell immediately what these signals meant.

[5] This was Richter and Company's *Anker Stone Blocks* Kits.

The special significance of the presence of this inquiry in this letter arises from the fact that Madame Victorica claimed to be acting under the auspices not only of the Foreign Office but also of the Navy. The secret letter, it will be noted, was dated more than two months before the declaration of war. Interests of the same general nature are discussed in April 1918, a year after the United States entered the war.

The reference of the next message is brief but clear. The whole message, dated February 20, reads:

> A messenger is coming there at the end of March, identified by Agatit Oscar Nt. 89, and brings our plans regarding transfer of our material. You can put full confidence in him. Put him into touch with W (i.e. Wessels).
> Fels.

In the interval of seventeen days between this and the preceding message, the naval secret service had apparently worked out a tentative plan for the U-boat bases and was sending it for consideration. The reference to Oscar taken in connection with later evidence shows that the messenger referred to was the man known as Eberhardt,[6] who did not arrive until toward the end of May.

This secret message—No. 3—was obviously regarded as very important for it was contained in invisible ink in for identical letters dated February 25 and addressed to Madame Victorica. Two of these were delivered by hand at an unknown date; one was mailed in New York City, May 17; the other at Hoboken on June 4. The clear text of all three letters reads as follows:

> My dear friend,
> Since I wrote you last, I have nothing received from you. Did you receive my remittance in the meantime?
> I got a cable message from our mutual friend Hudson, that his business is going well and that he requires money. I hope very much that you could arrange everything with him. As it is a long time since I had any news from you, since you arrived in New York, I send this letter with several steamers, hoping that one will reach you.
> - My Director was very pleased to have the news that Hudson's business is going on alright. Kindest regards from your friend,
> Alfredo.

The artificial character of this letter would have subjected it at once to the suspicion of any experienced examiner. Most persons think that it would be a very easy matter to compose a thoroughly innocent letter to be used as the medium for carrying a letter in secret ink, but this is far from being true. Certainly all these letters written by the Germans were entirely lacking in genuineness of tone and meaning. The difficulty, of course, in writing a letter that would sound real was that the writer had really nothing to say to his correspondent except in connection with the subjects of his activity. The moment he began to try to invent subject matter, he fell into the snare of artificiality, and this was especially true if he attempted at the same time to give a double meaning to the sentences of his clear text.[7]

[6]This is most likely J. Willard Robinson, who did not arrive back in the United States from the Netherlands until June.

[7]In the copy of this article in the Friedman Collection, William Friedman leaves the following comment, "Stupid letter writing. Needed a course in business correspondence."

It hardly needs explanation that "our mutual friend Hudson" refers to Wessels or Rodiger, to use the name by which he was known in New York. The money that Hudson had asked for was obviously the remittance of $10,000, which arrived on March 6.

References

Kahn, David. 2004. *The Reader of Gentlemen's Mail: Herbert O. Yardley and the Birth of American Codebreaking*. New Haven: Yale University Press.

Macrakis, Kristie. 2014. *Prisoners, Lovers, & Spies: The Story of Invisible Ink from Herodotus to Al-Qaeda*. New Haven, CT: Yale University Press.

Chapter 20
Madame Victorica: *Captured*!

John Matthews Manly

Abstract The final Madame Victorica article, numbered VII, focuses on a number of letters sent to Madame Victorica during 1917 and early 1918. Many of them are desperate efforts on the part of the German Secret Service to maintain contact with its agents in the United States after the American declaration of war in April 1917 and also letters from Madame Victorica trying to contact her superiors. The final part of the article details the arrest of Madame Victorica in late April 1918 and her subsequent statements to the US authorities. The Manly folder in the Friedman Collection also contains a letter to Manly and a memorandum from the US Mixed Claims Commission dated in 1932 that gives details of the interrogation of Madame Victorica after her arrest in 1918.

In the preceding articles we have followed Madame Victorica from the time when she entered the invisible ink laboratory of the German Secret Service in Berlin to receive instructions concerning her mission to the United States and training in the methods of communication with invisible inks. From the secret writing developed in the letters she received, we have learned that this mission involved such important and dangerous undertakings as the importation into America of explosives to be used in blowing up ships and the location of bases for U-boat supplies on the American coast. Certainly it would be difficult to devise a more ingenious plan, or one which apparently had a better chance of success, than her plan of having the ingredients for making the explosives enclosed in an altar and the sacred figures appropriate to it and then having them shipped by a firm in a neutral country to the address of some clergyman in the United States. The plan for the location of U-boat materials was apparently not conceived by her, but her last letter informed her that the plan had been tried successfully off the coast of Spain.

Even more important, however, than the messages of the letters already discussed is the long and detailed message written in invisible ink and enclosed in letters both to Madame Victorica and her principal associates. How important the Germans regarded it is shown by the fact that six copies of it exist incorporated in six different letters.

The secret message itself is designated as no. 7 and is dated March 15. After a long list of cover addresses in Europe and America, the message contains instructions concerning the ordering of iron columns and balustrades for some unexplained but obviously important purpose. It then repeats the instructions concerning the altar in the following terms:

J.F. Dooley, *Codes, Ciphers and Spies*, DOI 10.1007/978-3-319-29415-5_20

Order an altar with three figures of saints, four columns of two meters, style Renaissance-Baroque, painting in rustic style.

This order is explicitly entrusted to L.A. 3, that is, Madame Victorica.

The message then continues:

Has L.A.1 (Wessels) received $10,000? L.A. 3 (Madame Victorica) $35,000? Report cover addresses here. Report intermediaries for transmission of packages. L.A.1 is to place agents on American warships. Where and how can hundred ton sailboat land secret material on that side?

The matters thus far mentioned had been included in previous letters, but the next sentence of this message in invisible ink introduces another and very important subject. "What can be undertaken against the Panama Canal?"

This is the only allusion in the letters to this interesting subject, and we have no information concerning the reply that was made. Whether the question was addressed to Wessels or to Madame Victorica is perhaps a matter of no consequence, as the two were working in very close collaboration at this time, and the plans of one were apparently the plans of both.

The next sentence orders the recipient to "look for a chemist over there." It appears that there had been a plan to send over a chemist to aid in the construction of the bombs, but this plan had somehow fallen through, and it was thought that it would be best to secure someone on this side who could take his place.

The next sentence of the message reads, "Texts describe dice games." This apparently means that certain dice games imported into this country from abroad contained pamphlets in which secret messages had been incorporated.

Following this are two sentences dealing with secret inks—"L.A. 1 (Wessels) is not to write with S ink. 'Harry sends greetings' means L.A. (7?) has adopted a new ink....make known method of development in secret writing."

S ink was entirely different from F ink and would become visible only when an entirely different reagent was applied to it. For some reason, however—perhaps because the method of developing this ink had been discovered by the British and the French—it was regarded as dangerous to use it, and L. A. 1 (Wessels) was ordered to discontinue its use.

The sentence, "Harry sends greetings," is to be used as a warning that a new ink has been adopted by one of the other agents who was corresponding with these. It would, of course, appear in the plain text of a letter, and no one could possibly guess what significance it had been agreed to attach to it.

Following this is a sentence warning Madame Victorica and Wessels that in their letters, whether open or secret, they are never to use either their real names or the aliases under which they were known but were to make use of other names, the significance of which would be understood only by the correspondents. The sentence reads: "Never give names, only Hudson, Henderson-L.A.1; Clark, Carson – L.A.3; Octavio Evaristo, Harry Wood – L.S.7." Hudson and Clark are already known to us as pseudonyms for Wessels and Madame Victorica. The names for L.S.7 belong to the agent also referred to as Oscar who visited Wessels at the end of April.

Then follows a new set of numbers for use with the ABC Code and the warning that a reference to any private affair indicates that these code numbers are to be used.

This message was written in secret ink in three identical letters addressed to Wessels under different aliases and at different addresses, and three addressed to Madame Victorica under her own name at one and the same address. Although the secret message was dated March 17, the six letters containing it were all dated March 22. Perhaps copies had also been sent.

The letters to Madame Victorica read as follows:

> My dear friend,
>
> Have many thanks for your kind letters, dated 2 and 3 ult. which I received at the same time with a letter from your uncle. Everything was very interesting for me, but especially the details you have written about our mutual friend Hudson. I am pleased to hear that you agree with him. In the meantime I received your first letter, dated 25th of January, and I found that your writing was far better than in those dated 2 and 3.

This sentence is, perhaps, the only one in the letter that requires any explanation or comment. The writing referred to is the writing in invisible ink that all the letters contained. Fels is informing his correspondent that the writing in the first letter developed more easily and clearly than in the two later ones. It is possible that she used to dilute a solution in writing the letters of February 2 and 3, or it may be that she did not exercise so much care with her writing.

The letter continues with another reference to Kalium Iodatum—a subject that Fels never gets tired of discussing. Here, however, his purpose is specifically to say, "I am going to write to Wessels with the ink that is developed by this reagent."

> How do you like my specialty "Kalium Iodatum"? Are you satisfied? Please recommend it also to our mutual friend Hudson, it will be very useful to him; under all circumstances I write to him by the same mail, but as you will see him before my letter is in his hands, please give him all details you received here about using Kalium Iodatum, it is far better than his medicine.
>
> I hope that Hudson received my last remittance by wireless. Did you get my wireless regarding your reports? You will be pleased to hear that my cousin Oscar will call upon you on his trip to the South. Give him every assistance you can and I hope he will meet our mutual friend Hudson, too. By same mail I send the wanted documents to Hudson.
>
> With kindest regards
> I am yours truly,
> Fels.

The three identical letters to Wessels have been given in the preceding article.

It was not intended that these letters should contain anything else, but a curious accident happened to one of them. When examined in the laboratory of MI-8, it was discovered that one sheet contained writing which could be read only with the aid of a mirror. Obviously this sheet had been in contact with another letter that was not dry and had received what printers call an "offset"—that is, a reversed copy of the original writing. Only fragments of this letter could be read, and it is not known for whom it was intended or whether, in fact, it was ever sent to anyone.

The earlier sentences show fragmentary remarks about placing men on English warships, and the text then proceeds as follow:

Es können grosse Belohnungen versprochen werden. Wenn ------ Dreadnoughts in die Luft befördern konnen. Ist ihm eine Million in Hbg (?) zu Diensten (?).

This can be translated:

Great rewards can be promised if …. can blow up dreadnought. A million is at his service in Hamburg (?).

The message then proceeds thus:

To indicate that you have received and developed these messages accurately, cable to your Meisner cover address: 1. If you are able to develop the S ink; "can buy colony collection". 2. If you were not able to develop the S ink because you could not get any developer there: "can buy old French collection.

Such a cable message as "Can buy colony collection" or "can buy old French collection" would, of course, be passed by the cable censor without any question as an innocent business message, and no one who was ignorant of these special instructions could ever by any possibility discover or guess what such a message meant. This type of code communication has been mentioned several times before as one of the most effective means of secret communication used by the Germans. The only difficulty about its use is that an agreement has to be made beforehand concerning the secret meaning to be attached to such innocent words.

Although the entry of the United States into the war did not put an end to or materially change the plans of the secret service for Wessels and Madame Victorica, it greatly increased the difficulties under which they were operating. It was perhaps owing to it that Eberhardt, who had been expected at the end of March, did not arrive until April 29.[1] He then had to borrow $25 from the clerk at the Hotel Astor, but a few days later, he had $25,000, apparently derived from the sale of German securities. Although his coming had been announced, Wessels suspected him and when Eberhardt offered to furnish any amount of money up to $25,000 declared that he had been unsuccessful and did not need any.

Soon after this Madame Victorica concluded that she was being shadowed at the Netherlands Hotel, and she and Miss Sullivan[2] moved to the Hotel Nassau at Long Beach, where she remained nearly a year.

Meanwhile Wessels made a very suspicious visit to Lake Hopatcong. When questioned about it later, he declared that his sole purpose was rest and recreation, but it is significant that there were several powder plants near the hotel and that munitions experts sometimes came to the hotel for conferences with Hiram Maxim.[3]

On June 20, the messenger whom Madame Victorica had sent to Europe with reports from herself and Wessels, and inquiries and requests from O'Leary and

[1] Manly seems confused here. In the previous article, Eberhardt is to arrive at the end of May, while in this article he arrives in April.

[2] This is Madame Victorica's maid, Margaret Sullivan. See the footnote about her in Article V.

[3] Manly means Hudson Maxim, the inventor of smokeless gunpowder. Hudson Maxim lived the last 25 years of his life in Lake Hopatcong, NJ. His brother was Hiram Maxim, the inventor of the Maxim machine gun. (See Wikipedia entries, http://en.wikipedia.org/wiki/Hudson_Maxim and http://en.wikipedia.org/wiki/Hiram_Maxim accessed on 31 July 2014.)

Ryan, returned.[4] He reported himself very much dissatisfied because he had not been allowed to enter Germany but had been met in Amsterdam by Fels, who had with him an old clerk to take down the interviews in shorthand. Madame Victorica distrusted his statements and later felt that he had double-crossed her. She believed that although he was sent only as her messenger, he had arranged with Fels to act independently. This suspicion is confirmed by the later correspondence of the German Secret Service with this man.

He did not make a full report until July 5, when he had an interview at a Long Island restaurant with Madame Victorica and O'Leary. He told them that he brought the promise of the central powers to take up Ireland's cause when peace was negotiated. In reply to the request for a submarine to take the three Irishmen to Ireland, he reported that the German government believed that if a man like Sir Roger Casement could not succeed, no person of less influence could manage it.[5] Concerning the so-called third document, the reply of the Foreign Office was that it would be best for Ireland to wait for the Peace Conference, when Germany would back up the absolute freedom of Ireland—not home rule but independence.

This "third document" was, as we have seen, a proposal that Germany should lend troops to Ireland.[6]

The messenger then delivered to Madame Victorica four sheets of paper containing secret writing, one for herself and three for Wessels. They were the front and back fly leaves of a copy of the seaman's Bible—a very ingenious method of avoiding suspicion. Madame Victorica read her message, as it was written in an ink that she could develop, but Wessels was informed that the messages for him were written in a new ink which could be developed only with the aid of materials and instructions that were to be brought by a messenger who was then on his way.

Unfortunately for the hopes and plans of Wessels, this messenger was seized before he left Europe and consequently the materials and instructions never arrived.

---

[4] This was Willard J. Robinson. He had sailed from Halifax, Nova Scotia on March 24, 1917 on board the steamer *American* for Rotterdam, carrying messages from Wessels and Mme. Victorica for the German Secret Service, including messages and questions about the Imperial German government's support for Irish nationalism. While in Rotterdam he had met with German agents and was given new secret inks and instructions. He left for the United States on June 6, 1917 and arrived in New York on June 20, 1917 (New York Times 1919).

[5] Sir Roger Casement (1864–1916), the Irish patriot and revolutionary took a submarine from Germany to Ireland just before the Easter Rising of 1916. He landed in Tralee Bay on April 21, 1916 but was captured and arrested the next day. He was charged with treason, tried, convicted, and hanged on August 3, 1916.

[6] Roger Casement attempted to get the Germans to create an Irish Brigade of Irish prisoners of war that would help in the Irish War of Independence. He also attempted to get the Germans to supply weapons and German officers to train Irish volunteers in Ireland. Casement also proposed a German task force to land in the west of Ireland simultaneously with an uprising in the east. The Irish Brigade did not end up with many volunteers, and the Germans were unwilling to supply many weapons, no officers to train the Irish, and no task force. Only one ship, the *Aud*, was sent with arms destined for the Easter Rising. The Royal Navy intercepted the *Aud* off the Irish coast on April 21, 1916, and its captain scuttled it as it was being escorted into Cork Harbour (McNally and Dennis 2007).

For some time Wessels carried the three sheets of paper on his person, but owing to the passage of time or some chemical reaction, writing began to show on the sheets, and he became frightened and destroyed them. The markings were not clear enough to be read—so Wessels said later—but seem to have related to the construction of bombs, for parts of drawings could be distinguished similar to other drawings for bomb construction. But it is not clear that Wessels was telling the truth about his inability to read the message or that he gave up his plans for the construction and placing of bombs; for less than two weeks later, he made unusual efforts to secure $10,000, visiting Fricke at his country home in the night and declaring that the importance of his need was so great that it would justify Fricke in taking the money from the funds of the firm, a course which Fricke refused to pursue. The German-Americans were obviously becoming more and more embarrassed by their relations with the Secret Service agents.

About this time a disagreement developed between Ryan and O'Leary, and O'Leary soon had nothing more to do with the negotiations for German aid to Ireland. In consequence of the distrust aroused about the former messenger, Ryan offered to secure another one. He came in a car for Madame Victorica, and they drove "deep in the night" to a point near Dobbs' Ferry where the interview with the messenger was held. Madame Victorica advanced $300 for this tryout trip, but the messenger was not able to get off on the steamer he expected and was obliged to go via South America.

During the course of the summer, Madame Victorica was frightened by the appearance of a man who she thought was shadowing her and still more by the arrest of her former messenger on a charge of making seditious speeches. He was acquitted of this charge but she had many unhappy moments. Her troubles were increased by the death of the man to whom she had entrusted the care of her money. Instead of keeping it in a safety deposit box as she had requested, he had deposited it to the credit of his firm and had been furnishing her from time to time with such sums as she required. Upon his death, however, the accounts of the firm were tied up by his executors and by the Alien Property Custodian. Fricke lent her some money but was unable to supply the whole amount needed, and she was later obliged to borrow from Ryan.

The Ryan loans were supplied by methods worthy of the most romantic fiction. Ryan had agreed to let her have $1000 a month, and the arrangement was that at 6 o'clock on the appointed days Miss Sullivan, with a folded newspaper under her arm, should go to the cathedral and take a seat in pew 50. Shortly afterward, Ryan's messenger would come in carrying also a folded newspaper containing the sum promised and would take a seat either in pew 50 or in the immediate neighborhood of it. The two would then meet and arrange to exchange newspapers without being observed. This plan was successfully carried out on four different occasions. On each of the first three, the sum transferred was $1000. On the fourth occasion it was $1500. The plan was ingenious and entirely successful for it was never discovered and was known only from a later confession by Madame Victorica.

Meanwhile Madame Victorica had written to her uncle, Count Alvensleben, in Zurich, asking for money. He, of course, transmitted her letter to the German authorities, who began at once to make arrangements for sending it. It was undoubtedly for this reason that a messenger left Madrid about November 3 with $10,000 for Wessels. On November 5, the authorities here received from the British Secret Service a warning that he was coming, and shortly afterward a similar warning was sent by the French Secret Service. Whether this messenger was intercepted or not does not appear. In any event, he failed to arrive.

Madame Victorica knew that Ryan could not long continue to supply her with funds, and she seemed to be unable to get them directly from Germany. In her despair she turned to the German representatives in Mexico. Ryan agreed to aid her and recommended a man who was said to have had a great deal of successful experience in crossing the border with messages and contraband articles. Madame Victorica, therefore, promised him $2000 for making the trip and dispatched him with a letter requesting that she be furnished immediately with $20,000, partly for the expenses of herself and Wessels. She gave references in abundance to guarantee her right to ask for such a favor and urged that a cablegram be sent to Germany to secure the necessary authorization.

The message was written, Madame Victorica said, on six slips of thin paper. She turned these over to Ryan, who took them to a cigar maker whom he could trust and had them rolled up into a cigar. The messenger was instructed to deliver the money to Ryan, who, in turn, would transfer it to Madame Victorica.

Meanwhile the German Secret Service made repeated efforts to communicate with its agents in New York. Madame Victorica had given them as an emergency address the name of Miss Sullivan and two letters addressed to her but intended for Madame Victorica were mailed in Barcelona, Spain. These letters were signed "Henry." Madame Victorica thought that they contained secret writing but she was unable to develop it. In fact, she did not try to do so, for she knew that the old ink had been discarded and that the writing could be developed only by the aid of the messenger to Wessels, who never came. She said that the open writing was of no consequence—merely to the effects that her "uncle was worried at not hearing from her or Dillon," and there was some reference—which must have been very displeasing to her—to leaving the correspondence in the hands of the messenger whom she had sent to Amsterdam but whom she now believed to be attempting to act independently and thus displace her from the direct relations with the authorities which she had previously enjoyed.[7]

About the same time a Dutch sailor on the steamer *Nieuw Amsterdam* arrived with a letter which he had received six weeks earlier from a German agent in Rotterdam. He was to post the letter in New York, addressing it to Madame Victorica's former messenger to Europe. The letter shows very clearly that Madame Victorica was justified in suspecting that her messenger had undertaken to do business on his own account. The open text of this letter inquires "what is going on in our business in both the United States and Canada," whether the capital invested therein "actually pays," and furthermore whether all the friends, especially Marie and Dillon, "are in good health." It then suggests that he secure a confidential

[7] Once again, this was J. Willard Robinson. He went to Rotterdam, though, not Amsterdam.

courier, to call on the writer personally with "ample information both business and private matters." "I am quite willing to do business on a larger scale and place more money in it .... please do not mind the expense resulting from the employment of a special courier."

The most important part of this letter, the secret message contained in it, was never read. The German agents could not read it because it, like the rest of the letters they were receiving at this time, was written with an invisible ink that could be developed only with the aid of the messenger, who never came. Nor was it possible to develop it in our laboratory, probably because of the length of time that had elapsed and the careless usage to which the letter had been subjected or, possibly because in this case, the Germans had succeeded in inventing an ink for which we had not yet succeeded in finding a developer.

Whether the agents in the United States succeeded in letting the German authorities know that they were unable to deal with the new inks, we are not informed. Apparently they did not, for letters continued to come which the agents in the United States were unable to develop. Another letter to Madame Victorica's messenger, mailed in Denmark and postmarked February 19, 1918, renews the inquiries about "cousin Marie" (Madame Victorica) and her business. The letter refers also very clearly to the agent's proposal to establish an independent business with the writer as a sleeping partner. The writer, therefore, requests a more definite statement in regard to the particulars of the business and indicates his willingness to go into it although he has doubts of the present outlook. In this letter again the secret writing, which it undoubtedly contained, remained unread.

What the business referred to in these two letters also becomes clear from a letter dated Stockholm, April 23, which on the face of it submits a list of goods for which the writer has applied for licenses from the Stockholm representative of the War Trade Board. But the open letter is obviously only a cover for the secret writing, developed in the laboratory of MI-8, which reads as follows:

> No. 10. With reference to my last letter No. 9, I would point out once more that I am taking great interest in U.S.A. and Canada business, all the more as there is apparently not much to be done in England, where industry and trade are seriously hampered by the wars (?).
>
> Shipbuilding and shipping trade as also mercury plants should deserve our special attention for investing money. I am looking for some more American Firms of high standard with whom I should like to take up business. Perhaps you could bring me in touch with some good firms, by preference pure American ones. Please hurry forward a special messenger whom you can trust with a verbal report about the actual state of our affairs and prospects for the future. Furthermore, I expect you to send me a detailed report by letter at your earliest convenience. Some of our mutual friends have asked me as to how Dillon is getting on in business; perhaps he can drop them a line directly which they would appreciate very much; their present addresses are as follows:.............. Personally I am alright; the H—medicine recommended to me by one of our good specialists had done me a lot of good, I can only advise you to take it regularly.
>
> With kindest regards to you and Dillon, I remain, cordially yours
>
> Frank.

Particularly interesting in this letter are the allusions to investments in the shipping trade and in mercury plants. That these references apply to efforts to interfere with ships and with the manufacture of munitions needs no argument in view of the general nature of the correspondence we are discussing. The proposal to induce "good American firms" to engage in these undertakings is equally clear in its meaning; and every reader will now recognize that the H medicine is a developer for the new H ink.

If further evidence were needed, it would be furnished by what is perhaps the most famous and interesting letter of the whole series.

Although addressed to Mrs. Chaudière, this letter was clearly intended for Madame Victorica. A translation of the secret writing in it reads in part as follows:

> Please examine both sides of the sheet of paper for secret writing. I confirm my letter No. 7 of October sent in several copies. You and Dillon now are free to take up your business affairs in South America entirely or to invest capital in the great war industries, docks and navigation as you judge best. The works for obtaining quicksilver in the West are particularly recommended to me by well informed persons.
>
> In the view of the enormous shipbuilding program of the United States, capital should preferably be invested in the docks over there, but the firm must not become known to the banks as a stockholder in shipbuilding companies. Your Irish friends will surely not lose the opportunity of speculating in such a good thing. It will, therefore, be all the easier for these friends to play dummy for my firm in this affair. Remittances are on the way. Furthermore, sufficient credit has been opened for you in the South American Branch companies. Therefore, get into communication with South America.
>
> In Argentina business is to stop until further notice. On the other hand, Brazil is now a very good place to invest capital in, to which I call the special attention of the Branch offices in Brazil. Mexico naturally does not interest me on account of the absolutely confused political conditions now prevalent there. You must leave no stone unturned to get a good neutral, or better still, American (not German-American) cover address and let me know it at the first opportunity.
>
> It must under all circumstances be avoided that possible losses or unlucky speculations should lead to the break-down of my whole enterprise over there. Therefore a second firm must be established entirely independent of the present one, which, pursuant to the commercial law there, cannot in any way be made responsible for the operations of the present firm, and which will have no internal or external connection with the present firm, and which will be in a position to deal with me directly.[8]

This letter arrived in January 1918. A steward on a Swedish steamship had brought in into this country together with another letter of the same general character. In New York he got a friend to put them in new envelopes and send them by registering mail. Through some accident one of the letters was not delivered but was returned to the sender's address where agents of the Department of Justice seized it.[9]

This letter, as will be seen, renews the insistence upon investments in shipping and in quicksilver plants and adds the suggestion that aid may be expected from

[8] This is the famous "Maud" letter that was decrypted by Dr. Emmett Carver at the MI-8 secret ink laboratory in New York in 1918 (Yardley 1931, pp. 96–100).

[9] This steamship steward may have been Benjamin E. Benson, an American importer who had lived in Stockholm and had offices there for a number of years. Benson was approached by the Germans to act as a courier for pay. Instead, he went to the American embassy in Stockholm and laid out all the details of the proposed arrangement. He was told to accept the proposal. The Germans gave him clothing impregnated with secret ink and developer, and he allegedly delivered the materials to Willard Robinson in New York. (Cornell Daily Sun 1919).

South America. The indications that Mexico is not to be considered as a safe base for operations are also interesting and important.

But perhaps the most interesting part of the letter is the last paragraph which so strongly urges the organization of separate and unconnected groups of agents so that if the operations of one should be discovered or come under suspicion, the others could continue without hindrance. It was now, however, too late for these excellent suggestions to be put into operation.

A few more letters arrived but the recipients were still unable to read the secret writings contained in them, and meanwhile evidence of various sorts was accumulating against these agents. On April 17 Madame Victorica's messenger became alarmed and fled. On April 27, she and Miss Sullivan were arrested.

That Madame Victorica's arrest should have occurred on April 27 is a curious instance of the irony of fate. It will be remembered that the messenger whom she sent to Mexico for $20,000 was to deliver the money to Ryan to be transmitted to Madame Victorica. This messenger had started on his errand early in March. On April 6, Ryan met Miss Sullivan at the cathedral, reported that the messenger had not returned, and agreed to meet her again on the twentieth. On the twentieth Miss Sullivan reported that she had visited the cathedral but had missed Ryan, and shortly after Ryan wrote asking why she had not met him and agreed to meet her on the following Saturday. Madame Victorica herself always believed that the messenger returned with the money and that Ryan would have given it to Miss Sullivan if they could have met on this fateful day.[10]

Wessels was arrested on May 8.[11] Finally on June 8, a federal grand jury indicted Madame Victorica and Wessels on a charge of conspiracy to commit espionage and at the same time several of the American citizens who had been associated with them on a charge of conspiracy to commit treason.[12]

The evidence against Madame Victorica is indicated in the documents that have been presented. What she accomplished, it does not lie within the scope of our problem to inquire. Suffice it is to say that broken in spirits and in health, she made a full confession of the purposes and plans with which she had come to America, but

[10]However, the messenger apparently did deliver the money. The story of the arrest of Mme. Victorica on April 27, 1918 begins with a botched delivery to Victorica of $20,000 in one thousand dollar bills folded into a newspaper at the Nassau Hotel in Long Beach, New York, on April 16, 1918. Margaret Sullivan entered St. Patrick's Cathedral in the late afternoon of April 16 and casually left a newspaper on a pew. Moments later a mysterious man with his own newspaper sat in the same pew, swapped newspapers, and left. He then took a train out to Long Beach, NY, and sat for a while in the lobby of the Nassau Hotel reading everything but his newspaper. Getting up, he left the paper on his seat, which was promptly occupied by Marie de Victorica. Unfortunately for all of these people, federal agents had watched the entire process (Yardley 1931, pp. 115–117; Macrakis 2014, pp. 168–169).

[11]Wessels was arrested on May 1, 1918 (New York Times 1918a).

[12]The complete list of those indicted on June 8, 1918 is Marie de Victorica, Herman Wessels, Jeremiah O'Leary, John T. Ryan, Willard Robinson, Emil Kipper, and Albert Paul Fricke (New York Times 1918b).

the case was postponed from time to time, and finally, because she was a woman broken in health and spirits and because the armistice had intervened and peace had been declared, the charges against her were dropped.[13]

References

Cornell Daily Sun. 1919. "Robinson Guilty Asserts Benson." May 23, News.

Macrakis, Kristie. 2014. *Prisoners, Lovers, & Spies: The Story of Invisible Ink from Herodotus to Al-Qaeda*. New Haven, CT: Yale University Press.

McNally, Michael, and Peter Dennis. 2007. *Easter Rising 1916: Birth of the Irish Republic*. Dublin, Ireland: Osprey Publishing.

New York Times. 1918a. "Take Teuton Sailor As Chief Spy Here." May 1, News.

New York Times. 1918b. "Seven Indicted as German Spies Face Execution." June 8, News. http://query.nytimes.com/mem/archive-free/pdf?res=9A01E1D8163EE433A2575BC0A9609C946996D6CF.

New York Times. 1919. "Indict O'Leary's Man For Treason." January 21, News.

Yardley, Herbert O. 1931. *The American Black Chamber*. Indianapolis: Bobbs-Merrill.

[13] After testifying at the trials of Jeremiah O'Leary and Willard Robinson, Marie de Victorica was released from custody in September 1919 but was not allowed to leave the United States. The indictments against her were not dropped until 1922. She went to live in a Catholic convent in New York and died in a sanitarium at the age of 42 on August 12, 1920 from a mysterious case of pneumonia. She is buried at Gate of Heaven Cemetery in Hawthorne, New York.

Part IV
Epilogue

Chapter 21
Epilogue

Abstract The end of the war saw the dissolution of most of MI-8 and G2-A6. Fragments of both organizations, notably Herbert Yardley's joint War-State Department Cipher Bureau and William Friedman's two-person organization in the US Army Signal Corps, survived. Friedman's Signal Intelligence Service, formed in 1930, provided the nucleus of American military intelligence for World War II. Overall, Manly's articles give us a good picture of both the AEF and domestic intelligence during the war.

The end of World War I saw the almost immediate dissolution of most of MI-8 and G2-A6. With that demobilization in 1919, the United States began to head down the road that ignored military and especially signal intelligence again. Luckily, this time there were people in place to stop the attrition. In the summer of 1919, Herbert Yardley convinced the War and State Departments to keep the Code and Cipher Solution Section of MI-8 as the new joint War-State Department Cipher Bureau. This bureau, that Yardley would later call the *American Black Chamber*, lasted until its dissolution by Secretary of State Henry Stimson in October 1929 (Yardley 1931). Yardley, however, did not continue all the different operations of MI-8. In particular he dropped or ignored Code and Cipher Creation, Training, and Printing and Distribution of codebooks. Code printing and distribution were assigned to the General Staff Adjutant General's Office, while the secret ink laboratories were just dismantled and the shorthand section completely disbanded. Radio interception and traffic analysis, such a crucial part of G2-A6's work in France, were also completely shut down and would not be revived until the 1930s. After writing a history of the G2-A6 code section and a treatise on German codes, William Friedman was demobilized and returned to the United States in the spring of 1919. He and his wife Elizebeth Smith Friedman reluctantly returned to the Riverbank Laboratories that fall. William Friedman finally left the Riverbank Laboratories to join the US Army Signal Corps as a civilian employee in 1921 and, except for one clerk, was the sole employee of the Code and Cipher Creation and Training Sections for a decade before the creation of the Signal Intelligence Service (SIS) in 1930. While Yardley's Cipher Bureau operation would be shut down in 1929, Friedman and the SIS would pick up the torch of code and cipher solution and create the nucleus of the military intelligence organization that would carry the United States into World War II. The Navy, after outsourcing all their cryptanalytic work during World War I to MI-8,

J.F. Dooley, *Codes, Ciphers and Spies*, DOI 10.1007/978-3-319-29415-5_21

resurrected their own Naval Intelligence Unit in 1922 and placed it under then Lt. Laurance Safford in 1924. Safford would build this organization into an efficient and effective intelligence group. However, cooperation between the Army and Navy intelligence organizations would be very hesitant and intermittent until the advent of World War II.

Other nations also preserved a core of their intelligence agencies during the interwar period: The British Admiralty's Room 40 transformed into the *Government Code and Cypher School* (GCCS), the French preserved the *Deuxième Bureau* (their cryptanalytic branch of military intelligence), and the Germans surreptitiously created a military intelligence agency, the *Abwehr*, in 1920 in violation of the Versailles Treaty. GCCS, while kept as a very small organization throughout the 1920s and transferred from the Admiralty to the Foreign Office, would nonetheless become the core around which the 10,000-person operation at Bletchley Park would grow during World War II.

World War I was the last war where paper and pencil cryptanalysis was the predominant form of code breaking. Wireless telegraphy changed everything in the world of military cryptology. From a few interceptions of messengers or mailbags, the cryptanalytic organizations began to intercept hundreds and, at the end, thousands of wireless messages. This forced the cryptanalytic groups to grow from a handful of intense, focused cryptanalysts hunched over their desks to scores of clerks, translators, and cryptanalysts. There would be no more Georges Painvins breaking complex cipher systems purely by dint of brains and unceasing labor. Cryptanalysis itself changed, moving from a primarily linguistic effort to a statistical and mathematical science, a move that would continue through to the present day. Beginning in the early 1930s, the Poles became the first military intelligence organization to begin hiring mathematicians on a systematic basis. In 1922, William Friedman published his masterwork, *The Index of Coincidence and Its Applications in Cryptography* (Friedman 1922), putting cryptanalysis on a firm mathematical footing and changing how cryptanalysis is done forever.

The advent of machines to do encryption and decryption also spelled the death knell for codebooks. With the development of small, light machines that enabled operators to create cipher messages as fast as they could type, there was no need for distributing hundreds of codebooks or complicated cipher systems. Nor was there a need for extensive operator training. Operators in the field could be trained to use cipher machines in a very short period of time.

Cryptology has always been a competition between the cryptographers—those who create new ciphers and codes and the cryptanalysts who break them. By the end of World War I, the cryptanalysts had the upper hand. Nearly all of the code and cipher systems created during the war were broken by cryptanalysts during the war or shortly thereafter. The next round of the competition would give the cryptographers the upper hand once again.

The first cipher machines were patented in the early 1920s. The first generation of these machines typically was electromechanical and used a set of wired rotors organized in a sequence to scramble the input plaintext letters and produce ciphertext letters as output. A rotor was composed of two plates each with 26 electrical contacts on them. Sandwiched between them was an insulating material. The 26 contacts on one plate (the input or plaintext side) were connected randomly to the 26 contacts on the other plate (the output or ciphertext side) via wires. When an electric current is applied to one contact on the input side, the current exits at a different, random, contact on the other side. If we attach a keyboard to the input side, and a set of lamps to the output side, each time we type a key, say d, a lamp, say R, will light up. At this point what we have is a monoalphabetic substitution cipher using a mixed alphabet, which is not a very secure system. But, in these machines when a key is typed, the rotor steps forward one letter so that the first time we type a d, we will get an R, but the rotor then steps forward, and so typing a second input d will now result in an output of, say Q. Because we stepped the rotor forward, we have used a different mixed alphabet. This still just gives us 26 mixed alphabets, and so we now have a 26-alphabet polyalphabetic substitution cipher. This is still not terribly secure. But what if we add a second rotor to get a second encryption of the input letter and arrange it so that the second rotor will advance one step only every time the first rotor goes through an entire revolution of 26 alphabets? In this case we then get $26 \times 26 = 676$ different mixed cipher alphabets. Adding a third rotor gets us $26 \times 26 \times 26 = 17{,}576$ different cipher alphabets, four rotors get us 456,976 alphabets, etc. This is the key to the new rotor-based cipher machines. There are $26! = 403{,}291{,}461{,}126{,}605{,}635{,}584{,}000{,}000$ possible mixed alphabets of 26 letters. The more of them you can use in a ciphertext, the more difficult it will be for the cryptanalyst to decipher the cryptogram. With three rotors, one would have to have a message longer than 17,576 letters before one would repeat a mixed alphabet!

The most famous of the new rotor machines was the *Enigma* used by the German armed forces during World War II. However, the Enigma and its cousins all started life immediately after World War I. The first Enigma patent was applied for in Germany by Arthur Scherbius in 1918 and granted in 1925 (German patent no. 416,219). A US patent for the Enigma (US patent no. 1,657,411) was granted to Scherbius in January 1928. The German Navy adopted the three-rotor Enigma (Model C) in 1926, followed by the Army (Enigma Model G) in 1928. German Enigmas used either three or four rotors, chosen out of a set of either five or eight. An American, Edward Hebern, was designing a rotor-based electromechanical cipher machine as early as 1917. Hebern applied for his patent in 1921 and had it granted in 1924 (US patent no. 1,510,441). As with all things cryptologic, the United States was slower to adopt the electromechanical cipher machines than other countries, but William Friedman's organization, the Signal Intelligence Service, was using, analyzing, and recommending cipher machines by the mid-1930s (Kahn 1991).

The US Army adopted a purely mechanical device, the M-94 cipher cylinder, as its standard field cipher system, in 1922. The M-94 was designed by Parker Hitt and Joseph Mauborgne during World War I. Theirs was a design that had been reinvented independently several times since the Renaissance, including by Thomas Jefferson. It consisted of a set of 25 metal disks, each with a mixed alphabet inscribed around the outside edge. Each disk had a hole through the center, and the disks were placed on a rod that had wing nuts at each end that could be tightened to hold the disks in place. To use the disk, the operator put the disks on the rod in a particular order—the key for the day—and twisted the disks until the first 25 letters of his plaintext message appeared. The operator then read any other row of the disk as the ciphertext. If the message was longer than 25 letters, the operator simply repeated the steps until the entire message was encrypted.

William Friedman's SIS organization in the US Army Signal Corps succeeded in cryptanalyzing two Japanese cipher machines, code-named Red and Purple before the beginning of World War II. With help from the Navy, William Friedman and Frank Rowlett also designed their own electromechanical rotor-based cipher machine, the SIGABA, in the late 1930s. SIGABA, while similar in concept to the Enigma, overcame a weakness of the Enigma—the regular stepping of the rotors—by introducing a separate set of rotors to cause the cipher wheels to step irregularly. This irregular stepping added another puzzle to the decryption of any SIGABA message; the cryptanalyst not only had to figure out which rotors were used, the order in which they were placed in the machine, the starting place of each rotor, but now they also needed to figure out the pattern of the irregular stepping each time a key was pressed. SIGABA also used more rotors than the Enigma and has the distinction of never having been broken during World War II.

What do Manly's stories tell us about the Code and Cipher Solution Sections of the AEF and of MI-8? First we learn that they started from scratch. Second, we know that they caught up quickly. We also see that initially the Army field officers didn't take their work and intelligence data seriously. This has been a problem with intelligence officers since the beginning of intelligence work. Field officers only wanted to depend on human intelligence, particularly visual reports from cavalry patrols or spies. Interception of messengers and their messages, and later signals intelligence, was not given the cachet or importance of human intelligence. In every war, only a number of real successes—and failures resulting from ignored predictions—would change the field officers' opinions. This would not change until after World War II and the establishment of a real, permanent intelligence organization in the United States.

The Americans also suffered from the fact that their abilities and expertise weren't at first taken seriously by the Allies who had over 3 years real experience in the field. G2-A6 and MI-8 had to work with Allied cryptanalysts who were initially quite unwilling to share data but who began to trust their new American partners as the latter gained experience through 1918. By the armistice, the cooperation between G2-A6 and the French and British was excellent. Unfortunately, the fact that the Americans nearly totally dismantled their intelligence organization again after the war set back this cooperation between intelligence units for a quarter of a century. The dismantling of MI-8 in the United States also set back American intelligence by eliminating some functions and minimizing the rest.

What do Manly's articles tell us about domestic feelings during the war? By the time of the United States' declaration of war against Germany, a majority of Americans were in favor of entering the war on the side of the Allies. The United States was still basically a nativist and isolationist country that distrusted new immigrants and was very leery of any involvement in a European conflict.

The apprehension about being involved in World War I changed once men started being drafted and shipped to France. Patriotism and support for the government and the Army soared. The draft was embraced as over 24 million men signed up and nearly 3 million were drafted. Anti-German feelings skyrocketed, fueled by the press and semiofficial vigilante organizations like the American Protective League.[1] While there was little violence, German street and town names were changed, German language schools closed, historically German churches switched to English-only services, and some German immigrants or first-generation German-Americans even Anglicized their names. The letters, telegrams, and secret ink messages illustrated by Manly give us insight into the thoughts of prisoners of war, draftees, spies, and normal Americans who just want to keep secrets. They also remind us of the fact that during war many times privacy takes a back seat to security whether it is justified by circumstances or not.

References

Friedman, William F. 1922. *The Index of Coincidence and Its Application to Cryptography*. Geneva, IL: Riverbank Laboratories.

Kahn, David. 1991. *Seizing the Enigma*. Boston: Houghton Mifflin Company.

Yardley, Herbert O. 1931. *The American Black Chamber*. Indianapolis: Bobbs-Merrill.

[1] Founded in 1917 as a private organization, the American Protective League was approved by the United States Attorney General and given permission to work with the Justice Department's Bureau of Investigation to root out German spies, saboteurs, Bolsheviks, labor agitators, and anti-war leftists. At its high point it had 250,000 members. (See http://www.nypl.org/blog/2014/10/07/spies-among-us-wwi-apl.)

Bibliography and Further Reading

Ambruster, Howard Watson. 1947. *Treason's Peace: German Dyes and American Dupes.* New York: Beechhurst Press.

American Battle Monuments Commission. 1993. *American Armies and Battlefields in Europe (World War I): A History, Guide, and Reference Book,* Vol. 24. 24 vols. 23. Washington, DC: Center of Military History U.S. Army : U.S. G.P.O. Supt. of Docs. http://www.history.army.mil/catalog/browse/pubnum.html#23.

Anonymous. 1918. "Enemy Alien Property." *The Midland Druggist and Pharmaceutical Review* 52 (1): 520–22.

Anonymous. 1920. *The Federal Reporter Volume 259–260; Cases Argued and Determined in the Circuit and District Courts of the United States*. New York: West Publishing Company.

Anonymous. 1927. "Manly vs. Collier's. Facts." Item 811. George Marshall Foundation Research Library. William Friedman Collection.

Anonymous. 1967. "Notes on Manly Colliers Articles." 811. George Marshall Foundation Research Library. William Friedman Collection.

Anonymous. 2015. "National Counterintelligence Reader, Volume 1 – American Revolution to World War II: Chapter 3c – Imperial Germany's Sabotage Operations in the US." *CI Reader Volume 1 Chapter 3*. July 26. http://fas.org/irp/ops/ci/docs/ci1/ch3c.htm#imperial.

Anonymous. n.d. "Undated Handwritten Notes Relating to the Manly Collier's Articles (Probably by Elizebeth Friedman)." Item 811. Friedman Collection, George Marshall Foundation Research Library, Lexington, VA.

Ayres, Leonard P. 1919. *The War With Germany: A Statistical Summary*. Washington, DC: Government Printing Office. https://archive.org/details/warwithgermanyst00ayreuoft.

Barker, Wayne G. 1979. *The History of Codes and Ciphers in the United States During World War I*. Vol. 21. Laguna Beach, CA: Aegean Park Press.

Barton, George. 1919. *Celebrated Spies and Famous Mysteries of the Great War*. Boston, MA: The Page Company. http://books.google.com/books?id=D8QiAAAAMAAJ&printsec=frontcover&source=gbs_ge_summary_r&cad=0#v=onepage&q&f=false.

Bauer, Craig P. 2013. *Secret History: The Story of Cryptology*. Boca Raton, FL: CRC Press.

Blum, Howard. 2014. *Dark Invasion: 1915, Germany's Secret War and the Hunt for the First Terrorist Cell in America*. New York: Harper Collins.

Boghardt, Thomas. 2004. *Spies of the Kaiser: German Covert Operations in Great Britain During the First World War Era*. EBook. St. Antony's Series. Hampshire, UK: Palgrave Macmillan.

Bruckner, Hilmar-Detlef. 2005. "Germany's First Cryptanalysis on the Western Front: Decrypting British and French Naval Ciphers in World War I." *Cryptologia* 29 (1): 1–22. doi:10.1080/0161-110591893735.

J.F. Dooley, *Codes, Ciphers and Spies*, DOI 10.1007/978-3-319-29415-5

Center of Military History. 1988a. *American Expeditionary Forces*. 2 vols. Order of Battle of the United States Land Forces in the World War 23. Washington, DC: Center of Military History U.S. Army : U.S. G.P.O. Supt. of Docs. http://www.history.army.mil/catalog/browse/pubnum.html#23.

Center of Military History. 1988b. *United States Army in the World War 1917–1919*. 18 vols. Washington, DC: Center of Military History U.S. Army : U.S. G.P.O. Supt. of Docs. http://www.history.army.mil/catalog/browse/pubnum.html#23.

Center of Military History. 1988c. *Zone of the Interior*. 3 vols. Order of Battle of the United States Land Forces in the World War 23. Washington, DC: Center of Military History U.S. Army : U.S. G.P.O. Supt. of Docs. http://www.history.army.mil/catalog/browse/pubnum.html#23.

Chenery, William L. Letter to John M. Manly. 1927. "Letter to John M. Manly," September 16. William Friedman Collection, Item 811. George Marshall Foundation Research Library.

Childs, J. Rives. 1919a. *German Military Ciphers from February to November 1918*. 1016. Paris, France/College Park, MD: United States Army Expeditionary Force. Friedman Collection/National Archives.

Childs, J. Rives. 1919b. *The History and Principles of German Military Ciphers, 1914–1918*. Paris, France/College Park, MD: United States Army Expeditionary Force. Friedman Collection/National Archives.

Childs, J. Rives. 1932. *Before the Curtain Falls*. Hardcover. Indianapolis, IN: Bobbs-Merrill Company Publishers.

Childs, J. Rives. 1978. "My Recollections of G.2 A.6." *Cryptologia* 2 (3): 201–14. doi:10.1080/0161-117891853018.

Clark, Ronald. 1977. *The Man Who Broke Purple*. Boston, MA: Little, Brown and Company.

Clauson-Thue, William. 1901. *The Abc Universal Commercial Electric Telegraphic Code*. 5th edn. London, UK: Eden, Fisher & Company, Ltd. https://archive.org/details/abcuniversalcom00clau.

Coffman, Edward M. 1968. *The War to End All Wars: The American Military Experience in World War I*. New York: Oxford University Press.

Commandant of War Prisons. 1918, August 14. *Annual Report of the Commandant of War Prisons*. Barracks 2, Ft. Oglethorpe, GA: Adjutant General's Office. RG 407. National Archives, College Park, MD.

Cornell Daily Sun. 1919. "Robinson Guilty Asserts Benson." *Cornell Daily Sun*, May 23, sec. News.

Eisenhower, John S. D. 2001. *Yanks: The Epic Story of the American Army in World War I*. New York: The Free Press.

Ferris, John. 1988. "The British Army and Signals Intelligence in the Field During the First World War." *Intelligence and National Security* 3 (4): 23–48.

Finley, James P. 1995. *U.S. Army Military Intelligence History: A Source Book*. Fort Huachuca, AZ: U.S. Army Intelligence Center.

Friedman, William F. 1919. "Field Codes Used by the German Army During the World War." 209. Washington, DC/College Park, MD: War Department. Friedman Collection/National Archives.

Friedman, William F. 1939. *Military Cryptanalysis, Part III. Simpler Varieties of Aperiodic Substitution Systems*. Washington, DC: War Department, Office of the Chief Signal Officer.

Friedman, William F. 1942. *American Army Field Codes in the American Expeditionary Forces During the First World War*. Washington, DC: War Department, Office of the Chief Signal Officer. https://www.nsa.gov/public_info/_files/friedmanDocuments/Publications/FOLDER_267/41784809082383.pdf.

Friedman, William F. 2006. *The Friedman Legacy: A Tribute to William and Elizebeth Friedman*. Sources in Cryptologic History #3. Ft. George Meade, MD: National Security Agency: Center for Cryptologic History.

Friedman, William F. Letter to John M. Manly. 1931a. *Letter to John M. Manly*. June 30. John Matthews Manly Collection. University of Chicago Library.

Friedman, William F. Letter to John M. Manly. 1931b. *Letter to John M. Manly*. November 21. John Matthews Manly Collection. University of Chicago Library.

Friedman, William Frederick. 1977. *Solving German Codes in World War I*. Cryptographic Series #11. Laguna Hills, CA: Aegean Park Press.

Gaffney, T. St. John. 1930. *Breaking the Silence: England, Ireland, Wilson, and the War*. New York: Horace Liveright. https://ia600505.us.archive.org/33/items/breakingsilencee00thom/breakingsilencee00thom.pdf.

Gaines, Helen Fouché. 1956. *Cryptanalysis; a Study of Ciphers and Their Solution*. New York: Dover Publications.

Gilbert, James L. 2012. *World War I and the Origins of U.S. Military Intelligence*. Lanham, MD: Scarecrow Press, Inc.

Gilles, Sealy, and Sylvia Tomasch. 2005. "Professionalizing Chaucer: John Matthews Manly, Edith Rickert and the Canterbury Tales as Cultural Capital." In *Reading Medieval Culture: Essays in Honor of Robert W. Hanning*, edited by Robert M. Stein and Sandra Pierson Prior, 364–85. Notre Dame, IN: University of Notre Dame Press.

Golway, Terry. 1998. *Irish Rebel: John Devoy and America's Fight for Ireland's Freedom*. New York: St. Martin's Press.

Gylden, Yves. 1935. *The Contribution of the Cryptographic Bureaus in the World War*. Edited by William F. Friedman. Vol. 75–81. Reprinted from Signal Corps Bulletin. Washington, DC: U.S. Government Printing Office.

Hallas, James H. 2000. *Doughboy War: The American Expeditionary Force in World War I*. Boulder, CO: Lynne Rienner Publishers.

Hastedt, Glenn P., ed. 2011. *Spies, Wiretaps, and Secret Operations: An Encyclopedia of American Espionage*. Vol. 2. 2 vols. Santa Barbara, CA: ABC-CLIO, LLC.

Hatch, David A. 2007. "The Punitive Expedition Military Reform and Communications Intelligence." *Cryptologia* 31 (1): 38–45. doi:10.1080/01611190600964264.

Hieber, Hanne. 2005. "'Mademoiselle Docteur': The Life and Service of Imperial Germany's Only Female Intelligence Officer." *Journal of Intelligence History* 5 (2): 91–108. doi:http://dx.doi.org/10.1080/16161262.2005.10555119.

Hitt, Parker. 1916. *Manual for the Solution of Military Ciphers*. Fort Leavenworth, KS: Press of the Army Service Schools.

Jones, John Price. 1917. *The German Spy in America: The Secret Plotting of German Spies in the United States and the Inside Story of the Sinking of the Lusitania*. London: Hutchinson & Company, Ltd.

Kahn, David. 1967. *The Codebreakers; The Story of Secret Writing*. New York: Macmillan.

Kahn, David. 1991. *Seizing the Enigma*. Boston, MA: Houghton Mifflin Company.

Kahn, David. 2004. *The Reader of Gentlemen's Mail: Herbert O. Yardley and the Birth of American Codebreaking*. New Haven: Yale University Press.

Keegan, John. 1999. *The First World War*. New York: Alfred A. Knopf.

Kennedy, David M. 1980. *Over Here: The First World War and American Society*. New York: Oxford University Press.

Koenig, Robert. 2009. *The Fourth Horseman: One Man's Mission to Wage the Great War in America*. New York: Public Affairs/Perseus Group.

Landau, Captain Henry. 1937. *The Enemy Within: The Inside Story of German Sabotage in America*. New York: G. P. Putnam's Sons.

Lasry, George, Nils Kopal, and Arno Wacker. 2014. "Solving the Double Transposition Challenge with a Divide-and-Conquer Approach." *Cryptologia* 38 (3): 197–214. doi:10.1080/01611194.2014.915269.

Lescarboura, Austin C. 1923. "A Small Private Laboratory: Some General Impressions Gathered during a Visit to the Riverbank Laboratories." *Scientific American* 129 (3): 154, 201–4.

Lischke, Ute. 2000. *Lily Braun: 1865–1916, German Writer, Feminist, Socialist*. Rochester, NY: Camden House.

Literary Digest, The. 1919. "'Frank Richards' American Head of German Spies." *The Literary Digest*, March 1.

Lloyd, Nick. 2014. *Hundred Days: The Campaign That Ended World War I*. New York: Basic Books.

Macrakis, Kristie. 2014. *Prisoners, Lovers, & Spies: The Story of Invisible Ink from Herodotus to Al-Qaeda*. New Haven, CT: Yale University Press.

Manchester, Harland. 1939. "The Black Tom Case." *Harper's Monthly Magazine*, December. Harpers.org/archive.

Manly, John M. 1921a. "Roger Bacon's Cipher Manuscript." *The American Review of Reviews*, July, 105–6.

Manly, John M. 1921b. "The Most Mysterious Manuscript in the World: Did Roger Bacon Write It and Has the Key Been Found?" *Harper's Monthly Magazine*, July.

Manly, John M. 1927a. "Articles for Collier's Magazine." Item 811. Friedman Collection, Lexington, VA: George Marshall Foundation Research Library.

Manly, John M. 1927b. "Waberski." Item 811. Friedman Collection, Lexington, VA: George Marshall Foundation Research Library.

Manly, John M. 1931a. "Roger Bacon and the Voynich MS." *Speculum* 6 (3): 345–91. doi:10.2307/2848508.

Manly, John M. Letter to William F. Friedman. 1931b. "Letter to William F. Friedman," July 24. Friedman Collection, Lexington, VA: George Marshall Foundation Research Library.

Manly, John M. Letter to William F. Friedman. 1931c. "Letter to William F. Friedman," August 28. Friedman Collection, Lexington, VA: George Marshall Foundation Research Library.

Manly, John M. Letter to William F. Friedman. 1931d. "Letter to William F. Friedman," December 12. Friedman Collection, Lexington, VA: George Marshall Foundation Research Library.

Manly, John M., and Edith Rickert. 1940. *The Text of the Canterbury Tales*. Chicago: University of Chicago Press.

Mann, Charles C., and Mark L. Plummer. 1991. *The Aspirin Wars: Money, Medicine, and 100 Years of Rampant Competition*. Boston, MA: Harvard Business School Press.

Martin, H. H. Correspondence. 1932. *Martin-To-Manly Secret Ink Letter.* October 28. John Matthews Manly Collection. University of Chicago Library.

Mauborgne, J. O. 1914. "An Advanced Problem in Cryptography and Its Solution." Army Service Schools Press. William Friedman Collection. George Marshall Foundation Research Library. http://marshallfoundation.org/library/digital-archive/advanced-problem-cryptography-solution/.

McMaster, John Bach. 1918. *The United States in the World War*. New York: D. Appleton and Company.

McNally, Michael, and Peter Dennis. 2007. *Easter Rising 1916: Birth of the Irish Republic*. Dublin, Ireland: Osprey Publishing.

Mendelsohn, Charles. 1937. *Studies in German Diplomatic Codes Used During the World War*. Register No. 191. War Department, Washington, DC: Office of the Chief Signal Officer, Government Printing Office.

Messimer, Dwight R. 2015. *The Baltimore Sabotage Cell*. Annapolis, MD: Naval Institute Press.

Millman, Chad. 2006. *The Detonators: The Secret Plot to Destroy America and an Epic Hunt for Justice*. New York: Little, Brown and Company.

Mixed Claims Commission (United States and Germany). 1940. *Opinions and Decisions in the Sabotage Claims Handed Down June 15, 1939, and October 30, 1939 and Appendix*. Washington, DC: U. S. Government Printing Office. http://hdl.handle.net/2027/mdp.39015073384821.

Moorman, Frank. 1920a. *Final Report of the Radio Intelligence Section, General Headquarters, American Expeditionary Force, 1918–1919*. Washington, DC/College Park, MD: Government Printing Office. Friedman Collection/National Archives.

Moorman, Frank. 1920b. "Wireless Intelligence." Presented at the Meeting of Officers of the Military Intelligence Division, Washington, DC, February 13.

Moorman, Frank. 1920c. "Code and Cipher in France." *Infantry Journal* XVI (12): 1039–44.
Munson, Richard. 2013. *George Fabyan*. North Charleston, SC: CreateSpace Independent Publishing Platform.
Nelson, Timothy G. 2012. "The Explosion and the Testimony: The Wwi Sabotage Claims and An International Arbitral Tribunal's Power to Revise its Own Awards." *American Review of International Arbitration* 23 (2): 197–230.
Newbold, William Romaine. 1928. *The Cipher of Roger Bacon*. Edited by Roland Grubb Kent. Philadelphia, PA: University of Pennsylvania Press.
New York Times. 1918a. "Spy Net Yields 2 Women Here; Men Also Taken." *New York Times*, March 19, sec. News. http://query.nytimes.com/search/sitesearch/.
New York Times. 1918b. "President Orders Spies Deported." *New York Times*, March 20, sec. News.
New York Times. 1918c. "Take Teuton Sailor As Chief Spy Here." *New York Times*, May 1, sec. News.
New York Times. 1918d. "Seven Indicted as German Spies Face Execution." *New York Times*, June 8, sec. News.
New York Times. 1919. "Indict O'Leary's Man For Treason." *New York Times*, January 21, sec. News.
O'Donnell, Patrick K. 2004. *Operatives, Spies, and Saboteurs: The Unknown Story of the Men and Women of World War II's OSS*. New York: Free Press.
O'Leary, Jeremiah A., and Michael A. Kelly. 1920. *My Political Trial and Experiences*. New York: Ulan Press.
Pershing, John J. 1919. "Report of General John J. Pershing, Commander-in-Chief, American Expeditionary Forces." Washington, DC: War Department and U.S. Government Printing Office. https://ia802708.us.archive.org/29/items/finalreportofgen00unit/finalreportofgen00unit.pdf.
Poe, Edgar Allan. 1843. "The Gold-Bug." *The Dollar Newspaper*.
Pohlmann, Markus. 2005. "German Intelligence at War, 1914–1918." *Journal of Intelligence History* 5 (2): 25–54. doi:http://dx.doi.org/10.1080/16161262.2005.10555116.
Richmond Times-Dispatch. 1919. "Associate of O'Leary Charged with Treason." *Richmond Times-Dispatch*, January 21, sec. News.
Roberts, Frank E. 2004. *The American Foreign Legion: Black Soldiers of the 93d in World War I*. Annapolis, MD: Naval Institute Press.
Rowlett, Frank R., Solomon Kullback, and Abraham Sinkov. 1934. "General Solution for the ADFGVX Cipher." Washington, DC: U.S. Army Signal Intelligence Service.
Scala, Elizabeth. 2000. "Scandalous Assumptions: Edith Rickert and the Chicago Chaucer Project." *Medieval Feminist Forum* 30 (1): 27–37.
Schwartz, Austin. 2010. "Unprepared in the Face of Hell: Wartime Leadership, 1917–1918." *Western Illinois Historical Review* 2 (1): 115–43.
Scott, Emmett. 1919. *Scott's Official History of the American Negro in the World War*. Chicago: Homewood Press. http://net.lib.byu.edu/estu/wwi/comment/scott/ScottTC.htm#contents.
Shannon, Claude. 1949. "Communication Theory of Secrecy Systems." *Bell System Technical Journal* 28 (4): 656–715.
Sheldon, Rose Mary. 1999. *The Friedman Collection: An Analytical Guide*. Electronic. Lexington, VA: George Marshall Foundation Research Library. http://marshallfoundation.org/library/wp-content/uploads/sites/16/2014/09/Friedman_Collection_Guide_September_2014.pdf.
Simpson, Rebecca. 2014. "Secrets and Spies of the First World War." UK National Archives. *National Archives Blog*. April 11. http://blog.nationalarchives.gov.uk/blog/secrets-spies-first-world-war/.
Strother, French. 1918a. "Fighting Germany's Spies: Robert Fay and The Ship Bombs." In *The World's Work: A History of Our Time*, 35:663–69. New York: Doubleday, Page & Company.

Strother, French. 1918b. "Fighting Germany's Spies: The Inside Story of the Passport Frauds and The First Glimpse of Werner Horn." In *The World's Work: A History of Our Time*, 35:513–28. New York: Doubleday, Page & Company.

Strother, French. 1918c. "Fighting Germany's Spies: The Inside Story of Werner Horn and The First Glimpse of The Ship Bombs." In *The World's Work: A History of Our Time*, 35:652–63. New York: Doubleday, Page & Company.

Tuchman, Barbara W. 1958. *The Zimmermann Telegram*. New York: Macmillan Company.

Tunney, Thomas J., and Paul Merrick Hollister. 1919. *Throttled! The Detection of the German and Anarchist Bomb Plotters*. Boston, MA: Small, Maynard and Company. https://play.google.com/books/reader?id=bNcLAAAAYAAJ&printsec=frontcover&output=reader&authuser=0&hl=en&pg=GBS.PR8.

van Der Meulen, Michael. 1998. "The Road to German Diplomatic Ciphers, 1919–1945." *Cryptologia* 22 (2): 141–66. doi:10.1080/0161-119891886858.

Venzon, Anne Cipriano, ed. 1995. *The United States in the First World War: An Encyclopedia*. New York: Garland Publishing, Inc.

von der Goltz, Horst. 1918. *My Adventures as a German Secret Service Agent*. London: Cassell and Company, Ltd.

Warner, Michael. 2002. "The Kaiser Sows Destruction: Protecting the Homeland the First Time Around." *Studies in Intelligence* 46 (1): 6.

Washington Times. 1918. "Mme. Storch - Vampire and German Spy." *Washington Times*, July, sec. The American Weekly Magazine. Library of Congress. http://chroniclingamerica.loc.gov/lccn/sn84026749/1918-06-02/ed-1/seq-32/.

West, Nigel. 2013. *Historical Dictionary of World War I Intelligence. Historical Dictionaries of Intelligence and Counterintelligence*. New York: Scarecrow Press.

Weyl, Nathaniel. 1950. *Treason: The Story of Disloyalty and Betrayal in American History*. Washington, DC: Public Affairs Press.

Wheelwright, Julie. 1992. *The Fatal Lover: Mata Hari and the Myth of Women in Espionage*. London, England: Collins & Brown, Ltd.

Wilcox, Jennifer. 2012. *Revolutionary Secrets: Cryptology in the American Revolution*. History narrative. Ft. Meade, MD: Center for Cryptologic History, National Security Agency.

Wisconsin State Journal. 1917. "New Jersey Arms Plant Blown Up; Losses Are Heavy." *Wisconsin State Journal*, January 12. http://www3.gendisasters.com/new-jersey/14821/kingsland-nj-munitions-explosion-jan-1917.

Witcover, Jules. 1989. *Sabotage at Black Tom: Imperial Germany's Secret War in America – 1914–1917*. New York: Algonquin Books.

Yardley, Herbert O. 1930. "Letter to William F. Friedman," December 20. William Friedman Collection. George Marshall Foundation Research Library.

Yardley, Herbert O. 1931a. *The American Black Chamber*. Indianapolis: Bobbs-Merrill.

Yardley, Herbert O. 1931b. "Letter to William F. Friedman," February 1. William Friedman Collection. George Marshall Foundation Research Library.

Yardley, Herbert O. 1934a. *The Blonde Countess*. New York: A. L. Burt Company.

Yardley, Herbert O. 1934b. "H-27, The Blonde Woman from Antwerp." *Liberty Magazine*, April 21.

Index

J.F. Dooley, *Codes, Ciphers and Spies*, DOI 10.1007/978-3-319-29415-5

CPSIA information can be obtained
at www.ICGtesting.com
Printed in the USA
LVHW080631210922
728911LV00001B/1

9 783319 294148